23rd Congress of the International Commission for Optics 2014

(ICO 23)

Journal of Physics: Conference Series Volume 605

Santiago de Compostela, Spain
26 – 29 August 2014

Editors:

J. R. Salgueiro
M. T. Flores-Arias
J. B. Vazquez-Dorrio
A. Guzman
Y. Arakawa

ISBN: 978-1-5108-0265-0
ISSN: 1742-6588

Printed from e-media with permission by:

Curran Associates, Inc.
57 Morehouse Lane
Red Hook, NY 12571

Some format issues inherent in the e-media version may also appear in this print version.

Printed by Curran Associates, Inc. (2015)

For permission requests, please contact the Institute of Physics at the address below.

Institute of Physics
Dirac House, Temple Back
Bristol BS1 6BE UK

Phone: 44 1 17 929 7481
Fax: 44 1 17 920 0979

techtracking@iop.org

Additional copies of this publication are available from:

Curran Associates, Inc.
57 Morehouse Lane
Red Hook, NY 12571 USA
Phone: 845-758-0400
Fax: 845-758-2634
Email: curran@proceedings.com
Web: www.proceedings.com

TABLE OF CONTENTS

APPLIED OPTICS

COLOUR

OPTOMECHANICS

OPTICAL IMAGING

BIOPHOTONICS

NONLINEAR OPTICS

QUANTUM OPTICS

OPTICS IN EDUCATION

OPTICS FOR SUSTAINABLE DEVELOPMENT

ICO-23

-enlightening-the-future-

| 23RD CONGRESS | INTERNATIONAL COMMISION FOR OPTICS |

Santiago de Compostela 2014

Foreword

The 23rd Congress of the International Commission for Optics (ICO) was held in Santiago de Compostela (Spain) 26–29 August 2014, organized by the Universities of Vigo and Santiago de Compostela. Approximately 450 people attended the conference, sharing their knowledge in the cheerful, warm atmosphere of this lovely city. The conference was extremely successful in contributing to the mission of the ICO: to contribute worldwide, on an international basis, to the progress and diffusion of scientific and technological knowledge on optics and photonics.

Optics and photonics have reached a critical level of importance for the development of our societies and are present in a great many aspects of our technological progress, from communication systems supporting the Internet to the most modern techniques in medicine. Consistent with the conference slogan *Enlightening the Future*, the meeting stressed the importance of optical science as a key to technological progress in the coming years. UNESCOs designation of 2015 as the International Year of Light and Light-Based Technologies (`www.light2015.org`) acknowledges the importance of raising global awareness of how light and light-based technologies are present in a large fraction of todays advances and how they can address challenges in important areas such as energy, education, agriculture, and health.

The four-day conference highlighted eleven plenary talks by outstanding scientists working in important areas of optics and photonics. A. Aspect, T. Kippenberg (2013 ICO Prize awardee) and K. Razewski (2013 ICO Galileo Galilei Award) spoke on quantum optics; P. Russell and Yu. Kivshar lectured on topics related to optical processing devices as optical fibers and metamaterials for light shaping; N. X. Fang (2011 ICO Prize), U. Woggon, and A. Alú (2013 IUPAP Young Scientists Prize) discussed applications of optics to nanoscience; and K. Dholakia and J. Widjaja (2008 Galileo Galilei Award) presented in their plenaries applications of light to the tracking and manipulation of particles. Each day, in addition to plenary talks, nine or ten parallel sessions were held covering advances in both theory and applications of optics and photonics, ranging from quantum optics to technological applications. A total of 190 talks were given. Poster sessions allowed an additional 150 presentations. Congress participants came from every one of the five continents.

Included in the conference were three ceremonies for the granting of awards to plenary speakers noted above. Prizes awarded were the ICO Prize, the ICO Galileo Galilei Award, and the IUPAP young scientist Prize in Optics. Two additional prizes were awarded by OSA and SPIE to students outstanding poster presentations.

Summaries of the presentations were available to Congress attendees. In addition, a call was made for submission of elaborated papers to be peer reviewed and published in the Journal of Physics: Conference Series. Conference papers were presented on applied optics, including color science and optomechanics (7 papers), optical imaging (7 papers), biophotonics (2 papers), optical metrology (4 papers), lasers and laser processing (5 papers), optoelectronics and optical communications (6 papers), nonlinear optics (5 papers), quantum optics (4 papers) and education and sustainability (4 papers). We would like to express our gratitude to the Bureau of the International Commission for Optics, the members of the organizing and scientific advisory committees, and to the University of Vigo and the University of Santiago de Compostela for their support and help in the organization of the Conference. We finally want to acknowledge all the authors that have submitted their work to be included in this number and very specially thank all those collaborating persons who acted as reviewers of the papers included in this number for their excellent (and often little recognized) work.

The editors:

José Ramón Salgueiro, *Universidade de Vigo, Spain*
María Teresa Flores-Arias, *Universidade de Santiago de Compostela, Spain*
José Benito Vázquez-Dorrío, *Universidade de Vigo, Spain*
Ángela Guzmán, *CREOL, University of Central Florida, USA*
Yasuhiko Arakawa, *University of Tokyo, Japan*

Journal of Physics: Conference Series **605** (2015) 011001 doi:10.1088/1742-6596/605/1/011001

Local organising committee

Humberto Michinel, *Universidade de Vigo*

Carlos Gómez-Reino, *Universidade de Santiago de Compostela* (in memoriam)

José R. Salgueiro, *Universidade de Vigo*

Ángel Paredes, *Universidade de Vigo*

María T. Flores-Arias, *Universidade de Santiago de Compostela*

Benito Vázquez-Dorrío, *Universidade de Vigo*

Eva Cid, *Universidade de Vigo*

Local organising committee (sections)

P. Andrés, *Universitat de València*
(Applied Optics)

J. Arines, *Univ. de Santiago de Compostela*
(Biophotonics)

J. L. Nieves, *Universidad de Granada*
(Colour)

A. Yañez, *Universidade da Coruña*
(Laser Processing)

M. Nieto-Vesperinas, *CSIC*
(Nanophotonics)

J. P. Torres, *ICFO*
(Nonlinear Optics)

B. V. Dorrío, *Universidade de Vigo*
(Optical Metrology)

J. Capmany, *Univ. Politècnica de València*
(Optical Telecommunications)

M. L. Calvo, *Univ. Complutense de Madrid*
(Optics in education)

M. Flores, *Univ. de Santiago de Compostela*
(Optoelectronics)

C. Ferreira, *Universitat de València*
(Optical Imaging)

R. Arias-González, *IMDEA Nanociencia*
(Optomechanics)

G. J. de Valcárcel, *Universitat de València*
(Quantum Optics)

L. Roso, *CLPU*
(Ultrashort pulses)

International advisory committee

Y. Arakawa, *University of Tokyo, Japan*

Z. Ben Lakhdar, *University of Tunis, Tunis*

Z. Bingkun, *Tsinghua University, China*

M. L. Calvo,
Univ. Complutense de Madrid, Spain

C. Cisneros, *UNAM, Mexico*

A. Diaspro, *IIT, Italy*

Y. Ding, *Leigh University, USA*

A. Friberg, *KTH, Finland*

U. Gibson, *NTNU, Norway*

A. Guzmán, *Univ. of Central Florida, USA*

J. Harrington, *Rutgers University, USA*

H. P. Herzig, *EPFL, Switzerland*

F. Höller, *Carl Zeiss AG, Germany*

D. T. Moore, *Univ. of Rochester, USA*

M. Oron, *Kilolambda Technologies, Israel*

R. Ramponi, *Politecnico di Milano, Italy*

T. Szoplik, *Univ. of Warsaw, Poland*

S. Vallmitjana, *Univ. de Barcelona, Spain*

G. Von Bally, *Univ. of Münster, Germay*

A. Wagué,
Cheikh Anta Diop University, Senegal

M. Yzuel, *Univ. Autónoma de Barcelona, Spain*

Scientific committee

E. Acosta,
Univ. de Santiago de compostela, Spain

G. Agrawal, *Univ. of Rochester, USA*

P. Andrés, *Univ. de Valéncia, Spain*

G. Assanto, *University Roma III, Italy*

A. Asundi, *Nanyang Tech. Univ. Singapore*

V. Bagnato, *Univ. de São Carlos, Brazil*

R. Castañeda,
Univ. Nacional de Medellín, Colombia

I. Cirac, *MPQ Garching, Germany*

K. Chalasinka-Macukow,
Univ. of Warsaw, Poland

P. Chavel, *Institute d'Optique, France*

R. Corbalán,
Univ. Autónoma de Barcelona, Spain

C. Dainty,
National University of Ireland, Ireland

A. Desyatnikov,
The Australian National University, Australia

K. Dholakia,
National University of St. Andrews, U. K.

V. Fajer-Ávila, *SCF, Cuba*

M. F. Costa, *Univ. do Minho, Portugal*

J. Frejlich, *Univ. de Campinhas, Brazil*

J. W. Goodman, *Stanford University, USA*

J. Holdsworth,
The Univ. of Newcastle, Australia

G. Kaufmann, *CONICET, Argentina*

U. Keller, *ETHZ, Switzerland*

B. Y. Kim,
Korea Science Technology Center, Korea

doi:10.1088/1742-6596/605/1/011001

W. Krolikowski,
The Australian National University, Australia

M. Kujawinska, *Univ. of Warsaw, Poland*

N. Kundikova,
South Ural State University, Russia

F. Laurell, *KTH, Sweden*

L. Lilge, *Univ. of Toronto, Canada*

J. Liñares,
Univ. de Santiago de Compostela, Spain

S. Millán,
Univ. Politécnica de Catalunya, Spain

A. Papoyan,
National Academy of Sciences, Armenia

E. Rosas, *CIO, México*

G. Righini, *CNR, Italy*

P. Russell, *MPL, Germany*

P. Saari, *Univ. of Tartu, Estonia*

C. Saavedra, *Univ. de Concepción, Chile*

V. Sandoghdar, *MPL, Germany*

M. Segev, *Technion, Israel*

V. Sirutkaitis, *Vilnius University, Lithuania*

M. Trivi, *UNLP, Argentina*

E. Van Stryland, *CREOL, USA*

INTERNATIONAL COMMISSION FOR OPTICS
COMISSION INTERNATIONALE D'OPTIQUE

Peer review statement

All papers published in this volume of *Journal of Physics: Conference Series* have been peer reviewed through processes administered by the proceedings Editors. Reviews were conducted by expert referees to the professional and scientific standards expected of a proceedings journal published by IOP Publishing.

Journal of Physics: Conference Series **605** (2015) 012001 doi:10.1088/1742-6596/605/1/012001

Sintered silicon carbide mirror substrates surface finishing for embedded applications

A J Damião[1,2] and M Valentim Ribeiro dos Santos[2]
[1] Instituto Tecnológico de Aeronáutica, São José dos Campos, Sao Paulo, Brazil
[2] Instituto de Estudos Avançados, São José dos Campos, Sao Paulo, Brazil

E-mail: damiao@ieav.cta.br

Abstract. Stiffness is the main reason to use SiC as a mirror substrate for aerospace applications. Also the SiC mechanical and thermal properties are superior to the optical material that is normally used for those applications. In this work, diamond tools and pastes were applied to obtain flat small diameter mirror substrates. The goals were to obtain flatness of λ/10 and Ra roughness of λ/100, for visible light.

1. Introduction

In this work, diamond tools and pastes were applied to obtain flat small diameter mirror substrates. The goals were to obtain flatness of λ/10 and Ra roughness of λ/100, for visible light [1,2].

Light weight is a main concern in aerospace applications. Vitroceramics, Be, Safire are materials used in this case. If low specific mass material is not available, hollowed stiff material can be used. On mirror substrates, the idea is to use a honeycomb structure or some blind holes in the back part, without losing the front surface profile. Stiffness is then a major property. SiC is a good candidate and it has been used in aerospace applications.

Normally it is easier to polish a hard material than a soft one. Soft materials are easily scratched and it is difficult to keep the substrate front surface shape during the polishing process. But SiC is a very hard material and the polishing process must use diamond tools and pastes. The SiC removing rate is 1/35 of the fused silica and 1/50 of vitroceramics. That is why the surface finishing processes is longer and more expensive [1, 3-5]. In general, it is not easy to find detailed information about the finishing processes on SiC, due to commercial interests.

This work presents the procedures and techniques applied to obtain low roughness and high flatness SiC mirror (sintered) substrates.

2. Materials and methods

The H. C. Starck Grade BF17 SiC powder was the starting material for the sintering process, having YAG as sintering additive. The mixture was processed by: ball milling, drying and deagglomeration, sift, uniaxial (40 MPa) and isostatic (300 MPa) pressing, and, finally, argon atmosphere sintering at 1950°C, figure 1. Part of the samples were uniaxially (80 Mpa) pressed only. The raw surface presented after sintering is quite irregular when viewed by a confocal laser microscope (Figure 2).

Figure 1. Sample preparation: (a) ball milling; (b) prepared for drying; (c) sifting; (d) filling the mold; (e) uniaxial pressing; (f) uniaxial pressed sample; (g) ready for isostatic pressing; (h) isostatic pressing and (i) isostatic pressed sample.

Figure 2. Confocal laser microphotography from a Sic surface after sintering.

Samples were prepared in groups of 20 units or more in order to have enough samples to compound a blockade and a few for characterization: hardness, X-Ray, number and size of open and closed porous.

Journal of Physics: Conference Series **605** (2015) 012001 doi:10.1088/1742-6596/605/1/012001

Figure 3. (a) samples after sintering and (b) blockade of samples ready for grinding.

3. Surface (curve) generation

There are three main steps to obtain an optical surface: curve generation (profile), grinding and polishing. In this work, a Karl Zeiss Curve Generator was used. The main parameters were: 200 mm diameter diamond tool of 3750 rpm and work piece rotation of 0.67 rpm, cooled by ethylene glycol. The lower Ra roughness (15 nm) was obtained after 8 hours of surface shaping.

4. Polishing

Due to high quality of the surface generation, the grinding process was discarded and the samples were immediately taken to polish. A 1 µm diamond paste polishing compound diluted in glycerin was the abrasive used to polish on a polyurethane work plate, in LOH PM-250 polishing machine. The whole polishing process lasts 24 hours. The flatness was measured on a Zygo Mark IV phase interferometer and the roughness on a Taylor Hobson PGI 1000. Those are calibrated instruments, traceable to the Brazilian Metrology Institute (INMETRO).

5. Results and discussion

The flatness measurement of a surface right after curve generation ranged from 1.1 µm and 1.7 µm, after 8h machining. One result is presented in figure 4. This shape is characteristic in this curve generation process.

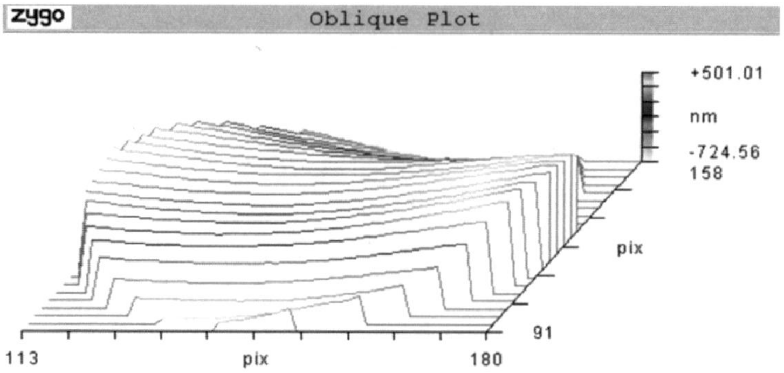

Figure 4. Flatness measurement after 8h machining on a curve generator.

After every 4 hours of polishing process, the flatness (and the roughness) was measured. The flatness results for sample number 6 are shown in figure 5.

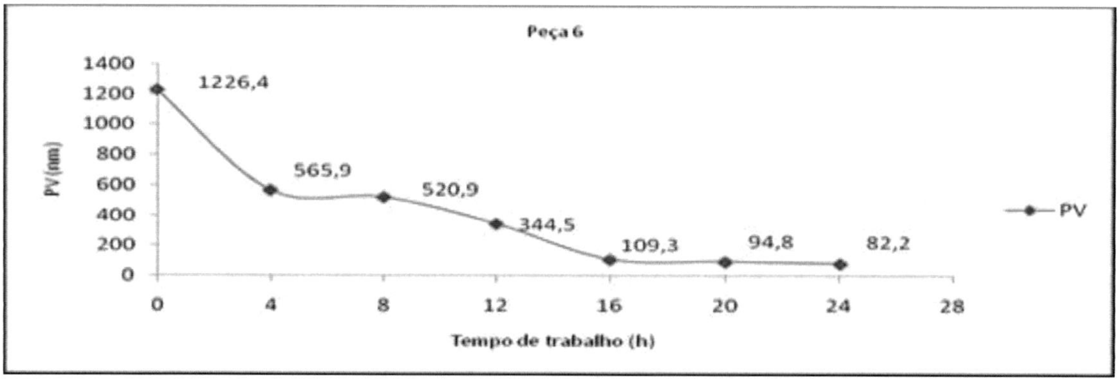

Figure 5. Flatness measurement of sample 6 during the polishing process.

The results for 6 samples polished in the same blockade are shown in table 1. It can be observed that the flatness range is a bit large for samples polished in the same blockade, but it is very hard material, having a very low material removal rate.

Table 1. The results for 6 samples polished in the same blockade.

Sample	Initial flatness	Flatness after 24 h polishing	Roughness (Ra) after 24 h polishing
1	1749 nm	133 nm	7,4 nm
2	1546 nm	157 nm	6,4 nm
3	1066 nm	152 nm	10,2 nm
4	1665 nm	95 nm	5,9 nm
5	1102 nm	226 nm	3,8 nm
6	1226 nm	82 nm	5,7 nm

6. Conclusion

The procedures and techniques applied to obtain low roughness and high flatness SiC mirror (sintered) substrates were presented. The flatness of $\lambda/6$ and Ra roughness of $\lambda/100$, for visible light was obtained. These results were obtained on sinterized Grade BF17 SiC powder.

Acknowledgments

To CNPq processes 559925/2010-7 and 310046/2012-2.

References

[1] Torres R V 2006 *Estudo de retificação de vidros ópticos* (São Carlos: Universidade de São Paulo) Mastering thesis 104 f

[2] Damião A J 2002 *Processo de deposição e propriedades de filmes de dióxido de titânio obtidos por Laser Ablation* (Campinas: Instituto de Física Gleb Wataghin/Universidade Estadual de Campinas) PhD thesis 111f

[3] Johnson J S and Grobsky K *Rapid fabrication of lightweight silicon carbide mirrors* Zygo Corporation Donald J Bray, POCO Graphite, Inc. www.zygo.com/library/papers/proc_4771_243.pdf

[4] Konaka M *et al* *Ultraprecision* Polishing of CVD-SiC Mirrors. http://www.spring8.or.jp/pdf/en/ann_rep/96/P215-217.pdf

[5] Hu H *et al* 2011 Research on reducing the edge effect in magnetorheological finishing *Appl. Opt.* **50** 1220-6. doi: 10.1364/AO.50.001220.

Journal of Physics: Conference Series **605** (2015) 012002 doi:10.1088/1742-6596/605/1/012002

Re-carbonized vitreous carbon substrates for optical applications

F Nascimento[1,2], A J Damião[1], M V R Santos[1], F D Origo[1], J F Reis[1,2], L Trevenzolli[3], A C M Gaspar[4] and J Baesso[4]

[1]Instituto de Estudos Avançados, São José dos Campos, Brazil
[2]Instituto Tecnológico de Aeronáutica, São José dos Campos, Brazil
[3]Faculdade de Engenharia Química, Universidade Estadual de Campinas, Campinas, Brazil
[4]Escola de Engenharia de Lorena, Universidade de São Paulo, Lorena, Brazil

E-mail: damiao@ieav.cta.br

Abstract. Imaging optical systems components for satellites must have low specific mass and high stiffness, as weight is a problem for payloads and stiffness is essential to keep the substrate front surface shape. In this work, Re-carbonized Vitreous Carbon (RVC) was tested as a substrate material. The process to obtain RVC is different from the traditional process to obtain the Monolithic Vitreous Carbon (MVC). It is essential to understand the process to evaluate the surface roughness data. This work describes the process to obtain RVC, as a candidate for optical component substrate, and the results of its surface roughness measurements.

1. Introduction

The traditional process to obtain the Monolithic Vitreous Carbon (MVC) is well described in the literature [1-4]. Vitreous (or glassy) carbon has low specific mass (1.5 g/cm^3) and good mechanical strength [5], combining glassy and ceramic properties [6]. In principle, it is a good material for optical substrates applications. Mainly in aerospace imaging reflective optical components low weight is essential. But the state-of-art thickness limit in this material is 10 mm [2] and could be a problem for many applications. This limit is caused by stress due to the exit of volatile gases formed during the poly (furfuryl alcohol) resin (PFA resin) carbonizations process. It is common to find 20% of the samples broken after carbonization [7]. In order to avoid the stress, a new route [8] was used to produce the substrates. The broken samples were crushed and ball milled (Retsch PM 100). The powder was classified by size. Powder metallurgy technique was applied, using the same resin as lubricant and humectants. The pressed (400 MPa) green samples were carbonizes at 1000 ºC. Samples till 30 mm thick were produced, with no breaks during carbonization. The sample surfaces had the surface finishing: curve generation, grinding and polishing. The flatness and roughness were monitored during the process.

2. Material and methods

One conventional route to obtain vitreous carbon [9] can be very shortly summarized as:

1. Furfuryl alcohol + sulfuric acid → PFA resin
2. PFA resin + p-toluenesulfonic acid → polymerized resin
3. Polymerized resin + low gradient heat → vitreous carbon

All the steps depend on very controlled temperature gradient and can last 10 days. The shortly resumed route used in this work can be described as:

a) Furfuryl alcohol + sulfuric acid → PFA resin
b) PFA resin + p-toluenesulfonic acid → polymerized resin
c) Polymerized resin + high gradient heat → partially carbonized resin (T=600 °C)
d) Partially carbonized resin + ball milling + size classification → partially carbonized resin grains
e) Partially carbonized resin grains + PFA resin + uniaxial press → green samples
f) Green samples + medium gradient heat → re-carbonized vitreous carbon (T=1000 °C)

As the partially carbonized resin will be milled, the temperature gradient in step "a" can be high (1°C/min) in opposition to 0.1°C/min in step 3 in the conventional route. In step "f" temperature gradient of 0.5 °C/min can be used, as the amount of volatiles is much lower and also the volatiles can leave of the sample among the grains during the (final) carbonization process. None of the samples broke at the end of the process, which lasted 5 days. The number of open and closed pores on the substrate depends on: the amount of catalyst acid added to furfurylic; the proportion of resin added to the partially carbonized resin; the powder grain size chosen in step "e". Smaller grains lead to smoother surfaces [10]. As the uniaxial pressing is done using one moving piston, the neutral line appears in one surface. Hardness measurements showed a factor of 3 between the front and back surfaces. The front part was polished and the back part was honeycomb milled, in order to lower the weight.

All the surface conditioning was done using classical generation (Karl Zeiss Curve Generator), grinding (Zeiss grinding) and polishing (Loh PM 2000 and PM 250) machines. The main aspect concerning light scattering is the surface roughness. Values as λ/10 for flatness and λ/100 for roughness were the targets.

During the polishing, the flatness was monitored by a Zygo Mark IV phase interferometer and the roughness by a Taylor Hobson PGI 1000. Both equipments belong to a metrological laboratory (LMSO) that is accredited by INMETRO (Brazilian NMI).

3. Results and Discussion

Catalysis of organic compounds highly depend on the amount of each reactant, temperature, among others. The catalysis of furfuryl alcohol is a good example of this [11]. The proportion (% m/m) of sulfuric acid in the furfuryl alcohol catalyses in order to produce furfuryl resin leads to visible differences after carbonization at 1000°C (Figure 1). The size of the pieces is smaller when 1 % m/m of sulfuric acid is used when compared to 1.5 % and 2 %. All samples were carbonized in the same batch. So, the whole process to produce the re-carbonized vitreous carbon can be directed in order to get smaller grain sizes, to increase the packaging.

The proportion resin/partially carbonized resin also plays an important role on the green sample texture. When less than 20% of resin is added, it is very difficult to manipulate the samples, because they are too dry. When more than 30% is added, they become pasty. A set of samples having 20, 25 and 30% of resin in partially carbonized (at 600 and 1000 °C) resin grains is shown in figure 2.

Figure 1. Vitreous carbon pieces after 1000 ºC carbonization. The resin had 2% (left), 1.5 % (midle) and 1 % (right) m/m of sulfuric acid catalyst.

Next, the surfaces of the samples had the surface finishing. The first step was the surface generation. Flat surfaces were generated in a traditional curve generator, using diamond tools. The removing rate was very slow (1 μm/h) and lasted for more than 8 h, after a leveling process (Figure 3).

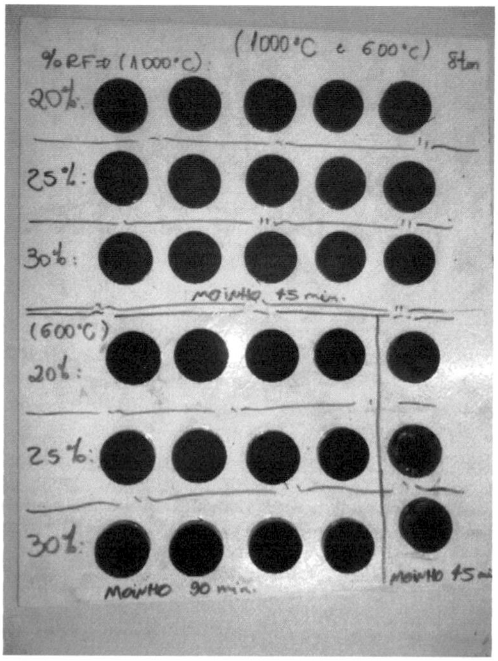

Figure 2. A set of samples having 20, 25 and 30% of resin and resin grains partially carbonized at 600 ºC and 1000 ºC.

Figure 3. Blocking having 5 samples during polishing process (flat).

Each blocking had 6 or 9 samples. Some samples went directly from generation to polishing. The flatness and the roughness were measured along the surface finishing process. The roughness measurements were performed on an area of 5 mm x 1 mm in the central sample of figure 3. The measurements attended the ISO 4288. The 3D image was obtained by 101 measurements on the 1 mm direction (Figure 4). Open pores can be observed and their origin can be attributed to tearing of small grains during grinding.

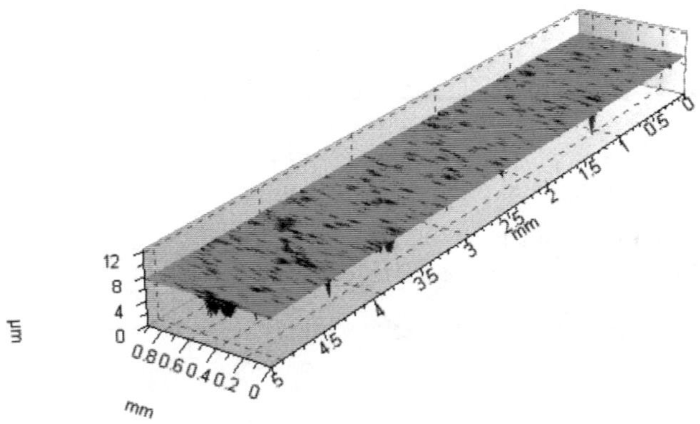

Figure 4. 3D virtual image of the surface of the central sample blocky during the polishing process.

The analysis results of the 3D surface showed a medium Ra value of 66 nm, having a high standard deviation. Both the standard deviation and the mean standard deviation are indicated in Table 1. This high deviation was caused by the presence of open pores on the surface.

Table 1. Main roughness parameter of 101 measurements performed on the central part of the sample.

		Mean	Std dev	Min	Max
Amplitude parameters – Roughness profile					
Rp	µm	0.139	0.0828	0.0461	0.492
Rv	µm	1.72	0.678	0.530	4.30
Rz	µm	1.86	0.726	0.576	4.55
Rc	µm	0.911	0.630	0.176	4.17
Rt	µm	4.15	1.91	0.832	10.3
Ra	µm	0.0656	0.0367	0.0168	0.180
Rq	µm	0.188	0.0908	0.0462	0.442
Rsk		-24.6	14.4	-65.8	-8.48
Rku		467	429	75.9	2187
Material Ratio parameters – Roughness profile					
		Mean	Std dev	Min	Max
Rmr	%	94.6	19.0	0.151	100
Rdc	µm	0.0471	0.0222	0.013	0.121

4. Conclusions

Re-carbonized Vitreous carbon can be considered as a good candidate as optical component substrate. It presented a low surface roughness, even with open pores on the surface. Probably the pores were caused by the removing of grains during the grinding or polishing process. Lower grain sizes will be used in order to get less open pores in the future.

Acknowledgments

To CNPq Projects 559925/2010-7 and 310046/2012-2.

References

[1] Jenkins G M and Kawamura K 1976 *Polymeric carbons – carbon fiber, glass and char* (New York: Cambridge Universtiy Press)

[2] Zarzycki J 1991 *Materials Science and Technology – a Comprehensive Treatment* vol 9, ed. R. W. Cahn, P. Haasen and E. J. Kramer, Weinheim (New York , Cambridge)

[3] Rezende M C 1989 Pore structure and mechanical property relationships in glassy carbon obtained from furfuryl and phenolicresin *Proc. of School of Materials Science (Bath, UK)*

[4] Edmonds T E 1985 Electroanalytical applications of carbon fibre electrodes *Analytica Chimica Acta* **175** 1-22

[5] Chekanova V D and Fialkov A S 1971 Vitreous Carbon (Preparation, Properties, and Application) *Russian Chemical Reviews* **40** 413-28

[6] Cowlard F C, Lewis J C 1967 Vitreous Carbon - A New Form of Carbon *Journal of Materials Science* **2** 507-12

[7] Dondeo F and Damiao A 2013 Vitreous carbon obtained from glassy carbon powder: advantages and disadvantages *Proc. of the Annual World Conference on Carbon (Rio de Janeiro, BR, 14-19 July 2013)* pp 82

[8] Patent was requested.

[9] Oishi S S, Rezende M C, Origo F D, Damiao A J and Botelho E C 2013 Viscosity, pH, and Moisture in the Porosity of Poly(furfuryl alcohol) *J. Appl. Polym. Sci.* **128** 1680-86

[10] Suryanarayana C 2001 Mechanical alloying and milling *Progress in Material Science* **46** 1-184

[11] Schmitt C R 1974 Polyfurfuryl alcohol resins *Polym Plast Technol. Eng* **3** 121-58

System for measuring the spatial reflectance distribution of material surfaces

L A Berni, M S Ribeiro, T F Paes and A F Beloto
Instituto Nacional de Pesquisas Espaciais (INPE), Laboratório Associado de Sensores e Materiais (LAS). Av. dos Astronautas, 1758, São José dos Campos/SP, Brazil 12227-010

E-mail: berni@las.inpe.br

Abstract. An automatic device for measuring the reflectance distribution from material surfaces was assembled in the laboratory. The mechanical setup employs two aluminum rotating arms driven by stepper motors: one for the light source and the second for the collecting optics. The two arms can rotate in the zenith direction from -90 to +90 degrees and the light collecting arm in the azimuth angle over 360 degrees, both with adjustable angular resolution. Measurements from black anodized aluminum, PTFE, graphite, porous silicon and a lambertian reference surface were performed to validate the system.

1. Introduction

The interaction of light with an object surface can be represented as a single function, called Bidirectional Reflection Distribution Function (BRDF)[1, 2]. That describes the pattern of light reflected from a material surface to all directions of incident light. The BRDF is defined as the ratio of reflected radiance (W m^{-2} sr^{-1}) to the incident irradiance (W m^{-2}) on the surface, see Eq. (1). It is a five dimensional function f(θ_i, φ_i, θ_r, φ_r, λ), where θ_i is the incident zenith angle, θ_r is the reflected zenith angle, φ_i and φ_r are the incident and reflected azimuth angles respectively, and λ the measured wavelength. For simplicity, the polarization dependence of light was left out of the equation as our system employs an unpolarized light source. A standard reflectance from GetSpec (GetReflex) was used to determine the irradiance on the samples surface for each incident angle of the light beam.

$$f(\theta_i, \varphi_i, \theta_r, \varphi_r, \lambda) = \frac{dL_r(\theta_r, \varphi_r, \lambda)}{dE_i(\theta_i, \varphi_i, \lambda)} \tag{1}$$

The characterization of the reflective properties of a surface by BRDF measurements and theoretical models are widely used in computer graphics to generate realistic images of virtual objects. The knowledge of these reflective properties can also be used in optical system designs in the analysis and reduction of stray light improving the signal to noise ratio. There are several analytical models that mathematically describe this interaction more or less complete, such as, Lambert model (ideal diffuse), Phong (Blinn-Phong) [3], Oren-Nayar [4] And Cook-Torrance [5].

The reflectance profiles measured in the laboratory can be used both to validate the models described above and also as input data in optical simulation programs. In an optical design which seeks to optimize the signal to noise ratio is essential to know how each material surface that makes up the system scatters radiation. Thus, it is possible to determine the critical regions and choose the best material and finishing for every situation. Normally, we use materials that have a high rate of absorption in the wavelength region of spurious radiation, but if the materials must display characteristics like specular, diffuse or intermediate depends on each situation. The development of the system described here was motivated by the study of stray light in optical systems [6], the input optics of radiometers [7] and porous silicon as coatings for solar cells [8].

2. System set up and specifications

The system was completely assembled on a structure of area of 1100mm x 800mm and height of 900mm. The mechanical setup employs two aluminum rotating arms, one for the 50W halogen light source and the second for the lens-fiber collecting optics. In the illumination arm there are two 40 mm diameter apertures to partially collimate the light beam. The rotation radius of this arm is 320 mm. The collecting light arm was assembled with one lens of 50 mm diameter and focusing length of 40 mm that projects the collecting area from the sample surface on a fiber bundle of 3.0 mm diameter. The rotation radius of this arm is 210 mm. The two arms can rotate in the zenith direction from -90 to +90 degrees and the light collecting arm in the azimuth angle over 360 degrees. All three axes are driven by three 18 kgf.cm stepper motors with adjustable angular resolution. The wavelength of the scattered collected light can be selected by a set of interference filters from 300nm to 1100nm with an average bandwidth of 8.0 nm. The signal is detected by a Si photodiode detector (Oriel 70336) connected to a lock-in amplifier (Oriel-Merlin) and after, the signal is transferred to the computer by a A/D converter (NI USB-6210). The sample to be measured is positioned on an adjustable table so that its surface is on the center of rotation of the optical system. Figure 1 shows the design of the assembled system.

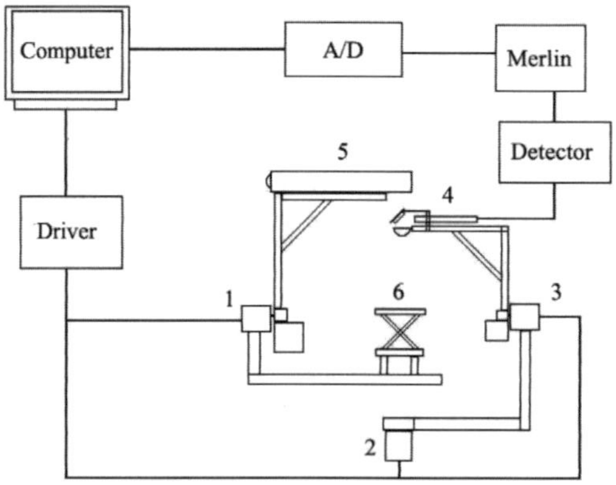

Figure 1. Schematic drawing of the system: 1, 2 and 3 - stepper motors; 4 – collecting scattered light arm; 5 – source light arm and 6 – sample adjustable table.

To characterize the system, it was measured the spectrum of the light source, the spatial irradiance distribution on the sample surface, the spot size of illumination and observation as a function of the incident light angle. Figure 2 (left) shows the irradiance on the center position on the surface table for an incident angle of 0° (the illumination is right above the sample surface). The wavelength dependence was measured with a compact CCD spectrometer from Getspec model 2048 from 250 nm to 1100 nm. As seen in figure 2 (left), the profile is a typical spectrum of a halogen dichroic lamp with the maximum irradiance at 640 nm. For the same incident angle, it was measured the spatial irradiance distribution over the sample table as shown in figure 2 (right). The maximum variation of the irradiance over the center region of 40 mm diameter is 4%. As the angle of the incident light increases from 0° to ± 90° the size of the illumination spot increases as shown in figure 3(left). The spot of collection optics was measured illuminating the fiber optic from the detector side with a HeNe laser. The projected area over the sample table was measured from -80° to 80° as shown in Fig. 3 (right). For the arm at 0° the collection area of the optics is about 10mm and increase up to 13.5mm at ± 40°. Since the observation spot size must be lesser than the sample diameter, this graph shows the limits of the zenith angle to be measured.

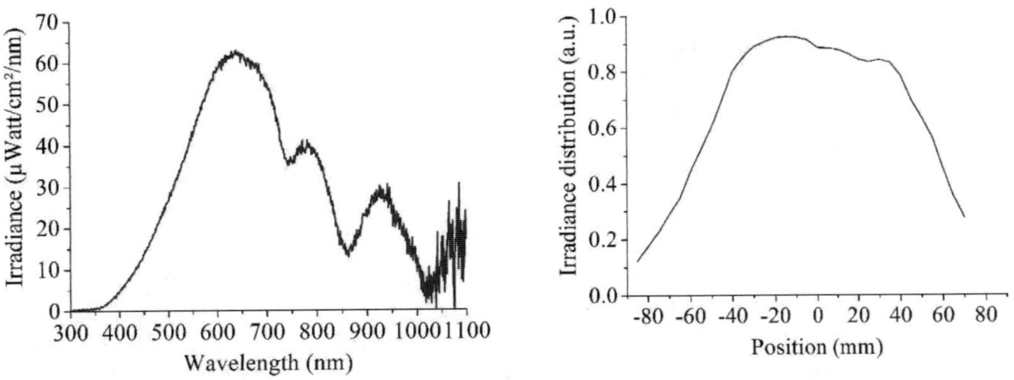

Figure 2. Left: Spectrum irradiance of the light source. Right: Spatial distribution of the irradiance on the sample position.

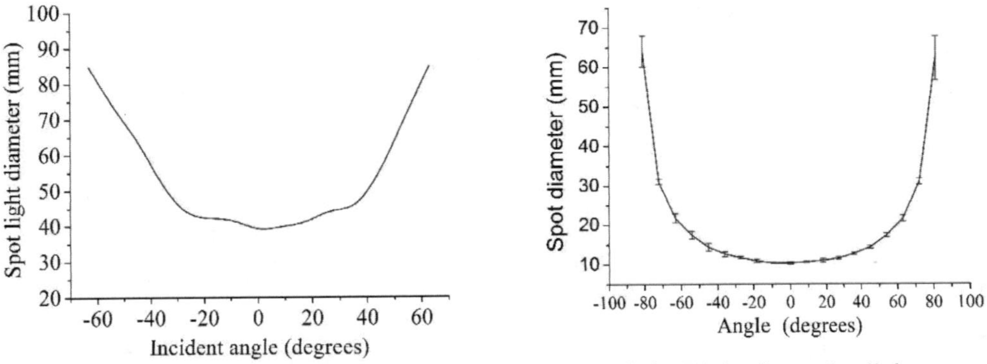

Figure 3. Spot diameter as a function of angle. Left: Incident light. Right: Scattering light.

3. Results

In order to verify the system, some surfaces usually employed in the laboratory were measured as shown in the following figures. In all the figures, the light beam is incident at the right side of the graph box with an angle of incidence measured from the vertical through the point (0,0). All these measurements were performed at the wavelength of 550nm. The scattered light was collected in the zenith plane from -80° to 80° and in the azimuth over 340° in steps of 10°. The face of the samples were positioned in the XY plane at position (0,0) and the relative scattered signal is indicated in the vertical Z axis. Figure 4 (left) shows the BRDF measured from our standard reflectance at 15° of incidence. Since this surface has a lambertian profile, this measurement was used to determine the incident irradiance for each angular position. In the figure there are some lost points due to the shadow of the detector arm at the positions where it passes in front of the light arm. Figure 4(right) shows the scattering distribution for a PTFE sample at the same incidence with step in the azimuth of 30°. PTFE is also used as a reference of reflectance surface, but in this case this sample presents a polished finishing surface that explain the increased signal at the specular reflection position. In the laboratory the PTFE is used as a diffuser for the entrance optics of radiometers.

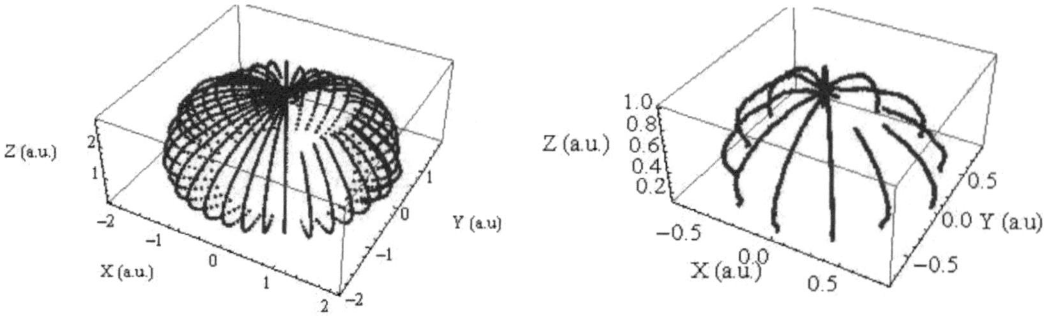

Figure 4. BRDF for standard reference (left) and PTFE (right) at 15° of incidence.

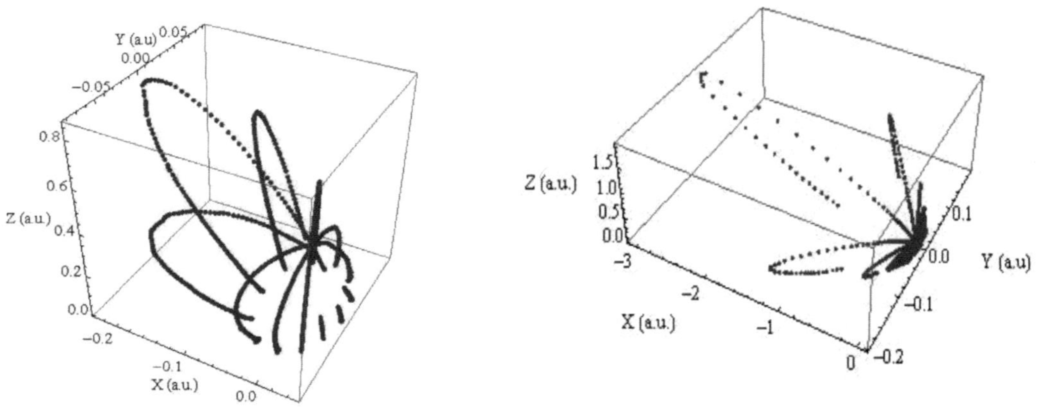

Figure 5. Left: BRDF for graphite sample at 15° of incidence and azimuth step of 30°. Right: graphite sample at 60° of incidence and azimuth step of 10°.

Figure 5 shows the scattering profile of a graphite sample that in some situations is used to control the stray light of optical systems [6]. For incidence at low angle, the sample shows a diffuse scattering profile which can complicate the control of stray light. For the same sample at 60°, the profile shows a lower diffuse signal and a high signal at the specular position. Figure 6 shows the profiles for black anodized aluminum at the same previous angles, which shows that it is possible to have a better control of the stray light at low angles with this sample. Figure 7 shows the BRDF for porous silicon samples at the same incidence angles. For these measurements, the zenith angle was limited from -60° to 60° due the small size of the samples and the azimuth step of 10°. These first measurements of this kind of sample, show profiles that could be explained by the complex structure of the surface of the porous silicon that will be the focus of the next study. Figure 8 shows BRDF measurements for graphite and black anodized aluminum samples at several angles of incident light for wavelengths of 550 nm and 900 nm. These measurements were performed at the same plane of incident light beam. The results show that the two samples present higher reflectance for near infrared wavelength and the reflectance signal increases with the incident angle. Comparing these results with the previous ones, it is possible to conclude that the black anodized aluminum sample is better for stray light mitigation with reflectance three times lesser than the graphite sample.

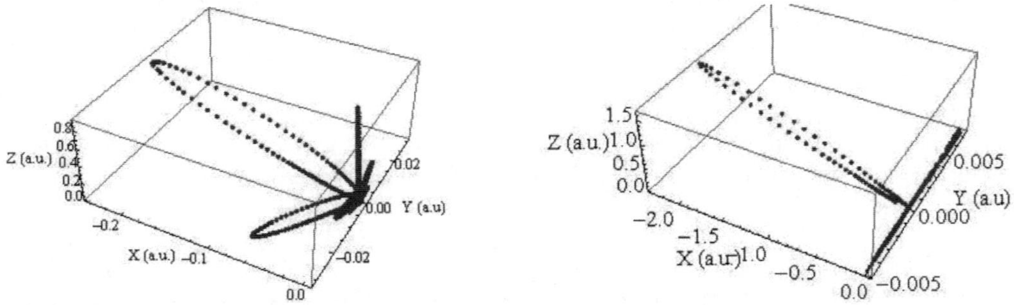

Figure 6. BRDF for black anodize aluminum sample at 15° of incidence (left) and 60° (right).

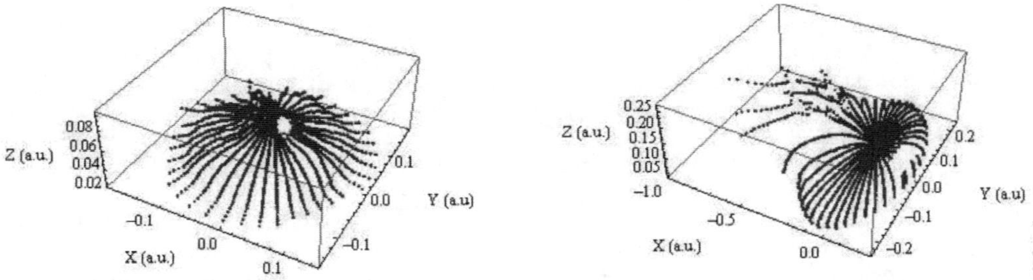

Figure 7. BRDF for porous silicon sample at 15° of incidence (left) and 60° (right).

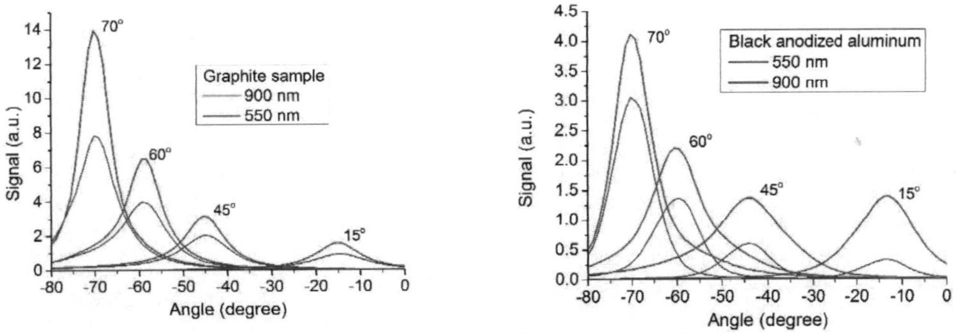

Figure 8. BRDF for several incident angles at 900nm and 550nm for graphite (left) and black anodized aluminum (right).

4. Discussion and conclusions

A system for measuring the BRDF from samples surface was implemented in the laboratory. The profiles of measured surfaces showed a clear dependence on surface finishing and with angle of incidence of radiation, which will be the focus of the next study as well as the dependence with the wavelength. Regarding to the porous silicon, we will attempt to verify the correlation between the profiles of the BRDF and the different porous structures of the material. To improve the irradiance on the sample surface, the light source will be changed by a QTH lamp powered by a radiometric power supply with a liquid light guide to drive the beam to the illumination arm. The spectral detection of scattered radiation will be changed by a compact CCD spectrometer to measure the wavelength radiation between 250 and 1100 nm at the same time.

Acknowledgments

The authors Mario S. Ribeiro and Tiago F. Paes are grateful to CNPq and CAPES, respectively, for the fellowships. The authors thank CAPES for the financial support to attend the ICO 23 congress.

References

[1] Nicodemus F E. Directional reflectance and emissivity of an opaque surface 1965 *Applied Optics* **4** No. 7.

[2] Martonchik J V, Bruegge C J and Strahler A H. A Review of reflectance nomenclature used in remote sensing 2000 *Remote Sensing Reviews* **19** pp. 9 – 20.

[3] Blinn J F. Models of light reflection for computer synthesized pictures 1977 *Proceedings of the 4th Annual Conference on Computer Graphics an Interactive Techniques* pp192-198.

[4] Oren M and Nayar S K. Generalization of Lambert`s reflectance model 1995 *International Journal on Computer Vision* **14** 3 pp227-251.

[5] Cook R L and Torrance K E. A reflectance model for computer graphics 1981 *Computer Graphics* **15** 3 pp307-316.

[6] Berni L A , Albuquerque B F C. Stray light analysis for the thomson scattering diagnostic of the ETE tokamak 2010 *Review of Scientific Instruments* 81, 123504.

[7] Berni L A, Vilela W A and Beloto A F. Otimização da ótica de entrada por traçado de raios no desenvolvimento de um radiômetro UV 2011 *Revista Brasileira de Energia Solar* **2**, série 1 ISSN 2178-9606.

[8] Ramizy A, Aziz W J, Hassan Z, Omar K and Ibrahim K. Improved performance of solar cell based on porous silicon surfaces 2011 *Optik* 122 2075 - 2077.

23rd Congress of the International Commission for Optics (ICO 23) IOP Publishing
Journal of Physics: Conference Series **605** (2015) 012004 doi:10.1088/1742-6596/605/1/012004

Experimental setup proposed to measure the quality of a spherical surface using deflectometry and a Hartmann screen

A Muñoz-Potosi [1], R Díaz-Uribe[2], M Campos-García[2], F Granados-Agustín[1]

[1] Coordinación de Óptica, Instituto Nacional de Astrofísica, Óptica y Electrónica, A.P. 51-216, Santa María Tonantzintla, C.P. 72000, Puebla, México

[2] Centro de Ciencias Aplicadas y Desarrollo Tecnológico, Universidad Nacional Autónoma de México, Circuíto Exterior S/N, Ciudad Universitaria, A.P. 70-186,Delegación Coyoacán, C.P. 04510 México D.F., México

E-mail: amunozpotosi@inaoep.mx

Abstract. Currently there are a variety of techniques to determine the quality of optical surfaces, which must provide quantitative information of the deformation or the shape of the surface under test. This paper proposes to use the deflectometry technique using a Hartmann screen to test a spherical surface of the diameter and radius of curvature known. For this, the proposed experimental setup is presented, highlighting the importance of alignment and the measures distances, location between the screen and image capture system, measured from the center of curvature of the analyzed surface. Subsequently, by using these distances, the positions of the reflected rays and the positions of location of points on the screen, you can find the equation of the line associated with these two points and given this equation, we can deduce the value of the normal surface. With this result of normals and using the general equation of deflectometry, we will obtain the shape of the surface under test.

1. Introduction

There are several methods for evaluating and testing the performance of optical elements that provide quantitative information about the quality of surfaces under test. Among these techniques, we can cite different methods, such as interferometry [1], deflectometry [2], Hartmann [3] and Ronchi [4] tests.

The deflectometry is a technique based on reflection that suffering light rays to impinge on a surface whose topography is desired to know. It consists on impinging a known ray on the surface to be characterized, measuring the ray reflected by it, we are able to determine the normal to the surface at each point [5, 6, 7, 8]. As the objective is to know the form of the entire surface, we have to obtain information from many points on it; for this, is necessary to scan all the surface. The fundamental equation in cartesian coordinates is given by:

$$\hat{n}.\vec{ds} = n_x dx + n_y dy + n_z dz = 0, \tag{1}$$

where $\hat{n}$ is the normal vector to the surface at the point of incidence of the light ray and $\vec{ds}$ is a tangent vector determined by the sweep that will be done on the surface defined as:

$$\hat{n} = n_x\hat{i} + n_y\hat{j} + n_z\hat{k} \tag{2}$$

and

$$\vec{ds} = dx\hat{i} + dy\hat{j} + dz\hat{k}. \tag{3}$$

In order to solve the equation 1, it is necessary to know the normal vector $\hat{n}$ at each sampling point because the vector $\vec{ds}$ is known from the path followed by the sampled rays.

2. Determination of the normal vector to the surface

To find the vector $\hat{n}$ we will use the law of reflection in vectorial form [9], where $\vec{S_1}$ is a vector along the direction of the incident ray, $\vec{S_2}$ is a vector along the reflected ray and $\hat{n}$ is a unit vector along the normal to the reflecting surface and I is the angle between the incident ray and the normal vector of the reflector surface at the point of incidence.

$$\vec{S_2} = \vec{S_1} + 2cosI|\vec{S_1}|\hat{n} \tag{4}$$

In this paper, we perform out the analysis for a concave spherical surface illuminated by a point source, where the incident ray, the reflected ray and the normal vector are parallel to each other, this means $cosI = 1$ and the equation 4 becomes:

$$\hat{n} = \frac{\vec{S_2} - \vec{S_1}}{2|\vec{S_1}|}, \tag{5}$$

Because the point source has an infinite number of rays and that the deflectometry measures the deviation that suffering one only ray, we propose to use a device to select some rays from the source, using a screen with holes that, in the literature, is known as Hartmann screen [10, 11, 12]. However, in this proposal the screen position is not in the exit pupil of the system, but can be located arbitrarily. In this case, it is located at a distance F between the surface and its center of curvature. With the device, the incident rays and its deviation are determined by using a CCD sensor located at a distance G from the center of curvature of the surface under study. The measurement of ray deflection is performed point by point until the entire surface is measured. With the largest number of holes in the Hartmann screen, the surface could be better sampled.

Because of the assumption that the normal vector is in the same direction of the incident and reflected rays, the deviation in the directions of these rays observed in the CCD sensor, will be due to the deformations of the surface of interest. The diagram of the proposed technique, where the dimensions of the spherical surface are exaggerated, is presented in Figure 1.

To perform the analysis, we define two vectors, $\vec{P}$ and $\vec{Q}$. Using the equation of the line, we obtained a vector $\vec{u}$, so that is parallel to the line L that connects the points P, Q and is in the same direction of the normal vector of the surface, as shown in equation 6.

$$\hat{n} = \frac{\vec{u}}{|\,\vec{u}\,|} = \frac{(x_0' - x_0'', y_0' - y_0'', G + F)}{\sqrt{(x_0' - x_0'')^2 + (y_0' - y_0'')^2 + (G + F)^2}}, \tag{6}$$

where (x_0'', y_0'') are the positions of the holes in the Hartmann screen and (x_0', y_0') are the positions of the reflected rays measured in the CCD.

The equation 6 is an approximation for calculating the normal vector for a spherical surface; however, this equation only holds in case the spherical surface is perfect. When the surface under test has local deformations, the incident and reflected rays at that point do not follow the same path, thus the normal vector changes. The theoretical analysis of this case is still in process.

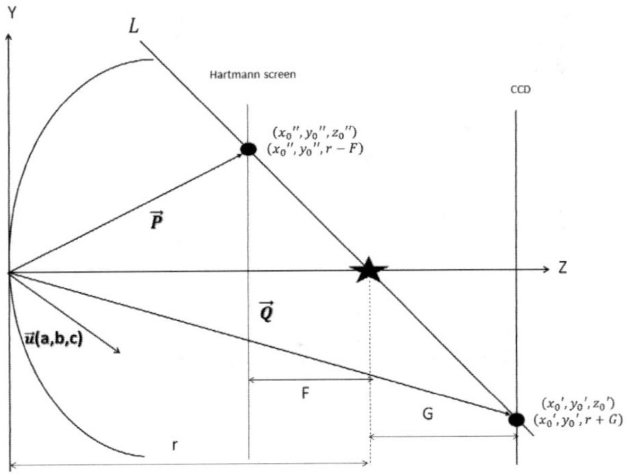

Figure 1. Scheme to determine the vector $\vec{u}$, parallel to the line L

3. Determination of vector $\vec{ds}$

To determine the vector $\vec{ds}$ is necessary taken into account the positions of the points on the Hartmann screen, which are known due to the manufacturing parameters. Subsequently the position of the rays passing over the screen after impinged on the surface being tested is determined. Figure 2, shows the geometric relationships taken into account in determining the vectors $\vec{r_0}$ and $\vec{r_1}$, considering only two points on the screen.

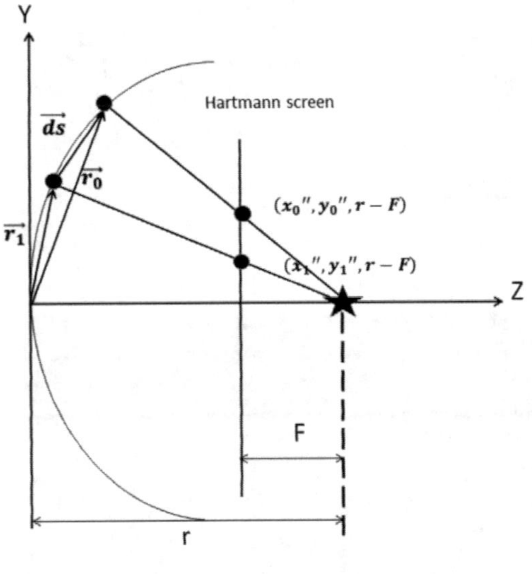

Figure 2. Vector position on the Hartmann screen

Once the position vectors $\vec{r_0}$ and $\vec{r_1}$ are known, the vector associated with the path is:

$$\vec{ds} = \vec{r_0} - \vec{r_1} = (r\sin\theta_0 - r\sin\theta_1, r\cos\theta_1 - r\cos\theta_0), \tag{7}$$

where r is the radius of curvature of the spherical surface and θ_0 and θ_1, are the angles between the z axis and the vectors $\vec{r_0}$ and $\vec{r_1}$, respectively. Considering equation 1, the coordinate z of the surface is given by:

$$z = -\int \frac{n_x}{n_z} dx - \int \frac{n_y}{n_z} dy \tag{8}$$

where dx and dy are determined by the positions where the sampling is performed.

Because sampling is proposed to the surface is done continuously, the equation 8 can be approximated numerically:

$$z = -\sum_{i=1}^{N} \frac{n_{xi}}{n_{zi}} \triangle xi - \sum_{i=1}^{N} \frac{n_{yi}}{n_{zi}} \triangle yi, \tag{9}$$

where N depends on the number of holes in the Hartmann screen, n_x, n_y, n_z are the components of the normal vector to the surface at each point and $\triangle x, \triangle y$ are determined with the number of holes in the screen.

4. Experimental setup proposed

The experimental setup proposed to test a spherical surface of diameter $215mm$ and center of curvature $767.6mm$, is presented in Figure 3. For this experimental setup, is necessary that the distance measured from the center of curvature of the spherical surface to the location of the Hartmann screen and the CCD sensor should be measured with major accuracy. In this case a Sony Alpha 37 camera was used.

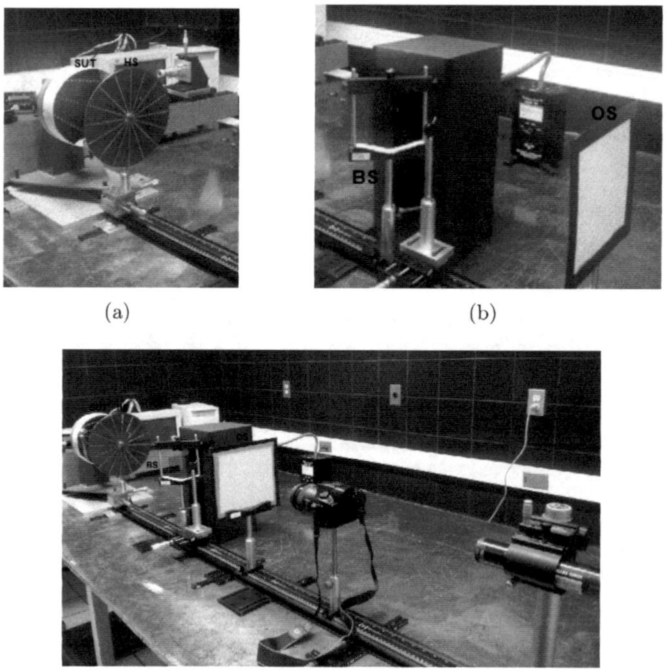

(a) (b) (c)

Figure 3. a) spherical surface under test (SUT) and Hartmann screen (HS), b) beamsplitter (BS) and observation screen (OS) and c) experimental setup complete

We use a He-Ne laser in order to ensure the best alignment of the experimental setup. This laser serves as a guide to determine the optical axis and the correct centering of all the elements of the setup, including the surface under study; this, considering the laser as incident at the center of the surface, the incident and reflected rays must follow the same path. In addition, a pair of rails are placed in order to have a guide for the location of the Hartmann screen, and the CCD. For the accurate location of the Hartmann screen, we use a micro-displacement transverse to the optical axis. This lets us adjust the optical axis of the system with the mechanical axis of the screen.

Then, using a laser distance measuring tool, brand Leica model DISTOD8, were performed the measures of the distances the location of the screen and the CCD sensor. These distances are: $F = 581.9mm$ for the Hartmann screen and $G = 272.2mm$ for the CCD sensor, both distances measured from the curvature center of the surface but in contrary directions.

Once the measurement of both distances F and G are completed, a thresholding and binarization of the image captured by the CCD sensor [13], showed in Figure 4, is necessary in order to determine the position of the centroids for each bright circle. These centroids will provide information about the vector normal to the surface at each sampling point. Therefore, is necessary an analysis of its accuracies because, when the quality of the surface under study is determined, the accuracy of the approximation considered in order to solve the equation of deflectometry could be obtained.

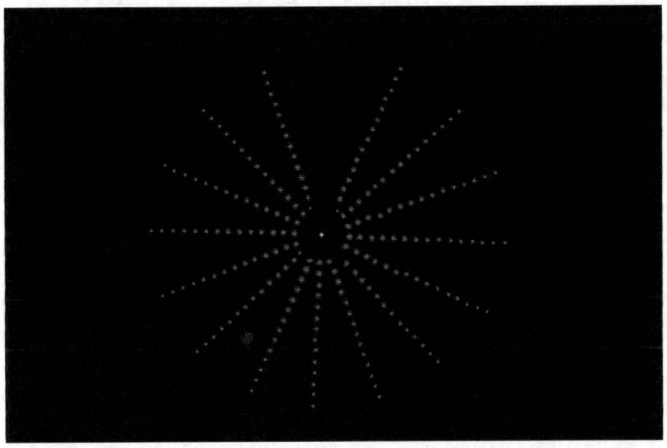

Figure 4. Centroid position for the image captured with the CCD

The evaluation process of the surface shape, considering a calculation of the normals and the vector tangent to the trajectory sampling is still in process.

5. Conclusions

The deflectometry as a method of measurement is proposed.

To determine the shape of the surface, is necessary to know the normal vector of it, in this work, an experimental setup for determining the normal vector from the distances F and G is proposed. The theoretical analysis of a surface with local deformations is still in process.

A method to determine the vector $\vec{ds}$, with which it can find the z coordinate of the surface under test is proposed.

An optimized experimental setup for alignment and image capture is proposed.

To calculate the normal to the surface under study and then determine the quality of spherical surface under test, it is necessary to analyze the measurement accuracy of the distances of the experimental setup. This, later, will be useful in order to know the quality of approximation by using the equation of the deflectometry.

Acknowledgments
Authors wishing to acknowledge the support by the CONACYT under grant 250427, assistance by Taller de Óptica INAOE and technical staff.

References

[1] Izazaga-Pérez R, Aguirre-Aguirre D, Granados-Agustín, F and Percino-Zacarías M 2014 *Optical Fabrication and Testing* **OM3C-6**
[2] Diaz-Uribe R and Granados-Agustin F 1999 *Optometry & Vision Science* **76** 40-49
[3] Díaz-Uribe R, Granados-Agustín F and Cornejo-Rodríguez A 2009 *Optics express* **17** 13959-13973
[4] Aguirre-Aguirre D, Izazaga-Pérez R, Percino-Zacarías M, Granados-Agustín F and Cornejo-Rodríguez A 2013 *Optical Review* **20** 271-276
[5] Granados Agustin F S 1993 *Evaluación de la topograía corneal por deflectometría láser* Universidad Nacional Autónoma de México
[6] Campos García M 2005 *Pruebas de superficies asféricas rápidas con pantallas nulas* Universidad Nacional Autónoma de México
[7] Rosete Aguilar M and Díaz-Uribe R 1993 *Applied Optics* **32** 4690-4697
[8] Canabal H, Alonso J and Bernabeu E 2001 *Optical Engineering* **40** 2517-2523
[9] Malacara D 2004 *Óptica Básica*(Fondo de Cultura Económica)
[10] Mejía Barbosa Y 2012 *Ciencia Tecnología Salud Visual Ocular* **10** 149-165 `http://revistas.lasalle.edu.co/index.php/sv/article/view/104/60`
[11] Malacara D 2007 *Optical Shop Testing* Ed 3 (New Jersey/Wiley)chapter 10 pp 361-397
[12] Díaz Uribe R,Granados Agustín F and Cornejo Rodríguez A 2009 *Optics Express* **17** 13959-13973
[13] Gonzalez R. C., Woods R. E. 2008 *Digital Image Processing*, Ed 3 (Prentice Hall)

Journal of Physics: Conference Series **605** (2015) 012005 doi:10.1088/1742-6596/605/1/012005

Analysis of a passive heat sink for temperature stabilization of high-power LED bulbs

Eduardo Balvís[1], Ricardo Bendaña[2], Humberto Michinel[21] Pedro Fernández de Córdoba[3] and Angel Paredes[2]

[1]Innebo Ingeniería, Lugar do Penedo, S. Cibrao das Viñas, Ourense, 32910 Spain
[2]Facultade de Ciencias, Universidade de Vigo, As Lagoas s/n, Ourense, 32004 Spain
[3]IUMPA, Universitat Politècnica de València, Camino de Vera 14, Valencia, 46022 Spain

E-mail: `hmichinel@uvigo.es`

Abstract. In this paper we present a numerical analysis and experimental measurements of the temperature stabilization of high-power LED chips that we have obtained by employing an aluminum passive heat sink, designed to be used in a compact light bulb configuration. We demonstrate that our system keeps the temperature of the LED chip well-below $70°C$ yielding long-term operation of the device. Our simulations have been performed for a low-cost device ready to install in public streetlights. The experimental measurements performed in different configurations show a nice agreement with the numerical calculations.

1. Introduction

The use of high-power light emitting diodes (HP-LED) for public illumination is an emerging subject, triggered by recent developments of different technologies including semiconductor materials[1, 2, 3], fluorescence techniques[4], driver electronics[5] or thermal control[6] among others[7, 8].

One of the key aspects concerning the performance and durability of HP-LED lighting systems is the adequate control of the temperature of the LED chip[9]. As it has been pointed by recent studies[10], LEDs have a high energy efficiency and long lifespan, however, a large amount of heat is dissipated during operation due to Joule effect; thus, cooling HP-LEDs is an important challenge in package designs, where a correct evacuation of the heat will substantially enlarge the lifetime of the device[11].

Besides the previous constraint, other practical aspects like a compact configuration, low cost, mass production or even esthetic considerations can play an important role in market-oriented products. Thus, in this paper we present a numerical analysis of the thermal stabilization of $30W - 50W$ LED chips attached to passive heat sinks, yielding a compact light bulb design that can be used for commercial purposes in the street lighting market. The system we propose keeps the temperature of the LED chip well-below 70^oC under realistic conditions, yielding long-term operation of the bulb, with the corresponding savings in energy consumption and maintenance.

After the numerical calculations performed, in order to compare the results of the computational simulations with experimental measurements taken in standard systems, a set of prototypes has been constructed in a compact and ready-to-install configuration. As we will

Journal of Physics: Conference Series **605** (2015) 012005 doi:10.1088/1742-6596/605/1/012005

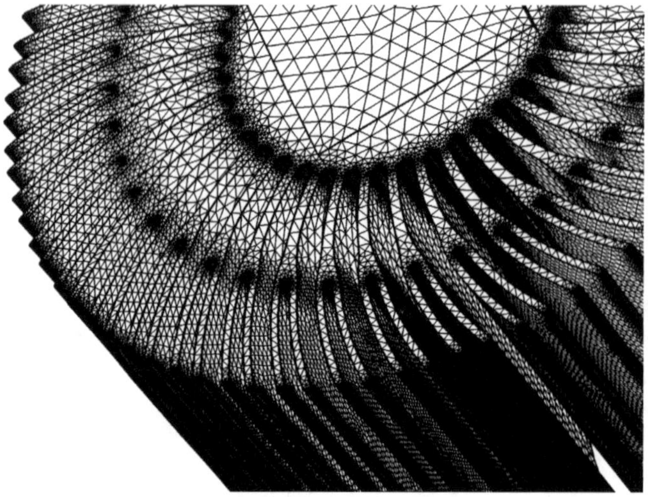

Figure 1. Detail of the computational grid used for the numerical simulations. The size of the real devices was $5cm$ and $10cm$ height and $9cm$ diameter. Details of the calculations are given in the text.

demonstrate, there is a nice agreement between the numerical simulations and the corresponding data obtained.

2. Numerical model

Our first aim is to calculate the steady-state temperature distribution over the surface of a heat sink with translational symmetry along one axis. This configuration is ideal for mass production at a very low cost via metal extrusion process. In our theoretical model, we have assumed that the heat sink is made of black anodized aluminum (Al 16061), which is surrounded by a laminar air flow of density (ρ) given by the ideal gas law. Thus, in the air side, the first expression that we formulate is the continuity equation:

$$\nabla \cdot (\rho \vec{v}) = 0, \tag{1}$$

being $\vec{v}$ the velocity of air. In addition we have the energy equation[12]:

$$\rho \frac{\partial \vec{v}}{\partial t} = -\nabla P + \mu \nabla^2 \vec{v} - \rho \vec{g}, \tag{2}$$

where t is the time, P the pressure and μ the dynamic viscosity of air. We assume that the acceleration of gravity $\vec{g}$, is parallel to the z-axis. Another formula to be added to the model is the moment equation:

$$\rho C_P \frac{\partial T}{\partial t} = \nabla \cdot (k \nabla T) + \frac{\partial P}{\partial t}, \tag{3}$$

being C_P the specific heat, T the absolute temperature and k the thermal conductivity. On the heat sink we have the condition $\nabla^2 T = 0$. At the interface, for a black anodized surface with high emissivity (≥ 0.8), we can neglect the effect of the incoming radiation heat flux (W/m^2) and thus the outgoing flux ($\dot{q}$) is given by the Stefan-Boltzmann law:

$$\dot{q} = \epsilon \sigma T^4, \tag{4}$$

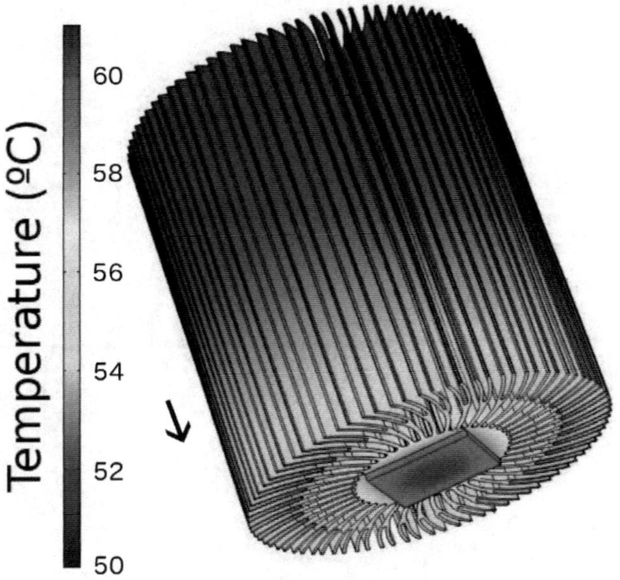

Figure 2. Numerical simulation of steady-state temperature distribution over the surface of a passive heat sink in vertical configuration corresponding to the geometry of Fig. 1. The color scale ranges from $50°C$ (blue) to $61°C$ (red). The arrow indicates the direction of the gravitational force, which in this case is parallel to the axis of symmetry of the heat sink. The system modeled is black anodized Al and the size used for the calculations was $10cm$ height and $9cm$ diameter. The LED power in this simulation is $50W$. Other details of the simulation are given in the text.

where ϵ is the emissivity of the aluminum wall, σ is the Stefan-Boltzmann constant and T is the absolute temperature of the heat sink.

We want to solve the previous problem for a heat sink made of black-anodized aluminum (Al6061) with the geometry shown in Fig. 1. To numerically integrate the previous set of equations, we have used COMSOL Multiphysics®, which is is a finite element analysis solver commercial package for various physics and engineering applications, especially coupled phenomena. In addition to conventional physics-based user interfaces, this software also allows for entering coupled systems of partial differential equations (PDEs). In particular, we have used the Heat Transfer Module which provides user interfaces for heat transfer by conduction, convection and radiation.

We have modeled the LED chip as a $1mm-$thick aluminum square plate which provides a constant heat flux at the base of the heat sink. For the chips under consideration the amount of waste heat can be estimated as 70% of the LED power[13]. The dependence on the grid density was investigated by changing the number of points. The final selection is shown in Fig. 1. The simulation corresponds to a heat sink $10cm$ height with a diameter of $9cm$. The diameter of the solid internal core is $4cm$.

The results of the numerical calculations are shown in Fig. 2 for a vertical configuration (i.e.: the symmetry axis of sink parallel to the direction of the acceleration of gravity $\vec{g}$) and simulating the effect of a $50W$ LED chip placed at the bottom of the heat sink. As it can be appreciated in the picture, the maximum of the temperature distribution is obviously located at the LED

Journal of Physics: Conference Series **605** (2015) 012005 doi:10.1088/1742-6596/605/1/012005

Figure 3. Front (left) and lateral (right) view of the real system composed of a black anodized aluminum structure in a compact LED bulb configuration. The height of the sink is $5cm$ and its diameter is $9cm$ the corresponding simulations and experiments for horizontal and vertical orientations with a 25W LED chip are shown in Fig.5 and Fig. 6, respectively.

chip and the values of T gradually diminish with the distance from the chip, showing a radially symmetric distribution around the axis of the cylinder. For an ambient temperature of $21°C$, the maximum of the resulting steady-state distribution calculated is $60.8°C$, well below the critical damage temperature provided by the LED manufacturer, providing thus a maximal lifetime of the device. As we will show below, this result is un good agreement with the experimental measurements performed in a real system. Of course, for higher ambient temperatures our predictions are still valid and in this case the gap with respect to the damage threshold is reduced accordingly[11] .

3. Experimental Setup

In order to check the validity of our numerical model, a series of experiments were made. Aluminum (Al6061) black anodized heat sinks with the same geometry as in Fig 1 and two different lengths ($5.0cm$ and $10.0cm$ were attached to a $25W$ and $50W$ LED chips respectively. In Fig. 3 we show a photo of one of the prototypes that we constructed to perform the experimental measurements.

In all the cases, to minimize the thermal contact resistance between the LED chip and the heat sink a graphite film of high thermal conductivity ($240Wm^{-1}K^{-1}$) was used. Finally, the light bulbs were mounted in several orientations in order to reproduce different operation conditions. The emissivity of the heat sink with the black-anodized surface treatment is 0.8. The geometric parameters of the experimental model are the same as in the numerical simulation described above and below. A temperature sensor with three different probe heads (DAQ-9172, NI9211), a power supply, a wattmeter, and a laptop were used in order to collect data.

The overall pattern for the air flow could be described as follows: the cooling air enters from the outer region of the heat sink and is heated while passing through the fins. The heated air rises upward in the inner regions of the heat sink due to the fact that the density of the air in

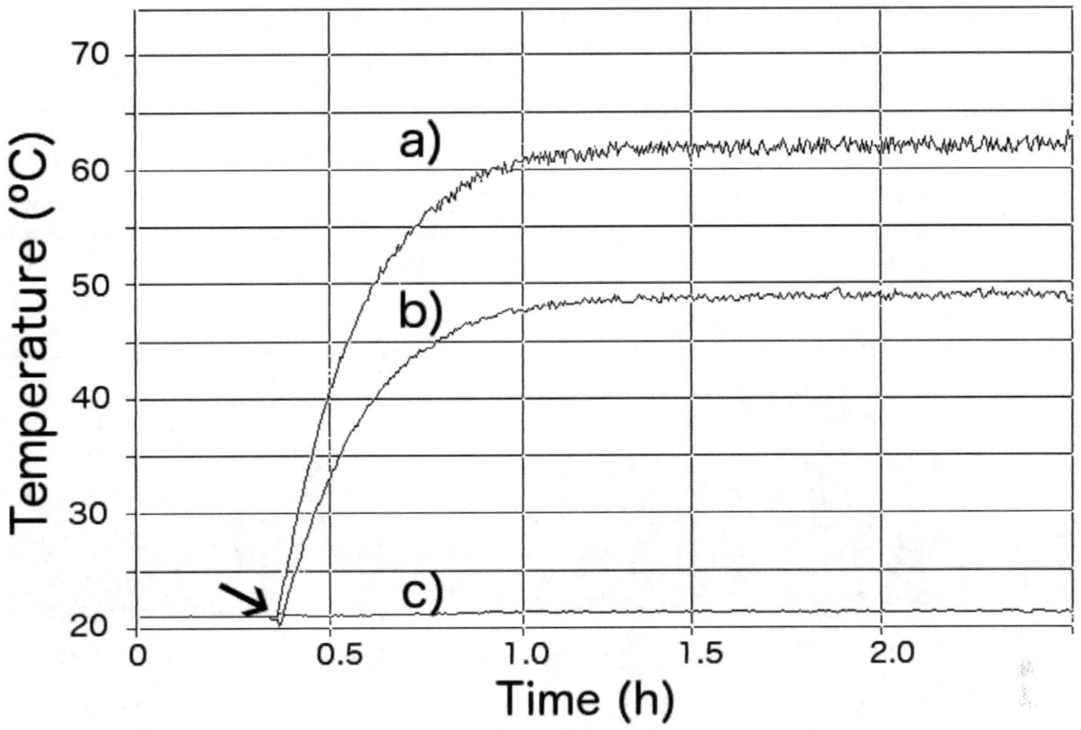

Figure 4. Experimental measurement of the temperature distribution in the vertical configuration (the axis of symmetry of the sink is parallel to the gravity force) corresponding to the numerical simulation of Fig.2. Curves a), b) and c) correspond respectively to the LED chip, the center of the top side of the heat sink and the ambient. Details of the experiment are given in the text. The arrow points to the instant when the power is switched on.

this zones became less than that of the surrounding air. In addition, a thermal boundary layer develops discontinuously, after some delay. Thus, a relatively high local heat transfer coefficient is expected in the inner regions of the heat sink.

The pin-fin heat sink will show uniform cooling performance in the case of natural convection considered. Repeated leading-edge effects will appear in the outer regions of the heat sink because the fins are arranged to keep the flow at a certain distance in the radial direction.

4. Results and discussion

In Fig. 4, we plot the experimental measurement of temperature values measured for a $50-W$ LED chip in vertical configuration (the axis of symmetry of the sink is parallel to the gravity force) corresponding to the numerical simulation of Fig.2 with the chip placed at the bottom of the heat-sink. Curves a), b) and c) correspond respectively to the temperatures measured at the LED chip, the center of the top side of the heat sink and the ambient.

As it can be appreciated in the curves, once the power is switched on (indicated by an arrow) the temperature increases until saturation after less than one hour. The maximum values measured in curves a) and b) are respectively $63°C$ and $49°C$ which are in very good agreement with the numerical simulation of Fig. 2.

To determine the optimum configuration and the precision of the numerical model, we have simulated and measured different orientations and lengths of the heat sink. The results of the numerical calculations can be seen in fig. 5 for a 25W HPLED and $5cm$ height heat sink. The results show the simulation of the steady-state temperature for vertical (a) and horizontal (b) configuration, meaning these names that the acceleration of gravity ($\vec{g}$) is parallel or perpendicular to the symmetry axis of the heat sink, respectively.

The comparison of the numerical simulations of Fig. 5 with experimental values is shown in Fig. 6. In this figure we plot the temperature measured at the chip (a) and at the center of the opposite side of the heat sink (b). Line c) shows the ambient temperature during the experiment.

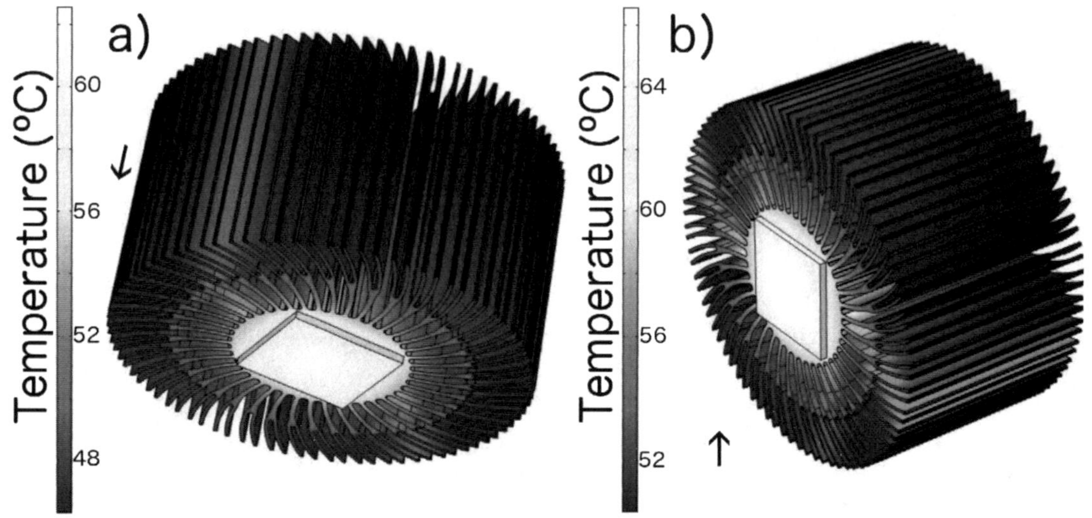

Figure 5. Numerical simulation of steady-state temperature distribution over the passive heat sink in horizontal (left) and vertical (right) configuration. The arrows indicate the direction of the gravitational force in each case. The color scale ranges are plotted in the left side of each simulation. The system modeled is black anodized Al as in Fig. 2. In this case the size used for the calculations was $5cm$ height and $9cm$ of diameter. The LED power is $25W$. Other details of the simulation are given in the text.

As it can be appreciated in the left non-shaded zone of Fig. 6, after less than one hour of operation of the LED, the system reaches a thermal steady state with a maximum temperaure in the chip of $63°C$, in excellent agreement with the simulations of Fig. 5-a. Once the temperature is stabilized, we rotate the system 90° which corresponds to a perpendicular orientation of the axis of symmetry of the heat sink with respect to the force of gravity (see Fig. 5-a). This situation is indicated by a dark-grey shading in the graph. It is obvious that this configuration is not the optimal one, yielding an increase of about $8°C$ in the chip temperature.

Then, we rotate the sink another 90° and thus, the axis of symmetry of the sink is aligned with the gravity force as in 5-b, however the chip is placed now at the top of the chip instead of the bottom. As it can be appreciated the temperature slightly decreases showing the important contribution of air convention along the pin-fins.

Finally, we restore the initial position and the temperature asymptotically recovers the steady-state value of the first zone of the graph. All the measurements made are in very good agreement with the numerical simulations shown in Fig. 5 for two different orientations.

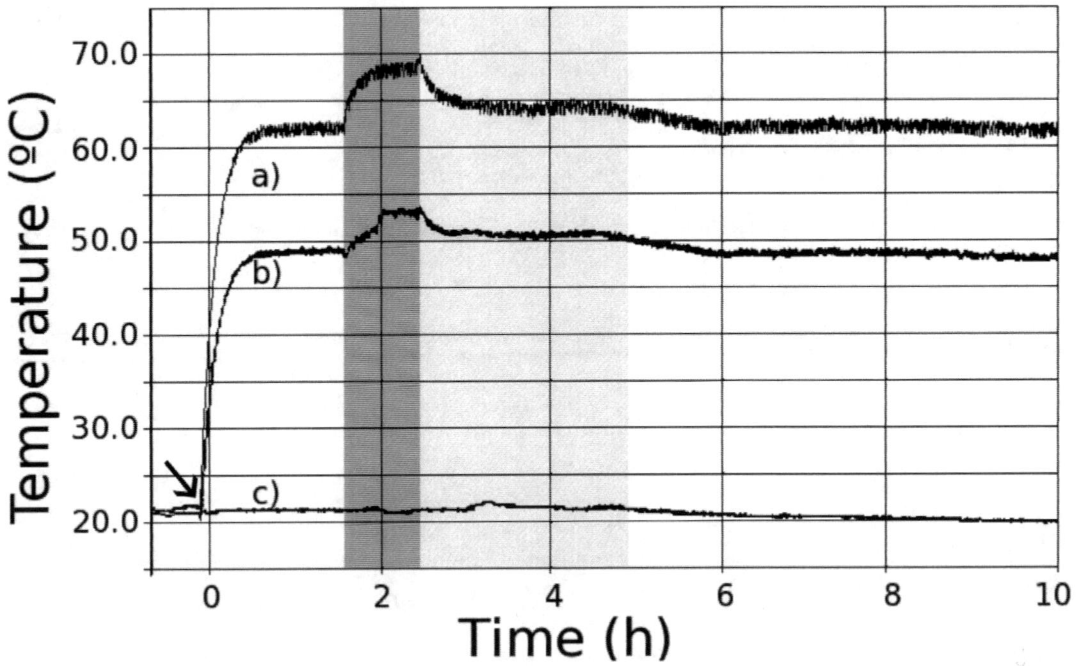

Figure 6. Experimental measurement of the temperature distribution measured at: a) the LED chip, b) the center of the back side of the heat sink, and c) the ambient. The non-shaded parts of the graph correspond to measurements made for parallel orientation of the symmetry axis of the heat sink with respect to the force of gravity and with the LED chip placed placed at the bottom of the heat sink corresponding to Fig. 5-a. The dark-gray zone displays the values obtained for perpendicular orientation (Fig. 5-b) and the light-gray region corresponds to the same orientation Fig. 5-a but with the LED chip place at the top of the sink. The rest of the parameters of the experiment are given in the text.

5. Conclusions

We have presented a numerical study of the steady-state temperature distribution of a realistic high-power light emitting diode (HP-LED) bulb. Our results have been compared with experimental data measured in a prototype fabricated under market considerations. A nice agreement has been found between the computer simulations and the measurements performed.

Therefore, from our analysis we can derive that aluminum low-cost passive heat sinks can be used to keep the temperature of $30W - 50W$ HP-LED chips below $70°C$, thus making it possible to reach life-times of $55.000h$ with the corresponding savings in energy consumption and maintenance.

Acknowledgments

HM thanks funding from projects K133131H64102 and K044131H64502 Xunta de Galicia. PF-C thanks Universitat Politècnica de València funding under project SP20120909.

References

[1] Mueller-Mach R, Mueller G, Krames M and Trottier T 2002 *Selected Topics in Quantum Electronics, IEEE Journal of* **8** 339 –345 ISSN 1077-260X

[2] Narukawa Y, Narita J, Sakamoto T, Deguchi K, Yamada T and Mukai T 2006 *Japanese Journal of Applied Physics* **45** L1084–L1086
[3] Krost A and Dadgar A 2002 *physica status solidi (a)* **194** 361–375 ISSN 1521-396X
[4] Luo H, Kim J K, Schubert E F, Cho J, Sone C and Park Y 2005 *Applied Physics Letters* **86** 243505 (pages 3)
[5] Chiu H J, Lo Y K, Chen J T, Cheng S J, Lin C Y and Mou S C 2010 *Industrial Electronics, IEEE Transactions on* **57** 735 –743 ISSN 0278-0046
[6] Efremov A, Bochkareva N, Gorbunov R, Lavrinovich D, Rebane Y, Tarkhin D and Shreter Y 2006 *Semiconductors* **40**(5) 605–610 ISSN 1063-7826
[7] Chen H Y, Lam W Y, Luo J D, Ho Y L, Tang B Z, Zhu D B, Wong M and Kwok H S 2002 *Applied Physics Letters* **81** 574–576
[8] Gardner N F, Kim J C, Wierer J J, Shen Y C and Krames M R 2005 *Applied Physics Letters* **86** 111101 (pages 3)
[9] Yu S H, Lee K S and Yook S J 2010 *International Journal of Heat and Mass Transfer* **53** 2935 – 2938 ISSN 0017-9310
[10] Narendran N and Gu Y 2005 *Display Technology, Journal of* **1** 167 – 171 ISSN 1551-319X
[11] Tan C M and Singh P 2014 *Device and Materials Reliability, IEEE Transactions on* **14** 742 – 750
[12] Jang D, Yu S H and Lee K S 2012 *International Journal of Heat and Mass Transfer* **55** 515 – 521 ISSN 0017-9310
[13] Arik M and Setlur A 2010 *International Journal of Energy Research* **34** 1195–1204 ISSN 1099-114X

Journal of Physics: Conference Series **605** (2015) 012006 doi:10.1088/1742-6596/605/1/012006

Measuring color differences in gonioapparent materials used in the automotive industry

M Melgosa[1], L Gómez-Robledo[1], G Cui[2], C Li[3], E Perales[4], F M Martínez-Verdú[4] and T Dauser[5]

[1] Department of Optics, University of Granada, 18071 Granada, Spain
[2] University of Wenzhou, Wenzhou 325035, China
[3] School of Electronic and Information Engineering, University of Science and Technology Liaoning, Anshan 114051, China
[4] Color and Vision Group, Department of Optics, Pharmacology and Anatomy, University of Alicante, 03690 Sant Vicent del Raspeig, Alicante, Spain
[5] AUDI AG, I/PG-C41, 85045 Ingolstadt, Germany

E-mail (corresponding author): mmelgosa@ugr.es

Abstract. This paper illustrates how to design a visual experiment to measure color differences in gonioapparent materials and how to assess the merits of different advanced color-difference formulas trying to predict the results of such experiment. Successful color-difference formulas are necessary for industrial quality control and artificial color-vision applications. A color-difference formula must be accurate under a wide variety of experimental conditions including the use of challenging materials like, for example, gonioapparent samples. Improving the experimental design in a previous paper [Melgosa et al., Optics Express 22, 3458-3467 (2014)], we have tested 11 advanced color-difference formulas from visual assessments performed by a panel of 11 observers with normal color vision using a set of 56 nearly achromatic color pairs of automotive gonioapparent samples. Best predictions of our experimental results were found for the AUDI2000 color-difference formula, followed by color-difference formulas based on the color appearance model CIECAM02. Parameters in the original weighting function for lightness in the AUDI2000 formula were optimized obtaining small improvements. However, a power function from results provided by the AUDI2000 formula considerably improved results, producing values close to the inter-observer variability in our visual experiment. Additional research is required to obtain a modified AUDI2000 color-difference formula significantly better than the current one.

1. Introduction

A color-difference formula just provides a number ΔE from instrumental color specifications of two color stimuli having a visually-perceived color difference ΔV [1]. Note that while ΔV is the subjective answer of the human visual system (unknown in many aspects), ΔE is an objective instrumental measurement. Industrial color-quality control requires successful color-difference formulas providing ΔE values which accurately predict subjective visual assessments of the color differences perceived by average observers under a wide variety of viewing conditions. For example, in the automotive industry, the different parts of cars are produced by different manufactures using different materials, and reliable color-difference formulas are necessary for automatic pass/fail color decisions. In the past years, a number of color-difference formulas have been proposed for industrial applications [2-10].

Currently, it is recommended to use the CIEDE2000 color-difference formula [7], which has been jointly proposed as a standard by the International Commission on Illumination (CIE) and the International Organization for Standardization (ISO) [11]. This recommendation mainly promotes uniformity of practice, enabling easier communication between manufacturers/users.

Gonioapparent materials or materials with flop effects represent a big challenge for color-difference evaluation in current automotive industry. The next definitions from the American Society for Testing and Materials (ASTM) [12] are considered relevant for this kind of materials: "Appearance" is "the aspect of visual experience by which things are recognized"; "goniochromatism" is the "change in any of all attributes of color of a specimen on change in angular illuminating-viewing conditions but without change in light source or observer"; "flop" is "a difference in appearance of a material viewed over two widely different aspecular angles"; and "aspecular angle" is the "viewing angle measured from the specular direction, in the illuminator plane unless otherwise specified".

In the middle of the 90's AUDI developed a tolerance formula for the approval of effect paint batches when colors with strong flop effects were important in the automotive sector [13]. Latter this formula was modified to predict color tolerances for solids (homogeneous) as well as for effect (goniochromatic) colors, leading to the AUDI2000 color-difference formula [14], currently employed by different manufacturers in the automotive industry. It is important to note that nowadays AUDI2000 is the only available color-difference formula considering flop effects. In previous works we have tested the performance of the AUDI2000 and other advanced color-difference formulas for different visual datasets with homogeneous colors [15, 16], as well as for a reduced dataset of 28 color pairs involving some gonioapparent samples [17]. The results found in this last experiment indicated that the AUDI2000 formula performs very well, but its weighting function for lightness (i.e. the dependence of perceived lightness-differences with flop) may be not optimal. Current paper reports new experimental results using color pairs with different controlled amounts of lightness flop. In a first step, our current results may be used to start the development of a modified AUDI2000 color-difference formula with improved performance with respect to the original one.

2. Materials and methods

We have performed a visual experiment where a panel of 11 experienced observers with normal color vision assessed a set of 56 color pairs with different amounts of lightness flop. These color pairs were selected from a set of 476 samples produced by AkzoNobel for the automotive industry, considering the next specific criteria:

1) The samples in all our color pairs were nearly achromatic ($C^*_{ab}<10$) and their total color differences were predominantly lightness differences. From our spectroradiometric measurements, the average color differences of our 56 color pairs was 3.1 CIELAB units, and the average percentage of lightness difference in the total CIELAB color difference was 79.8%.

2) Our 56 color pairs were spread around 10 color centers, in such a way that the two samples in all color pairs came from the same color center. Therefore, the two samples in all our color pairs had very similar textures (i.e. coarseness and sparkle), as well as very similar magnitudes of lightness flop. We measured the lightness flop for each color sample as $|\Delta L^*_{\gamma i+1} - \Delta L^*_{\gamma i}| / (\gamma i+1 - \gamma i)$, where $\gamma i = 15°$, 25°, 45°, 75° and $\gamma i+1$ is the next standard aspecular angle to the angle γi (e.g. $\gamma i+1 = 25°$ when $\gamma i = 15°$, or $\gamma i+1 = 110°$ when $\gamma i = 75°$). The lightness flop for each color pair was defined as the average of the lightness flops of the two samples in the color pair. The color pairs in our experiment appropriately covered a relatively wide range of lightness flop. Thus, from measurements performed with a BYK-mac multi-angle spectrophotometer (aperture 23 mm; illuminant D65; CIE 1964 standard colorimetric observer) the flop in our color pairs ranged from 0 to 3.8 units, with average and standard deviations values of 1.7 and 0.8 CIELAB units, respectively.

These selection criteria of our color pairs were adopted to achieve one of the main goals in our current paper: To test the performance of the weighting function for lightness in the AUDI2000 color-difference formula. In fact, in the design of our earlier experiment described in [17], the selected color pairs were nearly random and a gradual change in lightness flop in the assessed color pairs was not considered, in such a way that the results found in such experiment are not very appropriate to propose a new weighting function for lightness in the AUDI2000 color-difference formula. Other relevant differences between our earlier experiment in [17] and the current one are as follows: 1) Now, we have used color pairs with the highest available size (10x10 cm) against the previous size of 3.5x3.7 cm used in [17]. In this way the lightness flop is more evident to observers in the current experiment; 2) To have a reasonable number of visual measurements, allowing the whole experiment could be performed in a period of about 3 months, now we have not considered all 6 standard viewing geometries for each color pair as made in [17], which forced us to assess only a reduced number of 28 color pairs. Now, we assessed 56 color pairs selecting only specific viewing angles, in particular those close to the specular direction, because they are particularly important in gonioapparent materials. Specifically, the number of color pairs assessed at the different angles were 26 (15°), 26 (25°), 3 (45°), and 1 (75°).

Each one of the 56 selected color pairs was visually assessed using a byko-spectra effect light booth (BYK Additives and Instruments) placed in a dark room, as recommended by international recommendations and usually made in the automotive industry. The interested reader may find general information on how works this light booth in [18]. The distance between color pairs and observer's position in the byko-spectra effect light booth is 50 cm, approximately. After an adaptation time of around 3 min, each color pair was presented to each observer in a random order, and the task of the observer was to judge the magnitude of the perceived color difference in the color pair using the Gray Scale for Change in Colour of the Society of Dyers and Colourists [19], which was placed just above the color pairs. Each observer performed 3 non-sequential assessments of each color pair, in such a way that a total of 56 (pairs) x 11 (observers) x 3 (replications) = 1848 assessments were recorded. The visual sessions never lasted more than 20 min to avoid observers' fatigue. All our instrumental color measurements for next analyses were performed using a Konica Minolta CS2000 spectroradiometer placed on a tripod at the same position than the head of the observer, in such a way that the spectroradiometer just measured what the observer sees. The reference white for transformations to CIELAB and computations of color coordinates in other color spaces was a calibrated PTFE plaque placed at the positions of the samples and measured also with our spectroradiometer. Before computation of color differences, the tristimulus values of the two samples in each color pair were transformed using the chromatic adaptation transform CAT02 employed in CIECAM02 [20], considering as reference white the average of the reference whites placed in the positions of the two samples. This transformation was made to take into account slight differences (in some cases negligible) in the measured color of the reference white placed on the left and right samples of the test color pairs or on the 9 color pairs in the Gray Scale of the Society of Dyers and Colourists. Three independent spectroradiometric measurements were performed for the two samples in each of the 56 color pairs, and, on the average, the standard deviation of such measurements was 0.58 CIELAB units.

A detailed description of the AUDI2000 color-difference formula can be found in [17]. Here we will only indicate that AUDI2000 is a CIELAB-based color-difference formula (see Eq. (1)) with the structure of CIE94 [6]. The weighting function for lightness in the AUDI2000 color-difference formula is named S_{L,γ_i}, and it is given in Eq. (2):

$$\Delta E_{AUDI2000,\gamma_i} = \left[\left(\frac{\Delta L^*_{\gamma_i}}{S_{L,\gamma_i}} \right)^2 + \left(\frac{\Delta C^*_{ab,\gamma_i}}{S_{C,\gamma_i}} \right)^2 + \left(\frac{\Delta H^*_{ab,\gamma_i}}{S_{H,\gamma_i}} \right)^2 \right]^{1/2} \quad (1)$$

$$S_{L,\gamma_i} = a_1\left(\frac{\left|L^*_{\gamma_i+1} - L^*_{\gamma_i}\right|}{\gamma_{i+1} - \gamma_i}\right)^{a_2} + a_3 C^*_{ab,45^o} + a_4 = 1.0\left(\frac{\left|L^*_{\gamma_i+1} - L^*_{\gamma_i}\right|}{\gamma_{i+1} - \gamma_i}\right)^{2/3} + 0.002C^*_{ab,45^o} + 0.33 \quad (2)$$

The merit of a color-difference formula may be assessed by different indices measuring the strength of the relationship between ΔE_i and ΔV_i values (i.e. computed and perceived color differences for the i=1,…, 56 color pairs). In the past few years the *STRESS* index [21] became increasingly popular to assess the merit of color-difference formulas. *STRESS* values are in the range 0-100, low values indicating better performance of a color- difference formula. *STRESS* can be also used to measure intra- and inter-observer variability [22], as well as to know whether two different formulas are or not statistically significantly different with respect to the visual data [20]. Here we have used the *STRESS* index and a weighted *STRESS* index, named *WSTRESS* (see Eq. (3)). While the *STRESS* index gives the same weight to all color pairs (i.e. w_i =1 ; i=1,…, 56), the *WSTRESS* index assigns different weights to each color pair, those with higher accuracy (lower inter-observer variability) having a higher weight. Following Berns [23], here we have adopted as weight a reciprocal variance, defined as the square of the average visual difference of all observers divided by 4 times their standard deviation (both measured in CIELAB units), as shown in Eq. (3):

$$WSTRESS = \left(\frac{\sum_{i=1}^{56} w_i\left(\Delta E_i - F_1\Delta V_i\right)^2}{\sum_{i=1}^{56} w_i F_1^2 \Delta V_i^2}\right)^{1/2} \; ; \; F_1 = \frac{\sum_{i=1}^{56}\Delta E_i^2}{\sum_{i=1}^{56}\Delta E_i \Delta V_i} \; ; \; w_i = \left(\frac{Average\ \Delta V_i}{4\ St.\ Dev.\ \Delta V_i}\right)^2 \; (i = 1,\dots 56) \quad (3)$$

3. Results and discussion

The raw visual color-difference values reported by our observers were transformed into true visual differences, according to the procedure described in [17]. After this transformation, the *STRESS* values corresponding to the intra- and inter-observer variability in our experiment were 16.2 and 19.9 units, respectively. These values are lower than the 25.2 and 23.2 units found in our earlier experiment in [17] where some of the observers participating in the experiment were not experienced. However, in the current experiment most observers had considerable previous experience in visual color-difference experiments. Even so, a first analysis of the visual results in the current experiment indicated that the answers from some observers in some color pairs appeared as outliers. Accordingly, it was decided to remove in all further analyses the visual answers below Q1-1.5*(Q3-Q1) or above Q3+1.5(Q3-Q1), where Q1 and Q3 are the first and third quartiles in the population of all visual answers, respectively. This procedure resulted in 73 removed visual answers (i.e. about 4% of total answers) and a new lower inter-observer variability of 18.2 units. Therefore, any color-difference formula providing *STRESS* results below this value of 18.2 units may be considered satisfactory, because it can be considered as equivalent to an average individual observer participating in our experiment. After removing outliers, the standard deviation of the 3 visual assessments performed in each color pair was on the average 0.78 CIELAB units.

Table 1 shows *STRESS* and *WSTRESS* values for each one of the color-difference formulas tested in this paper. The parametric factors in all tested color-difference formulas were assumed as 1. For the CAM02-SCD and CAM02-UCS color-difference formulas, we adopted average surround, a neutral background with L^*=24.2 and illuminance on the plane of the samples equal to 1800 lx, according to the viewing conditions in our experiment. As shown in Table 1, the best predictions (lowest *STRESS* and *WSTRESS* values) of our visual experimental results were found for the AUDI2000 color-difference formula, followed by the CAM02-SCD, CAM02-UCS, CIEDE2000 and DIN99d color-difference formulas. In fact, from values in Table 1 it can be concluded that, at a confidence level of 95%, any of these four color-difference formulas is significantly different to the AUDI2000 formula.

Journal of Physics: Conference Series **605** (2015) 012006 doi:10.1088/1742-6596/605/1/012006

This result agrees with the one reported in our earlier experiment in [17], where AUDI2000, CAM02-SCD and CAM02-UCS were the best color-difference formulas without statistically significant differences among them. Anyway, it must be noted that the *STRESS* value of 27.3 units achieved by the AUDI2000 color-difference formula in our current experiment is considerably higher than the inter-observer variability (18.2 units), which means that the AUDI2000 formula must be improved to perform like a real average visual observer. Table 1 also shows that *WSTRESS* is always lower than *STRESS*, which means that the average similitude between perceived (ΔV) and computed (ΔE) color differences increases when a lower weight is assigned to the color pairs with higher uncertainty in observers' visual assessments.

Table 1. Performance of 11 advanced color-difference formulas against the visual results in our experiment, measured by the *STRESS* and *WSTRESS* indices. Lower values of these indices (always in the range 0-100) indicate better performance.

Color-Difference Formula	*STRESS*	*WSTRESS*
CIELUV [2]	33.4	29.7
CIELAB [3]	35.0	32.6
CMC [4]	40.8	33.8
BFD [5]	43.4	34.2
CIE94 [6]	36.2	34.2
CIEDE2000 [7, 11]	32.0	28.1
DIN99d [8]	31.5	28.4
AUDI2000 [14]	27.3	23.9
CAM02-SCD [9]	28.6	24.2
CAM02-UCS [9]	29.3	24.8
OSA-GP-Euclidean [10]	40.6	31.2

We have optimized independently the four coefficients (a_i; i=1,…, 4) in the weighting function for lightness of the AUDI2000 color-difference formula (Eq. (2)). Table 2 shows the original values of each one of these coefficients (second column) and the optimized ones, both, from results in our previous experiment in [17] (third column) and from results in current experiment (fourth column). Last column in Table 2 shows the *STRESS* values from the optimized coefficients in the current experiment. As we can see, all optimized coefficients except a_1 are enough different to the ones in the original AUDI2000 color-difference formula. However, taking into account that the original AUDI2000 color-difference formula achieved a *STRESS* value of 27.3 for the results in the current experiment (see Table 1), we can see that, unfortunately, in the best case the optimized coefficients in our experiment only produced a small decrease in *STRESS* values around 2 units (see Table 2, last column), which means a very small improvement of the original AUDI2000 formula. Likely, a simultaneous optimization of the four coefficients involved in Eq. (2) may produce a higher *STRESS* decrease. By comparing the third and fourth columns in Table 2, it can be also noted that the values of the coefficients minimizing *STRESS* in our earlier experiment in [17] were enough similar to those found in the current experiment, with the exception of the a_4 coefficient. In fact, the optimized value a_4=0.7 in the current experiment is much more convenient than the previous 0.01 value, because for highly achromatic homogeneous colors this last value may lead to values of the weighting function for lightness close to zero (see Eq. (2)), resulting in very high unacceptable/anomalous values of AUDI2000 color differences.

Table 2. Original and (two sets of) optimized coefficients (a_i; i=1,...4) in the AUDI2000 color-difference formula (Eq. (2)). The *STRESS* value for the original AUDI2000 formula was 27.3. Columns 3 and 4 in the last row show exponents α improving the original AUDI2000 formula (in the case of the two values marked with asterisks this improvement was only 0.02 *STRESS* units).

Coefficients	Original Values [14]	Optimized Values in [17]	Current Optimized Values	*STRESS* for Current Optimized Values
a_1	1.0	1.1	1.1	27.3*
a_2	0.67	0.5	0.3	25.2
a_3	0.002	0.01	0.01	27.3*
a_4	0.33	0.01	0.7	26.0
α	1.0	0.82	0.65	18.7

Finally, from last row in Table 2 it is worth to mention that an exponent α=0.65 in the original AUDI2000 color-difference formula produced a very important decrease of about 8.6 units in the *STRESS* value found in our current experiment. The use of this kind of power functions ($\Delta E' = \Delta E^{\alpha}$) in color-difference formulas was originally proposed by Attridge and Pointer [24], and it has been proved also useful in other previous experiments [25], including the one in [17] where the optimum value for the exponent α was 0.82. Interestingly, the value of 18.7 *STRESS* units found with α=0.65 (Table 2, last column) is very close value to the 18.2 *STRESS* units corresponding to the inter-observer variability in our current visual experiment.

4. Conclusions and future work

This paper illustrates how to design a visual experiment to measure color differences in gonioapparent materials, and how to assess the relative merits of different color-difference formulas trying to predict the visual results in such experiment. More specifically, we tested the weighting function for lightness in the AUDI2000 color-difference formula, complementing results found in another previous experiment [17]. Our current results confirm that AUDI2000 is a good color-difference formula while its weighting function for lightness seems to be not optimal. Beside some advances, we must conclude that the proposal of a modified weighting function for lightness in the AUDI2000 formula requires further experimental results with additional color pairs and corresponding analyses. These works are currently underway in our laboratories. In addition, an optimization of the whole AUDI2000 color-difference formula will require new visual assessments using color pairs with chroma and hue flops, because current color pairs and results are mainly related to lightness flop. The option of a power function as a second-step transform to improve the results achieved by the original or optimized AUDI2000 formulas seems very promising. The possibility of developing a total-difference formula involving coarseness and sparkle in addition to color, as previously suggested by other authors [26], must be also considered in future improvements of current AUDI2000 color-difference formula.

Acknowledgments

Authors are grateful to all volunteer observers at the Department of Optics of the University of Granada (Spain) participating in the visual experiment. Samples used in the current experiment were produced and kindly provided to us by AkzoNobel Automotive & Aerospace Coatings, http://www.akzonobel.com/aac/aboutus/who_we_are/ (Sassenheim, the Netherlands). The loan of the byko-spectra effect light booth used in our experiment by BYK Additives and Instruments, https://www.byk.com/en/instruments/products/color-measurement.html is also strongly acknowledged. This research was supported by the Ministry of Economy and Competitiveness of Spain, research projects FIS2013-40661-P and DPI2011-30090-C02, with European Research Development Fund (ERDF), as well as by the National Science Foundation of China (grant number 61178053).

References

[1] Melgosa M, Trémeau A and Cui G 2013 Colour difference evaluation *Advanced Color Image Processing and Analysis* ed E. Fernandez-Maloigne (Springer) chapter 3 pp 59-81

[2] Commission Internationale de l'Éclairage 2009 *CIE Colorimetry – Part 5: CIE 1976 L*u*v* Colour Space and u',v' Uniform Chromaticity Diagram* CIE S 014-5 E:2009
International Organization for Standardization 2009 *CIE Colorimetry – Part 5: CIE 1976 L*u*v* Colour Space and u',v' Uniform Chromaticity Diagram* ISO 11664-5:2009(E)

[3] Commission Internationale de l'Éclairage 2007 *CIE Colorimetry – Part 4: 1976 L*a*b* Colour Space* CIE S 014-4 E:2007
International Organization for Standardization 2008 *CIE Colorimetry – Part 4: 1976 L*a*b* Colour Space* ISO 11664-4:2008(E)

[4] Clarke F J J, McDonald R and Rigg B 1984 Modification to the JPC79 colour-difference formula *J. Soc. Dyers Colour.* **100** 128-32

[5] Luo M R and Rigg B 1987 BFD(l:c) colour-difference formula. Part I – Development of the formula *J. Soc. Dyers Colour.* **103** 86-94

[6] Commission Internationale de l'Éclairage 1995 *Industrial colour-difference evaluation* CIE Publication 116-1995

[7] Commission Internationale de l'Éclairage 2001 *Improvement to industrial colour-difference evaluation* CIE Publication 142-2001

[8] Cui G, Luo M R, Rigg B, Roesler G and Witt K 2002 Uniform colour spaces based on the DIN99 colour-difference formula *Color Res. Appl.* **27**, 282-90

[9] Luo M R, Cui G and Li C 2006 Uniform colour spaces based on CIECAM02 colour appearance model *Color Res. Appl.* **31**, 320-30

[10] Oleari C, Melgosa M and Huertas R 2009 Euclidean color-difference formula for small-medium color differences in log-compressed OSA-UCS space *J. Opt. Soc. Am. A* **26**, 121-34

[11] International Organization for Standardization - Commission Internationale de l'Éclairage 2014 *Colorimetry - Part 6: CIEDE2000 colour-difference formula* ISO/CIE 11664-6:2014(E)

[12] American Society for Testing and Materials 1995 *Standard terminology of appearance* ASTM E284 95a

[13] Deutsches Institut für Normung eV 2001-03 *Farbtoleranzen für Automobillackierungen – Teil 2: Effektlackierungen* DIN 6175-2

[14] Dauser T 2012 Color Management at AUDI (AUDI AG, I/PG-C41, 85045 Ingolstadt, Germany) http://www.detroitcc.org/files/Color%20Management%20at%20AUDI%20(DCC%20March%202012).pdf

[15] Melgosa M, Gómez-Robledo L, Martínez-García J, Perales E, Martínez-Verdú F M and Dauser T 2013 Testing a colour-difference formula for the automotive industry using the experimental visual datasets employed in CIEDE2000 development *Proc. CIE Centenary Conference* (Paris: CIE x038:2013 Publication) pp 465-69

[16] Martínez-García J, Melgosa M, Gómez-Robledo L, Li C, Huang M, Liu H, Cui G, Luo M R and Dauser T 2013 Testing the AUDI2000 colour-difference formula for solid colours using some visual datasets with usefulness to automotive industry *Proc. SPIE* vol 8785 (Porto, Portugal: 8th Iberoamerican Optics Meeting and 11th Latin American Meeting on Optics, Lasers and Applications) 87858P

[17] Melgosa M, Martínez-García J, Gómez-Robledo L, Perales E, Martínez-Verdú F M and Dauser T 2014 Measuring color differences in automotive samples with lightness flop: A test of the AUDI2000 color-difference formula *Opt. Express* **22**, 3458-67

[18] https://www.youtube.com/watch?v=7j6oBkDKnjE Experience the byko-spectra effect.

[19] International Organization for Standardization 1993 *Test for colour fastness – Part A02: Gray scale for assessing change in colour* ISO 105-A02:1993

[20] Commission Internationale de l'Éclairage 2004 *A colour appearance model for colour management systems: CIECAM02* CIE Publication 159-2004

Journal of Physics: Conference Series **605** (2015) 012006 doi:10.1088/1742-6596/605/1/012006

[21] García P A, Huertas R, Melgosa M and Cui G 2007 Measurement of the relationship between perceived and computed color differences *J. Opt. Soc. Am. A* **24**, 1823-29

[22] Melgosa M, García P A, Gómez-Robledo L, Shamey R, Hinks D, Cui G and Luo M R 2011 Notes on the application of the standardized residual sum of squares index for the assessment of intra- and inter-observer variability in color-difference experiments *J. Opt. Soc. Am. A* **28**, 949-53

[23] Berns R S and Hou B 2010 RIT-DuPont supra-threshold color-tolerance individual color-difference pair dataset *Color Res. Appl.* **35**, 274-83

[24] Attridge G G and Pointer M R 2000 Some aspects of the visual scaling of large colour differences – II *Color Res. Appl.* **25**, 116-22

[25] Huang M, Cui G, Melgosa M, Sánchez-Marañón M, Li C, Luo M R, Liu H 2015 Power functions improving the performance of color-difference formulas *Opt. Express* **23**, 597-610

[26] Huang Z, Xu H, Luo M R, Cui G, Feng H 2010 Assessing total differences for effective samples having variations in color, coarseness, and glint *Chin. Opt. Lett.* **8**, 717-20

Journal of Physics: Conference Series **605** (2015) 012007 doi:10.1088/1742-6596/605/1/012007

Capture of microscopic objects by contour optical traps formed by 4-channel liquid crystal modulator

A Korobtsov[1], S Kotova[1,2], N Losevsky[1], A Mayorova[1,2], S Samagin[1] and V Volostnikov[1,2]

[1]Lebedev Physical Institute of Russian Academy of Sciences
Novo-Sadovaya Str., Samara, 443011, Russia;
[2]Samara State Aerospace University (SSAU)
34, Moskovskoye Shosse, Samara, 443086, Russia.

E-mail: kotova@fian.smr.ru

Abstract. The efficiency of the contour optical traps generated by the 4-channel liquid crystal modulator for the manipulation with both transparent and absorbing particles has been experimentally studied. The maximum velocity of the substrate movement at which the captured particle is kept in the optical trap was used for the traps efficiency estimation. The obtained data show the ability of the effective capture of transparent objects with the ring traps even in the course of the objects boundary (periphery) being under the impact. The possibilities of a smoothly control of the size and the form of optical traps for increasing of the capture efficiency have been demonstrated.

1. Introduction

The object of this paper is an experimental study of the efficiency of the contour optical traps generated by a 4-channel liquid crystal (LC) modulator (LC focusing device). The LC focusing device is the spatial light modulator (SLM) characterized by the simplicity and compactness of the device and its control system, relatively low cost, sufficient energy efficiency and wide operating spectral range. The functionality of the 4-channel LC modulator is significantly lower as compared to the multi-pixel analogs. However, in problems of optical manipulation it can be effectively used for the generation of point traps, capable of capturing the transparent microobjects and their moving along a predetermined path within the focusing device aperture [1]. Besides, with the use of the focusing device the line segment traps can be formed applicable for the trapping and rotation of several particles or an elongated particle [1]. On taking into account a high practical interest to the capture and manipulation of absorbing microobjects including those of biological nature, we have considered the possibility of the formation and real-time control of the so-called contour traps [2] in the form of a ring, ellipse, and C-shaped.

Experiments on the optical manipulation by the traps in the form of light rings are described, for example, in works [3-6]. In [3] the behavior of silica microspheres moving under the influence of the periodic optical field is studied theoretically and experimentally. The axicon is used to form the Bessel beams with the intensity distributions in the form of concentric rings in the plane perpendicular to the direction of the beam propagation. The authors of [4] carried out experiments on the capture of absorbing and transparent microparticles with the use of hollow annular light beams. An array of such beams was formed by using a phase diffractive optical element (DOE), realized with the use of photolithography technique. The use of the stationary DOE is justified in researches where the dynamic changes of the beam are not required. And LC SLMs are the most promising in the studies where the control of light beams in real time is needed. The results of our experiments on the formation of optical traps of various types (including ring traps) with the use of the LC modulator

HOLOEYE HEO-1080P have been presented in [5]. In particular, the laser trapping of the 1.2 μm-diameter latex particles was fulfilled with the use of the ring-shaped optical trap that subsequently changed its shape from a ring to an ellipse. The experiments on the particles movement by the vortex optical trap in the shape of two concentric circulars have been also carried out.

By using the light modulator Holoeye LC - 2500R, the authors [6] have formed the Bessel beams of different orders and with different size of rings that allowed them to explore the behavior of transparent particles in such fields depending on the ratio of the particle radius and ring size, as well as the topological charge of the beam.

The 4-channel LC modulator application, on the one hand, allows to control the optical traps (their shape, dimensions) in real time, and on the other side – to create a compact and relatively inexpensive scheme of optical tweezers. The results of the implementation of this scheme and experiments on optical manipulation of both transparent and opaque objects with the use of contour traps are given in the article.

2. Formation of optical traps with a four-channel liquid crystal modulator

The LC focusing device is based on the integrated structure of crossed substrates of two cylindrical modal LC lenses. The nematic LC-layer is sandwiched between the two glass substrates covered with the transparent high-resistance coatings (the surface resistance ranges from 100 kOhm/square to several MOhm/square) and low-resistance non-transparent strip-shaped contacts. The substrates are oriented so that their contact electrodes are perpendicular to each other. The resulting device has a rectangular aperture. The LC-layer thickness is determined by the spacers and the initial planar orientation is ensured by the orienting coatings deposited onto the substrates. A detail description of the device principle of operation, its mathematical model and main operating regimes are given in [6-8]. In particular it has been shown that the frequency effect on the voltage distribution becomes negligibly small at the operating regime with a small modal parameter and as a result, the control of the voltage distribution is realized by means of the potentials amplitude and phase. In this case the equipotential lines of voltage distribution can only be of an elliptic and parabolic shape. The voltage distributions in the form of ellipses, circles and parallel straight lines generate the phase profiles of elliptic and circular truncated cones (due to the threshold nature of the voltage-phase dependence of LC-layer), as well as of cylindrical surface. While changing the amplitude and/or phase of the voltage fed to the contacts, it is possible to control both the location of the centers of circular and elliptic cones bases and the orientation of the elliptic cone axes and its eccentricity.

The optical transparency with the considered phase transmissions affects the plane uniform light wave in the following way. The light field with the central point maximum and concentric rings secondary maxima in the intensity distribution at the plane perpendicular to the light direction is formed in the far-field region when the phase delay has the shape of circle cone [7, 8]. For the case of the phase delay in the form of the cylindrical lens surface, the light in the focal plane is focused into the light segment with secondary maxima, and the central maximum intensity is substantially higher than secondary ones [9]. Thus it is possible to form point optical traps with the controlled position in the manipulation plane, as well as traps in the form of a segment with a given orientation [1]. In the field of the Fresnel diffraction at a short distance from the LC focusing device the following distribution of the light field is formed: the points of the maximum intensity are located on a curve similar to the shape of the lines of constant phase profile of the phase delay. Thus the contour optical traps in the form of rings and ellipses can be generated. The examples of possible optical traps formed with the use of the LC focusing device are presented in figure 1.

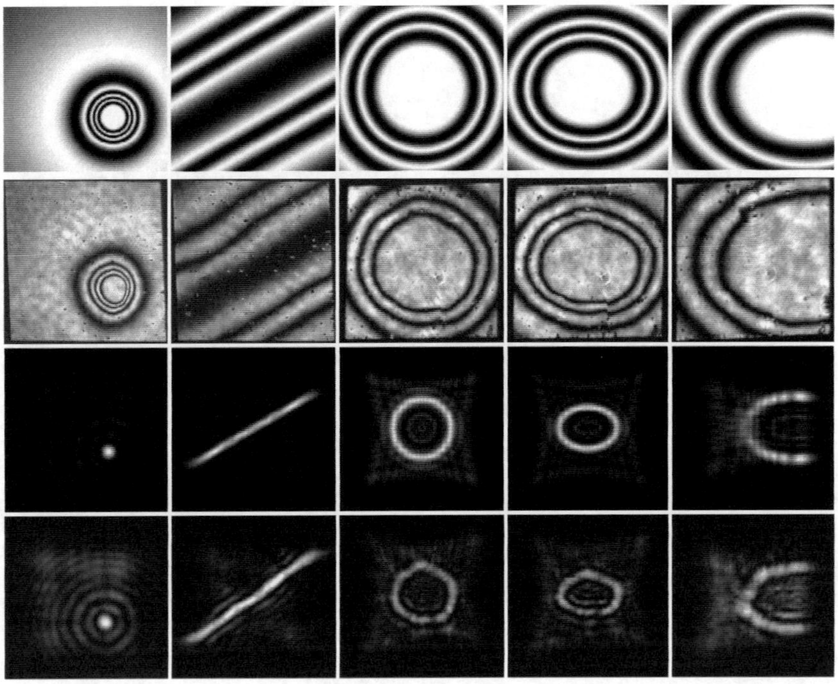

Figure 1. Different types of optical traps. The simulated (upper row) and experimentally obtained (second row) polarization interferograms; simulated (third row) and experimentally obtained (lower row) intensity distributions.

3. Experiments on optical capture

3.1. Experimental technique

In figure 2 the scheme of the laser manipulator with the integrated 4-channel LC modulator developed by the authors is shown. A solid-state diode-pumped laser with the emission wavelength of 0.53 μm and maximal output power of 500 mW was used to form the optical traps. The laser beam was directed through an interface collimator onto the LC focusing device. The LC focusing device and the lens form the required light field distribution in the specified plane. Then the beam was introduced into the 100x objective of an upgraded microscope XSP-104 and the result was a reduced intensity distribution reproduced in the manipulation plane. In order to visualize the working area of the experiment, an illuminator was placed under the cell with the microscopic objects. The imaging of the experimental procedure was made with the ocular digital camera DCIM-130 connected with the computer. The focusing device parameters were controlled via the graphical user interface that allowed specifying of the control voltage amplitude and phase in each of the channels.

The transparent microobjects (latex spheres of various diameters), nontransparent absorbing microobjects (particles consisting of a mixture of aluminum oxide and metal aluminum, silver particles, and their conglomerates) as well as microobjects of biological nature (e.coli and yeast *Saccharomy cescerevisiae*) were used for our experiments. The microparticles were suspended in water.

The efficiency of the traps was estimated by the maximum velocity of the substrate movement at which the captured particle is held in the optical trap. Microscopic objects were captured by the trap (either by the light ring boundary for transparent objects, or into a dark area for absorbing ones), then the microscopic stage was moved by a predetermined path with acceleration, and the velocity at which the particle was escaped from the trap was measured. It is evident that the particles separation from the trap occurs at the moment when the optical trap force is equal to the viscous force.

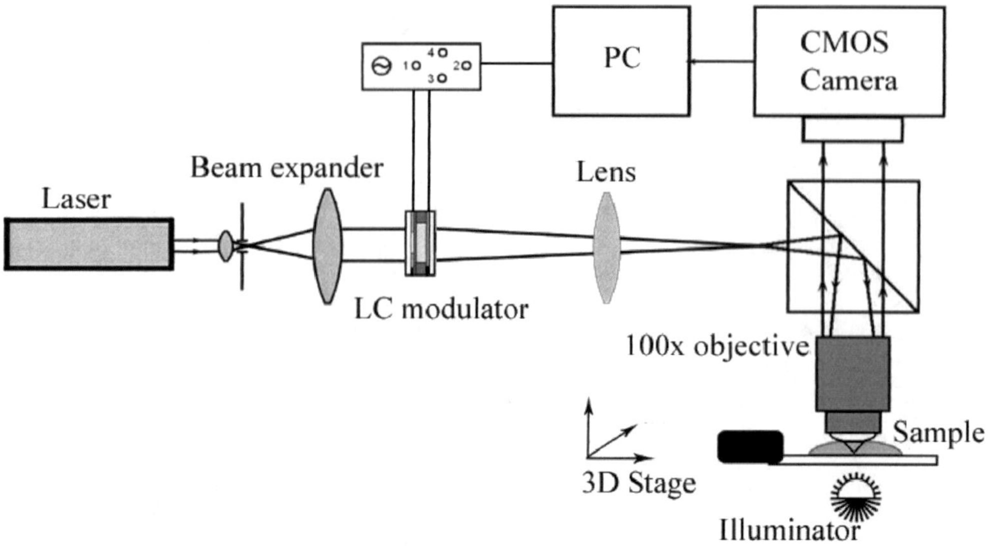

Figure 2. Scheme of the experimental set-up.

3.2. Capture and confinement of transparent objects

Transparent (non-absorbing) objects are trapped at the intensity maximum, i.e. they are captured by the light contour. The results of experiments on the capture and confinement of the latex microparticles with various diameters (d) by means of the ring trap with 5 μm diameter and 0.6±0.1 μm contour thickness are presented in table 1. The radiation power was 5 mW. The contact voltage was 6 V. The maximal velocity of microscopic stage relocation *(V_{max})* at which the captured particle of a specified size is held in the optical trap are presented. One can see that the velocity is maximal for the particle of 2 μm diameter, i.e. the formed optical trap is more efficient for the objects of this size.

Table 1. The experimental results of the study of the efficiency of the ring optical trap with 5 μm diameter and 0.6±0.1 μm contour thickness for the capture of latex spheres of various diameters (d).

d, μm	d/ contour thickness	V_{max}, μm/sec
0.9	1.5	3.5±0.4
1.2	2.0	8.5±0.9
2.0	3.3	25.0±2.5
2.9	4.8	13.9±1.4
5.0	8.3	11.2±1.1
6.1	10.2	1.9±0.2

The capture of transparent objects of the size substantially smaller than the diameter of the ring trap, is visually illustrated in figure 3. It is seen that the latex spheres of 2 μm-size and e.coli of the 0.7-0.9 μm-size are captured. The particles are arranged along the light contour (boundary) of the ring. The number of the particles captured by the optical trap is determined by their size and radiation power in the optical trap.

A qualitatively different picture for the capture is observed when the particle sizes are comparable to the size of the dark area of the ring (the diameter of the ring trap). In this case it is possible to capture the micro-object into the center of the trap. When the microscopic stage is moved with low speed the object is held in the trap center. The increase of the velocity results in the shift of the trapped object. The capture and confinement of the transparent objects of the size comparable with the diameter of the ring trap is illustrated in figure 4.

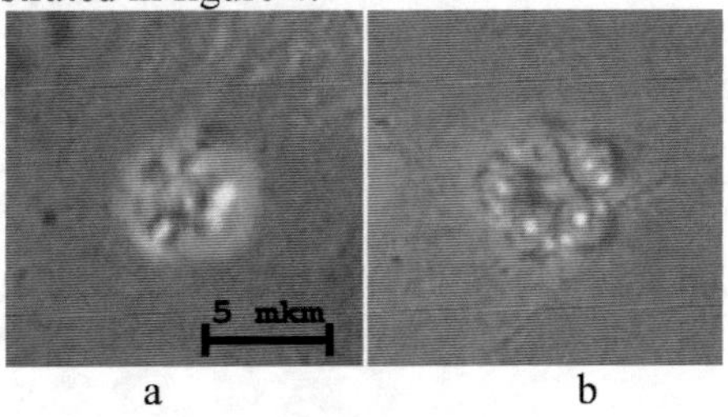

Figure 3. The capture of transparent objects of a size substantially smaller than the diameter of the ring trap a) e.coli; b) latex spheres.

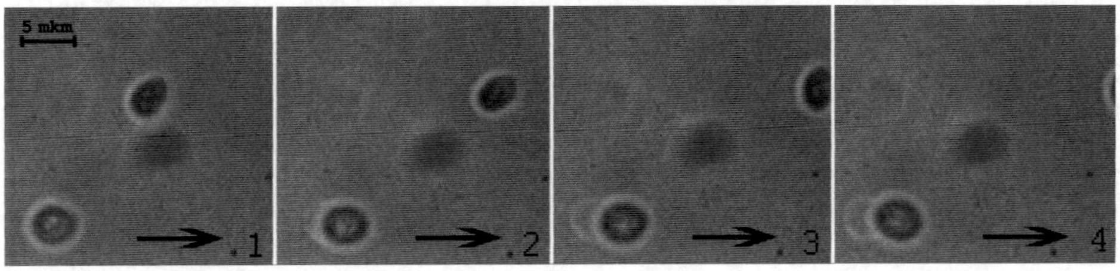

Figure 4. The capture and confinement of transparent object of the size comparable to that of the optical trap. Frames of video: different stages of capture (the arrow shows the direction of movement of the cell with particles).

The yeast *Saccharomy cescerevisiae* are used as the objects having the size comparable to the size of the dark region (intensity minimum) of the ring. The yeast is a weakly absorbing object at the used wavelength of 0.53 µm. The experiments were carried out at the radiation power of 3 mW. The dependence of the maximum velocity of the substrate relocation when the microobjects (with dimensions of 5.7 X 4.6 µm and 4.7 X 3.7 µm) were kept in a ring trap, on the change of its diameter (D) is presented in tables 2 and 3, respectively.

Table 2. The experimental results of the study on the efficiency of the ring optical trap for the capture of yeast cells with size 5.7 X 4.6 µm (d_{mean}= 5.15µm).

D µm	d_{mean}/D	$S_{particle}/S_{ring}$	V_{max}, µm /sec
3.1	1.66	2.70	9.0±0.9
3.3	1.56	2.40	6.9±0.7
3.6	1.43	2.02	5.7±0.6
4.1	1.25	1.56	4.7±0.5
4.4	1.17	1.35	4.7±0.5

Table 3. The experimental results of the study of the efficiency of the ring optical trap for the capture of yeast cells with size 4.7 X 3.7 µm (d_{mean}= 4.2µm).

D µm	d_{mean}/D	$S_{particle}/S_{ring}$	V_{max}, µm/sec
3.1	1.35	1.81	3.7±0.4
3.3	1.27	1.60	3.5±0.4
3.6	1.17	1.34	3.5±0.4
4.1	1.02	1.03	3.5±0.4
4.4	0.95	0.89	3.5±0.4

It is seen from table 2 that the maximal velocity of the microscopic stage motion at which the microscopic objects escaped from the optical trap is 9.0±0.9 μm/sec for the 5.7 X 4.6 yeast cell. This value is significantly lower than the corresponding velocity for the microobjects capture by means of the point optical trap. The maximal velocity is 27±2.7 μm/sec for the point optical trap (with beam waist 0.7-0.8 μm) with power radiation 3 mW. However, the obtained data show that the traps in the form of a ring can be effectively used for the capturing of transparent objects. The advantage of such traps as compared to the point traps for manipulation of biological objects consists in their ability to minimize the impact of the laser radiation on biological object by reducing of the power density, and impact on the periphery of the object. The relevance of the problem under consideration is discussed in [10], for example.

The efficient capturing under the influence on the objects periphery is confirmed by the last two lines of table 2. It is seen that the velocity values do not change and are 4.7±0.5 μm/s at the decrease of the ratio of particle diameter to the diameter of the trap from 1.25 to 1.17. The ability of a further increase of the trap diameter without having to reconfigure the optical scheme was limited by the aperture of the LC focusing device. However, the experiments with the microobject of a smaller size (see table 3) proved the possibility of the capturing and retention under the impact actually on the border of the object: $d_{mean}/D=1.02$ and $d_{mean}/D=0.95$. The maximum moving speed of the substrate for these cases was 3.5±0.45 μm/s. A similar result was obtained by the authors [4] for the capture of transparent (non-absorbent) particles in a dark area of light rings formed with the use of a stationary diffraction optical element. It is observed that for the effective capturing the size of the particles should exceed the radius of the ring.

You can also see from table 2 that V_{max} depends on the diameter of the trap provided that this diameter is sufficiently large (for our experiments $d_{mean}/D=1.43\div1.66$). Thus the reducing of the trap diameter from 4.4 μm to 3.1 μm, during the capture of the yeast cells 5.7 X 4.6 μm resulted in double increase of V_{max} (from 4.7 μm/sec to 9.0 μm/sec). So we can suppose that there exist the possibility to control the capture efficiency by means of the optical trap size.

The capture efficiency depends not only on the dimensions of the contour trap, but also on its form. This is confirmed by the experiments presented by movie frames in figure 5.

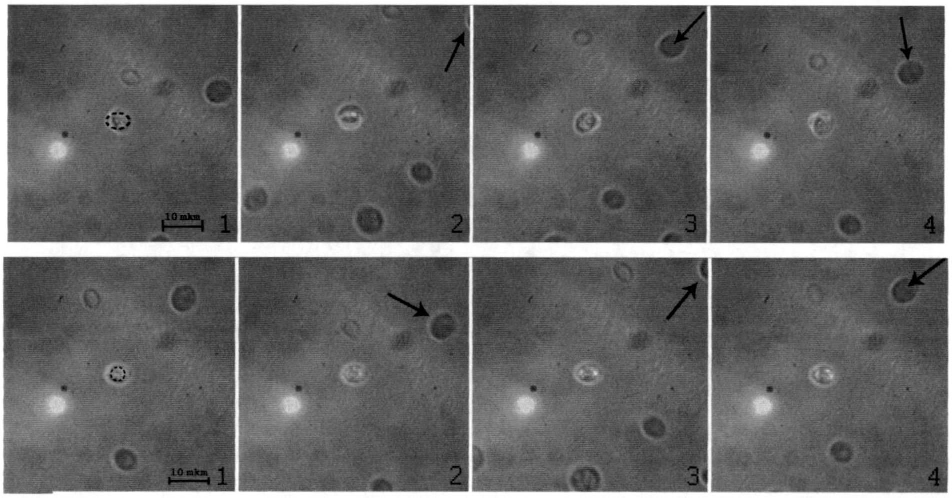

Figure 5. Behavior of the particle in the trap depending on the shape of the trap. Upper row: the ellipse trap – the object is moving along the trap contour. Lower row: the ring trap – the microobject is stably held by the trap. The arrow shows the instant direction of movement of the cell with particles The black dashed lines at the frames 1 are similar to the form of the intensity maximum of contour traps.

Yeast cells were used as micro objects. The microobject was captured by the contour trap and the microscope stage was moved along an arbitrary trajectory defined by the user. The shape of the trap was transformed from ellipse to ring during the experiment. You can see (upper line of figure.5) that for the ellipse trap the microscopic stage moving resulted in the relocation of the particle along a path (the particle does not leave the limits of the optical trap, but inside the trap it is moved by the fluid flow). Then the trap was transformed to the ring shape. In this case the microobject was kept by the traps and its relocation was not observed (see lower line of figure.5). Thus, the capture by the ring trap is more stable as compared to the trap in the form of an ellipse.

3.3.Experiments with absorbing objects

The traps with a minimum intensity inside (in the form of rings, ellipses, and so on) are the most interesting for the capture of absorbing objects into the dark region. In our experiments as the absorbing microobjects we used suspended in water microparticles (of 0.5÷1.5 μm size) and their conglomerates (of 5-6 μm size) consisting of a mixture of aluminum oxide Al_2O_3 and metallic aluminum obtained in the process of electrical explosion of aluminum wires and also suspended particles of silver (Ag) obtained by the laser ablation. To capture the absorbing object it is necessary to move the trap closer to the object to overlap the radiation and open it again after the objects appears in the center of the optical trap. The shutter was used for radiation overlapping and the objects were delivered to the virtual marker on the screen that indicates the position of the center of the light ring.

The capture of particles by a trap in the form of a crescent (C-shaped optical trap) with its subsequent transformation into the ring is more easy-to-use for a number of tasks. The LC 4-channel modulator allows to form C-shaped optical trap and to transform it into the ring trap due to the change of applied voltage. For the formation of traps in the form of circle or ellipse arcs it is need that the lines of constant phase profile of the phase delays would not be closed within the aperture (that is, had the shape of an arc of a circle or ellipse). Since the light field distribution with the intensity maxima located along the contour of similar curve is generated in the near-field region of LC modulator, it is possible to form the corresponding C-shaped traps. The line of the constant phase will be not closed provided that the phase delay distribution is formed in the shape of a cone with the circle or ellipse base with dimensions slightly larger than the aperture or the generated distribution is shifted to the edge of the aperture.

The experiment on the capture and confinement of microparticles in this way is illustrated in figure.6. Microobjects are moved to the C-shaped trap formed by the focusing device (frame 1). The trap is transformed into the circle and the particles are captured with the trap (frame 2) and kept in it during the substrate motion (frames 3-4).

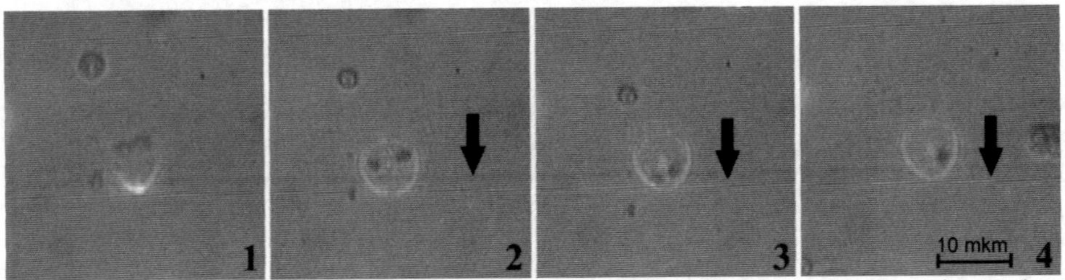

Figure 6. The capture and confinement of aluminum particles (of 1.5 μm diameter) by C-shaped trap transformed into the ring trap. The arrow shows the direction of movement of the microscopic stage.

The maximum velocity of the substrate relocation ensuring the confinement of the aluminum particle (with diameter 1.5 μm) in the ring trap of 8 μm diameter and 0.9 μm contour thickness at radiation power 10 mW was 2.5±0.3 μm/sec. This diameter of the trap was achieved due to the optical scheme reconfiguration which provided the input of the generated intensity distribution in the working area of the microscope.

The frames of the movie illustrating the argentum (Ag) particle confinement by the ring optical trap are presented in the figure 7. The outside diameter of the ring was 3.2 μm, the contour thickness was 0.9 μm and the Ag particle size was 2.7 X 1.1 μm.

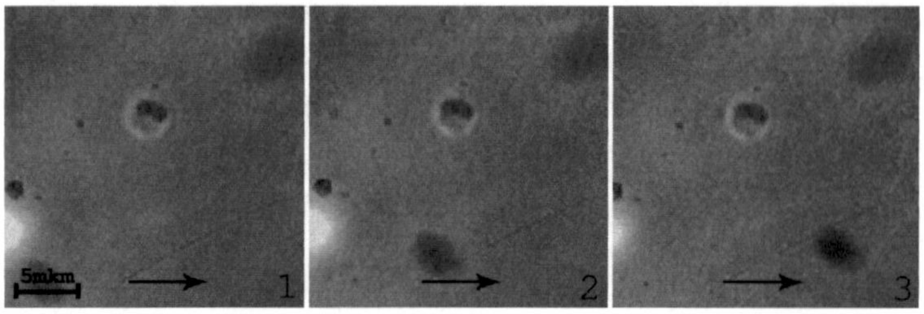

Figure 7. Ag particle (2.7 X 1.1 μm) confinement by the ring optical trap (diameter - 3.2 μm, the contour thickness - 0.9 μm) during the substrate motion. The arrow shows the direction of movement of the microscopic stage.

The maximal velocity of the microscopic stage relocation (V_{max}) when the argentum particles of various sizes are still kept in the optical trap are given in table 4.

Table 4. The experimental results of the study on the efficiency of the ring optical trap with 3.2 μm diameter and 0.9±0.1 μm contour thickness for the capture of absorbing Ar particles of various size.

Size, μm	*V max*, μm/sec	P, mW
4.3 X 2.5	1.9±0.2	3
2.7 X 1.1	1.9±0.2	3
2.7 X 3.7	5.7±0.5	2

The results of capture the Ag particle with the size 2.7 X 1.1 μm depending on the change of the trap size (ring diameter) are presented in table 5. It is seen that for the ring diameter equal to 3.5 μm the maximal velocity of the substrate movement was achieved. For the mentioned parameters of the particle and trap this velocity was 8.1 μm under the radiation power 2 mW.

Table 5. The experimental results of the study on the efficiency of the ring optical trap for the capture of absorbing Ar particle with the size 2.7 X 3.7 μm (d_{mean}= 3.2 μm).

D, μm	d_{mean} /D	$S_{particle}/S_{ring}$	V_{max}, μm/sec
3.2	1.00	0.98	5.7±0.5
3.5	0.91	0.83	8.1±0.8
4.2	0.76	0.56	5.7±0.5
4.4	0.73	0.52	5.7±0.5

Our experimental results show that there exists an optimal size of the ring traps, when the capture efficiency is maximum. This size is determined by the parameters of the captured micro-object (its size, shape, refractive index, absorption coefficient and so on) and this dependence is quite complicated even for transparent objects [3, 6, 11]. So it can be concluded that from a practical viewpoint the use of the 4-channel LC modulator for the absorbing microobjects capturing is reasonable and convenient since it allows to ensure a smooth control of the shape and size of the optical trap owing to the control voltage parameters change.

Journal of Physics: Conference Series **605** (2015) 012007 doi:10.1088/1742-6596/605/1/012007

4. Conclusion

The experiments on the optical manipulation of transparent and absorbing particles by the contour traps generated with the use of the 4-channel LC light modulator have been carried out. The capture efficiency was estimated by the maximum velocity of the substrate that ensures the particles confinement in the trap. This value depends both on the parameters of traps and the geometrical and physical properties of micro objects. It has been shown that the ability of the LC focusing device to smoothly control the shape and size of the optical trap in real time allows to optimize the size of the trap to the size of the captured particles thus improving the capture efficiency. This can be useful, for instance for manipulation of the objects of biological nature (both transparent and absorbing), because of the variations in size, even within a single sample, attributable to them.

Acknowledgments

The research project was supported by the Russian Foundation for Basic Research (grant No. 14- 02-31376) and the Program for Basic Researchers of the Physical Sciences Department of RAS "Fundamental aspects of physics and technology of semi-conductor lasers as basic elements of photonics and quantum electronic".

5. References

[1] Korobtsov A V, Kotova S P, Losevsky N N, Mayorova A M, Patlan V V and Samagin S A 2014 Optical trap formation with a four-channelliquid crystal light modulator *Journal of Optics*, **16**, 035704

[2] Korobtsov A V, Kotova S P, Losevsky N N, Mayorova A M and Samagin S A 2014 Formation of contour optical traps with a four-channel liquid crystal focusator *Quantum Electron.(in press)*

[3] Milne M, Dholakia K, McGloin D, Volke-Sepulveda K, and Zemánek P 2007 Transverse particle dynamics in a Bessel beam *Opt. Express* **15** 13972-13987

[4] Porfirev A P and Skidanov R V 2013 Generation of an array of optical bottle beams using a superposition of Bessel beams *Applied Optics* **52** 6230-6238

[5] Afanasiev K N, Korobtsov A V, Kotova S P, Losevsky N N, Mayorova A M, Patlan V V and Volostnikov V G 2013 Further Development of the Laser Tweezers Technique for Biomedical Applications *Journal of Physics: Conf. Ser.* **414** 012017

[6] Šiler M, Jákl P, Brzobohatý O and Zemánek P 2012 Optical forces induced behavior of a particle in a non-diffracting vortex beam *Opt. Express* **20** 24304-24319

[7] Kotova S P, Patlan V V and Samagin S A 2011 Tunable liquid-crystal focusing device. I. Theory *Quantum Electron.* **41 (1)** 58-64

[8] Kotova S P, Patlan V V and Samagin S A 2011 Tunable liquid-crystal focusing device. II. Experiment *Quantum Electron.* **41 (1)** 65-70

[9] Kotova S P, Patlan V V and Samagin S A 2013 Focusing light into a line segment of arbitrary orientation using a four-channel liquid crystal light modulator *Journal of Optics*, **15(3)**, 035706

[10] Skidanov R, Rykov M, Iannacchione G and Krivoshlykov S 2012 The modification of laser beam for optimization of optical trap force characteristic *Computer optics* **36** 377-386

[11] Muradoglu M and Wah Ng T 2013 Optical trapping map of dielectric spheres *Appl. Opt.* **52** 3500-3509

Journal of Physics: Conference Series **605** (2015) 012008 doi:10.1088/1742-6596/605/1/012008

Mueller matrix microscopy on a *Morpho* butterfly

Oriol Arteaga *, Ertan Kuntman, Joan Antó, Esther Pascual, Adolf Canillas and Enric Bertran

Dep. Física Aplicada i Òptica, IN2UB, Universitat de Barcelona, 08028 Barcelona, Spain

E-mail: oarteaga@ub.edu

Abstract. The brilliant iridescent colouring in male Morpho butterflies is due to the microstrutures and nanostructures present in the wing scales, rather than pigments. In this work Mueller matrix microscopy is used to investigate the polarization properties of butterfly wing scales in reflection and transmission. It is found that the top layer of more transparent scales (cover scales) have very different polarimetric properties from the ground iridescent scales. Images with high spatial resolution showing the retarding and diattenuating optical properties for both types of scales are provided.

1. Introduction

Structural colors arise from the interaction of light with biological nanostructures [1]. They rely exclusively on the shape of the material and not its chemical properties. Pigments and dyes that produce colors by the selective absorption of wavelength fade over time but structural color does not age. One of the most well-known examples of animal with structural color are the tropical *Morpho* butterflies. Most of these butterflies are colored in iridescent shades of blues and greens that arise in the microscopic scales covering the dorsal side of Morpho's wings. The structures present in the scales selectively cancel out certain colors through wavelength interference while reflecting others. The photophysics of structural color in Morpho butterfly wings has been one of the most extensively studied subjects in optics, however it is still not fully understood, as its uniform reflection in a surprisingly broad angular range is difficult to understand using simple models of multilayer interference [2].

The scales have a characteristic dimensions $\sim 100\mu m$ and some of its features are easily visible in an optical microscope. The scales have micron or sub-micron periodic ridges that form a diffraction-like grating. Electron microscopy reveals that every ridge contain a finer structure consisting of 6-12 lamellae overlapping one to another and forming a nanosized tree-like structure. Recently it has been shown that the lamellae distribution of the tree-like structure is the determining factor for producing a low incident angle dependence [2]. Despite all the efforts to understand the optics of Morpho butterflies, comparatively few works have considered how the butterfly wings respond to polarized light [3]. Maybe this omission comes from the fact that if the butterflies are illuminated with different types of linear or circular polarized light, it is not possible to perceive substantial differences in the coloration or brightness of the wing. This is different to what happens in several species of scarab beetles, that also feature structural coloration, but that are well-known for reflecting only left-handed circular polarized light while absorbing right-handed [4]. This response is produced by the helical arrangement of chiting-based multilayer structures present in the cuticle of the beetle forming a Bouligand-type architecture. Despite it is a more subtle effect than in the case of the beetle, we have found that the optical response of the scales of a Morpho butterfly is sensitive to the polarization of light and we have studied it with Mueller matrix microscopy.

The Mueller matrix microscope [5] uses monochromatic light together with a dual rotating compensator setup to carry out quantitative measurements of optical anisotropy of a sample in transmission (trans-illumination) or in reflection (epi-illumination). Results are quantitative and offer information about all the polarization optical effects present in the sample. In comparison, measurements made with standard polarization microscopes typically use white light and the specimen is observed in between crossed polarizers. This leads to colored interference patterns in the sample image that are related to the different amounts of retardation introduced by the sample.

2. Material and methods

2.1. Mueller matrix microscope

The Mueller matrix microscope produces 16 images of a sample in around one minute. Each one of these images corresponds to a different element of the 4×4 Mueller matrix, providing the complete quantitative analysis of the modification of the polarization as the light beam is transmitted in or reflected from the sample. False color is used to denote different magnitudes and signs through the Mueller matrix elements. The Mueller matrix microscope is able to measure the magnitude and orientation of linear diattenuation and linear retardation, the circular retardation, the circular diattenaution and the depolarization. The Mueller matrix microscope works by doing a frequency analysis of the images continuously captured by a Point Grey Flea3 CMOS camera while two rotating compensators modulate the polarization of light. As the intensity detected by every individual pixel of the camera is frequency analyzed by digital demodulation, the determination of the optical properties keeps the same high lateral resolution of standard microscopy measurements.

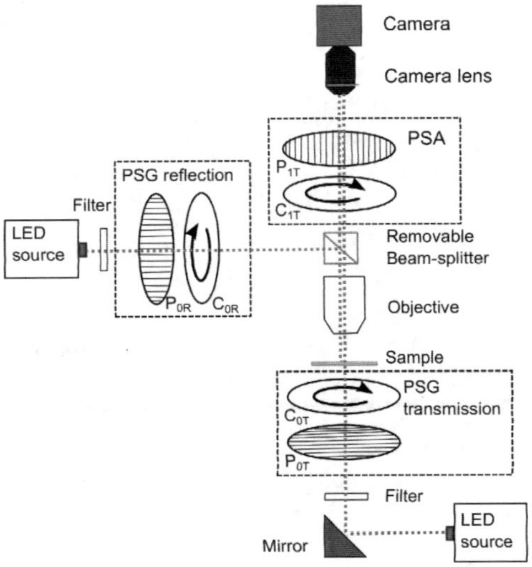

Figure 1. Basic scheme of the instrument. The green and red dotted lines respectively show the light path for transmission and reflection configurations respectively. Both configurations share the same PSA but use a different PSG.

The instrument contains two polarization state generators (PSG) and one polarization state analyzer (PSA). Only one of the two PSGs works at a time in combination with the PSA, depending if it is a transmission or reflection measurement. Fig. 1 shows the basic scheme of the instrument, identifying the position of the transmission and reflection PSGs together with the unique PSA. The image is formed using a infinite-corrected long-working distance objective which is free from strains. Every PSG or PSA is composed of a polarizer and a rotating compensator that has its own time dependence because their speed of rotation can be adjusted with a motor controller. In our most usual configuration, the camera acquires images at 50 frames per second (fps), the compensators in the PSGs rotate at $\omega_0 = 5°/s$ (0.0138 Hz), while the compensator in the PSA moves five times faster, at $\omega_1 = 25°/s$ (0.0694 Hz). This relation of angular speeds is not arbitrary since, as we have previously shown [5], it provides optimum conditions for the frequency demodulation . The number of collected frames is adjusted so that, at least, a complete turn of the slowest compensator is recorded (and therefore 5 complete turns of the fastest one). This means acquiring 3600 frames of data and, as the camera works at 50 fps, a complete measurements is done in about 72 seconds.

More details about the Mueller matrix microscope system are given in [5].

2.2. Polarization microscope

The butterfly scales were also examined using a JENAVAL polarization microscope (Carl Zeiss, Jena, Germany). All the measurements with this microscope were done with dark field epi-illumination (in between crossed polarizers). The images were captured with a Deltapix Invenio 3S camera.

2.3. The Morpho menelaus sample

We have studied the dorsal scales of the *Morpho menelaus* butterfly shown in Figure 2. All the optical measurements were done on a small ($\sim 2 \times 2$ mm) fragment clipped from the left wing The wings of this butterfly are composed of two independently formed membranes which cleave together in the nymphal stage. Both sides of the wing are covered with scales. The underside of the wing is covered with brownish scales that block light. These dark scales show no iridescence, but they seem to increase the contrast of the iridescence in the other side of the wing. The top of the wing shows an ordered arrangement of iridescent scales. In this part we can clearly distinguish two types of scales: blue (ground scales) and translucent (cover scales). Sometimes the cover scales are also called glass scales. Both sets of scales contain ridges that are aligned in the same direction. However the separation between adjacent ridges is different for both types scales. In the cover scale is around 1.5-2 μm but the ridges in the ground scale are difficult to observe well in an optical microscope as their separation is below 1 μm.

3. Results

Normal incidence back-reflection Mueller matrix microscopy measurements were done on the top of the butterfly wing at a wavelength of 535nm. The measured Mueller matrix image (Fig. 3) reveals that, in this configuration, the wing behaves mostly as a retarder (Mueller matrix elements m_{01}, m_{02}, m_{03}, m_{10}, m_{20}, m_{30} are close to zero). The fact that the diattenuating terms of the Mueller matrix are close to zero indicates that the reflectivity of the scale does not depend on the polarization of light. The total retardation (that is usually referred as birefringence when it is measured in transmission) can be calculated with the analytic method given in [6, 7]. The retardation calculated from the Mueller matrix in Fig. 3 is given in Fig. 4. For this calculation we have subtracted the default phase retardation of $\pm\pi$ radians in the linear retardation term that is intrinsic to the normal-incidence back-reflection configuration [8] (e.g. an ideal mirror at normal incidence back-reflection acts like a half-wave retarder).

Fig. 4b and Fig. 4d clearly show that the highest values of the retardation (close to $\pi/2$ radians) are obtained in the areas where the ground scale is not covered by the glass scale. This fact would be in agreement with some reports that show that this second layer of scales

Figure 2. The *Morpho menelaus* butterfly seen at different levels of magnification. a) Photo of the studied specimen. (b) Dark scales with wavy edges present in the underside of the wing. (c) The two types of scales (ground and cover) present in the top of the wing. (d) Detail of the iridescent ridges in a ground scale.

acts as a diffuser of radiation, spreading the angle over which incident light is reflected [9, 10]. It is also surprising that areas with cover scales appear to have quite higher reflectivity under the microscope, showing that the cover-ground interaction enhances the iridescence. Common retarders are based on the anisotropic property of birefringence exhibited by certain crystals (e. g., quartz) or on total internal reflection (the Fresnel rhomb). However, strong reflection retarders at normal incidence are quite rare since they are only possible in materials or structures that efficiently extinguish one direction of light polarization as it is discussed in Ref. [8]. In the case of Morpho ground scales, light that is polarized in a direction perpendicular to the direction of the ridges is greatly extinguished. We are not aware of any other biostructure causing these high levels of retardation. Our Mueller matrix microscope uses a green LED light source for epi-illuminiscence measurements, therefore we could not quantify the retardation in the blue side of the spectrum ($\sim$ 450 nm) but we would expect it has to be larger than in the green.

The levels of retardation exhibited by the scales are strong enough to be easily visible with a conventional polarization microscope. Fig. 5 shows the optical micrographs for two different orientations of the scales obtained with a Jenaval microscope with dark field illumination. When the orientation of the ridges coincides with any of the two crossed polarizers (Fig. 5a) the image is dark, but a rotation of the scales by 45° (Fig. 5b) returns the bright iridescence to the scales, indicating that the retardation in the scales has induced some high degree ellipticity to the reflected light.

The cover scales are quite transparent and they can be easily studied with transmission Mueller matrix microscopy. Ground scales are much less transparent but, still, they allow some

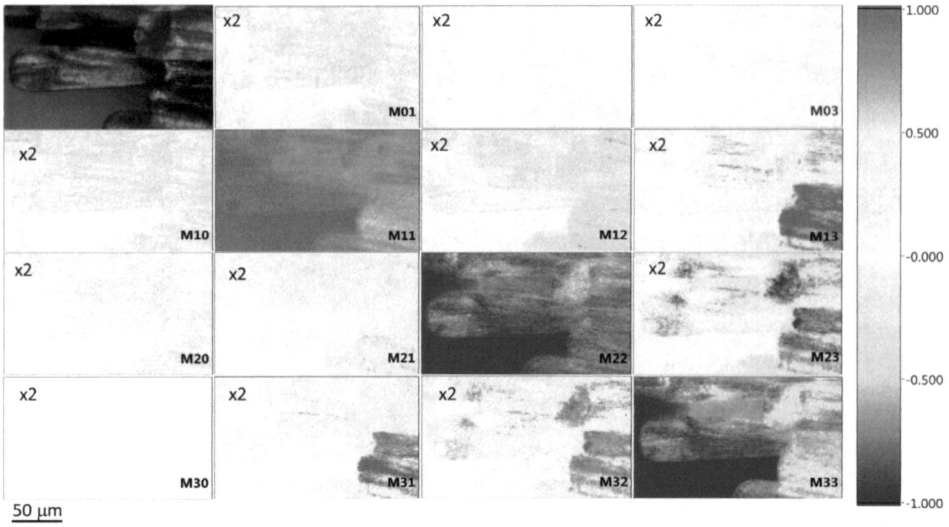

Figure 3. Normalized Mueller matrix measurement on the dorsal side of a section of a *Morpho* butterfly wing. The wavelength of measurement is 535 nm. The off-diagonal matrix elements have been multiplied by the factors indicated in the image to improve the visualization of the image. The top left image shows the unpolarized reflectivity, that corresponds to the Mueller matrix element m_{00}.

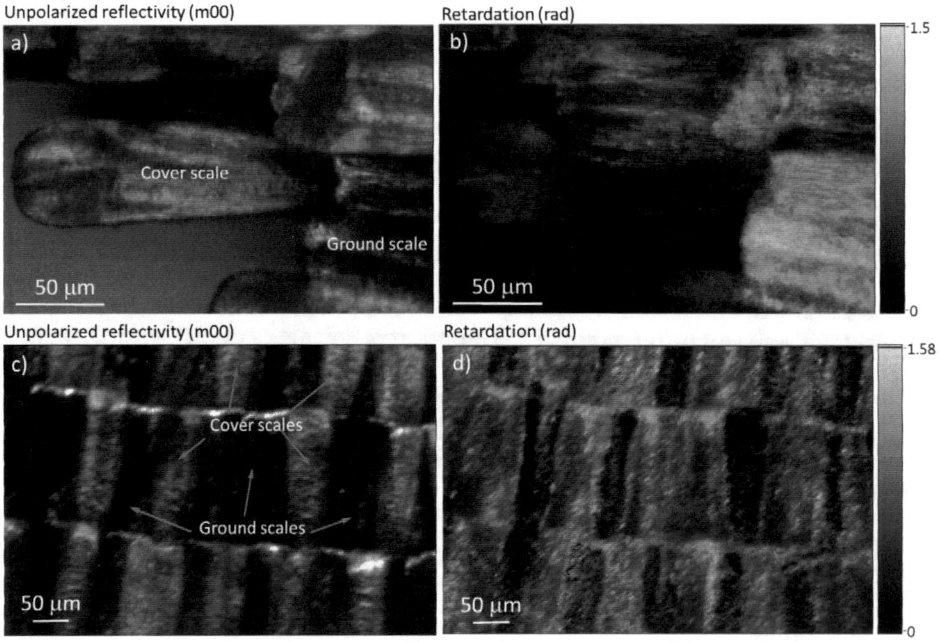

Figure 4. Measured retardation in different parts the butterfly wing. (a) and (b) are, respectively, the unpolarized reflectivity and the retardation corresponding to the Mueller matrix given in Fig. 3. (c) and (d) are, respectively, the unpolarized reflectivity and the retardation in a different part of the wing.

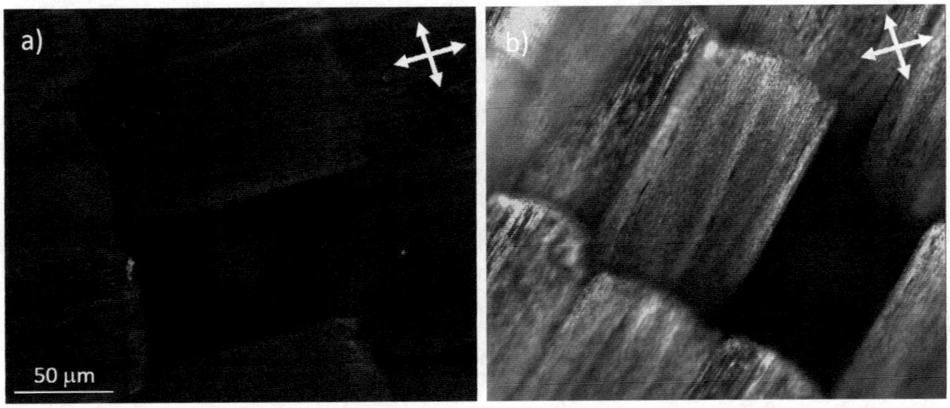

Figure 5. Optical micrographs as seen under crossed polarizers for two different orientations of the ground scales of the *Morpho* butterfly. The white arrows indicate the extinction axes of the polarizers. The camera was kept with the same exposure values for both orientations.

light to go through. To study the transmissive optical properties of both types of scales we focused our attention to a portion of the edge of the wing fragment that was under study. In this area we could find ground and cover scales that were not overlapped one to another and that were not protected by the dark scales of the underside of the wing. The transmission Mueller matrix image measured at 630 nm for this area is given in Fig. 6.

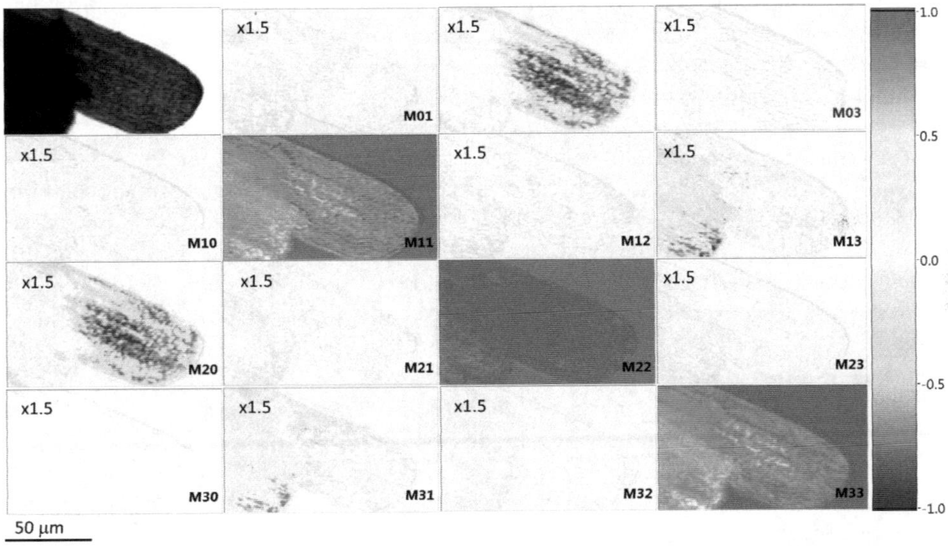

Figure 6. Normalized transmission Mueller matrix measurement on the scales of the dorsal side of a *Morpho* butterfly wing. The wavelength of measurement is 630 nm. The off-diagonal matrix elements have been multiplied by the factors indicated in the image to improve the visualization of the image. The top left image shows the unpolarized transmissivity, that corresponds to the Mueller matrix element m_{00}.

As indicated previously, using the analytic inversion method [7], it is possible to calculate the magnitude and orientation of the (structural) linear birefringence and linear dichroism present in the scales. The results have been summarized in Fig. 7 for two measurements made in the same area but at two different wavelengths (630 nm and 420 nm). With red light, the cover scale has a remarkable linear dichroism and a weak linear birefringence. Its optical response is very different from the ground scale, that has weak linear dichroism and greatest retardation. As the wavelength decreases to the blue-violet region all the transmissive effects appear to become weaker. It is remarkable, though, that for the cover scale the direction of greater extinction switches by 90° from violet to red (Fig. 7). This property is probably helping to enhance the extinction of red light as it first passes through the cover scale, is reflected in the ground scale (suffering attenuation and retardation) and passes again through the cover scale. A more detailed spectral analysis of the transmission and reflection processes would be required to fully understand the complex polarimetric relation between ground and cover scales. More work in this direction is now in progress.

In all the measurements we have reported, the orientation of micron-sized grating-like ridges present in the scales coincides with the axes of linear polarization properties. However, as the scales do not keep the same orientation through all the butterfly wing span, the global effect of the wing to the polarization of light is spatially averaged and, globally, at a macroscopic level, it seems weaker than it really is under the microscope. Only by observing the wing with an instrument with high spatial resolution it is possible to evaluate the real polarimetric response of the individual butterfly scales.

4. Conclusions

We have studied the polarimetric response of the cover and ground scales of a *Morpho menelaus* butterfly with a Mueller matrix microscope. The polarimetric optical properties that we have determined in transmission and in reflection are mainly due to *structural anisotropy*. This means that the anisotropic optical response do not come from the intrinsic anisotropy of the materials that form the scale (i.e. chitin) but from the complex ridged nanostructure of the wing.

The two types of scales (ground and cover) present in the dorsal side of the butterfly have very different polarimetric properties. At normal-incidence reflection, the ground scales induce a remarkable anisotropic retardation to the reflected optical wave but, at the same time, they do not have different reflectivity as a function of the polarization of light. In contrast, we have seen that the cover scales serve to attenuate this retardation effect and they also enhance the reflectivity of the scale as seen under the microscope. In transmission the cover scales show higher polarization-depending effects than the ground scales, and this is particularly evident for red light, for which the cover scales exhibit a rather strong linear dichroism.

Acknowledgements

The authors thank the *Museu de Ciències Naturals de Barcelona* for the loan of the *Morpho menelaus* butterfly. This work was partially supported by the project FIS2012-38244-C02-02 of the Spanish Ministry of Economy and Competitivity. O. A. acknowledges a Marie Curie Fellowship within the European Union Seventh Framework Programme (FP7/2007-2013) under REA grant agreement no. 330513

References

[1] Vukusic P and Sambles J R 2003 *Nature* **424** 852–855 ISSN 0028-0836 URL `http://dx.doi.org/10.1038/nature01941`

[2] Pris A D, Utturkar Y, Surman C, Morris W G, Vert A, Zalyubovskiy S, Deng T, Ghiradella H T and Potyrailo R A 2012 *Nature Photonics* **6** 195–200 ISSN 1749-4885 URL `http://dx.doi.org/10.1038/nphoton.2011.355`

[3] Vidal B 2011 *Micron* **42** 801–807 ISSN 09684328 URL `http://dx.doi.org/10.1016/j.micron.2011.04.006`

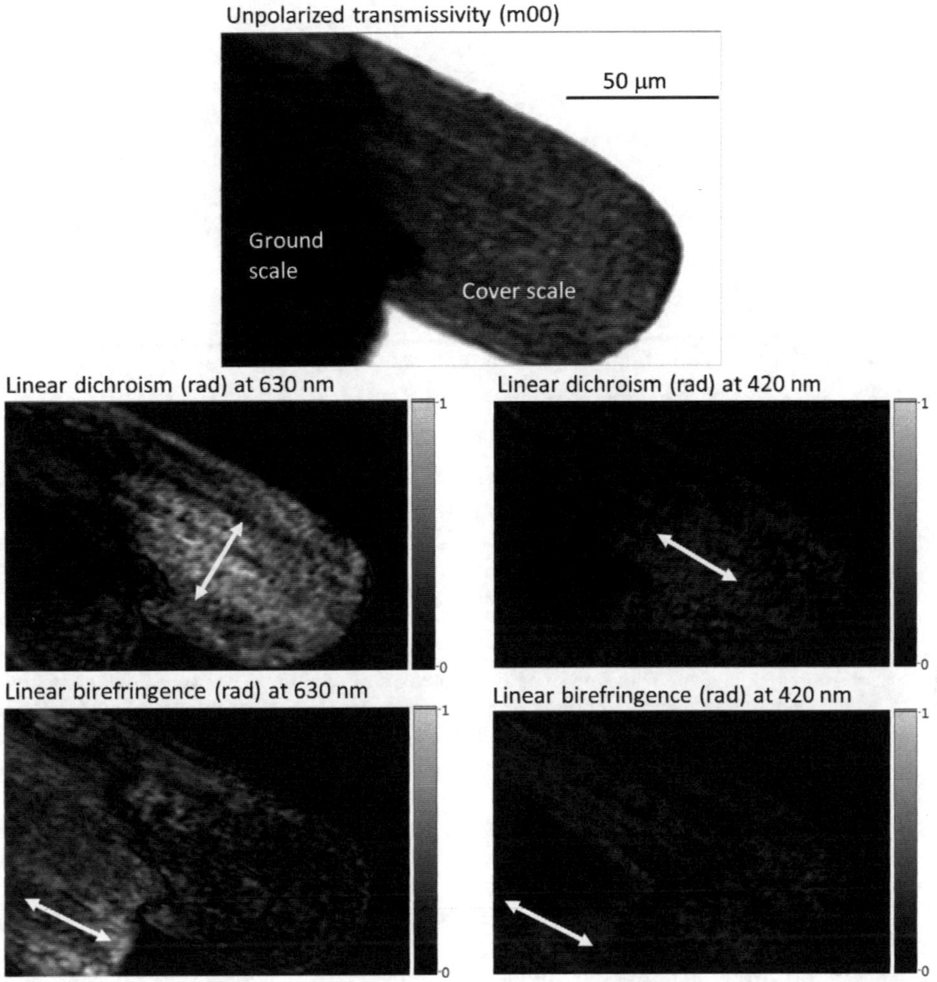

Figure 7. Magnitudes of the linear dichroism and linear birefringence properties of the ground and cover scales at two different wavelengths: 630 nm and 420 nm. In the linear dichroism plots the white arrows indicate the direction with highest extinction, while in the linear birefringence plots the white arrows the direction with lowest refractive index.

[4] Arwin H, Magnusson R, Landin J and Järrendahl K 2012 *Phil. Mag.* **92** 1583–1599 URL `http://dx.doi.org/10.1080/14786435.2011.648228`
[5] Arteaga O, Baldrís M, Antó J, Canillas A, Pascual E and Bertran E 2014 *Appl. Opt.* **53** 2236–2245 URL `http://dx.doi.org/10.1364/ao.53.002236`
[6] Arteaga O 2010 *Mueller matrix polarimetry of anisotropic chiral media* PhD Thesis
[7] Arteaga O and Canillas A 2010 *Opt. Lett.* **35** 559–561 URL `http://dx.doi.org/10.1364/ol.35.000559`
[8] Arteaga O, Garcia-Caurel E and Ossikovski R. 2014 *Opt. Lett* **39** 6050–6053 URL `http://dx.doi.org/10.1364/ol.39.006050`
[9] Vukusic P, Sambles J R, Lawrence C R and Wootton R J 1999 *Proc. Roy. Soc. Lond. Series B: Biol. Sc.* **266** 1403–1411 ISSN 1471-2954 URL `http://dx.doi.org/10.1098/rspb.1999.0794`
[10] Yoshioka S and Kinoshita S 2004 *Proc. Roy. Soc. Lond. Series B: Biol. Sc.* **271** 581–587 ISSN 1471-2954 URL `http://dx.doi.org/10.1098/rspb.2003.2618`

Super resolved imaging via variable pinholes array and time multiplexed object's coding

A Schwarz[1], A Shemer[2] , J Wang[1] , B Javidi[1] , and Z Zalevsky[2]
[1] Department of Electrical and Computer Engineering, University of Connecticut, Unit 4157, Storrs, Connecticut 06269, USA
[2] Faculty of Engineering, Bar Ilan University, Ramat Gan 5290002, Israel

E-mail: zalevsz@biu.ac.il

Abstract. In this paper, we describe a novel super resolution method for pinhole optics with variable pinholes array. The imaging system is based on super resolved time multiplexing method using variable or moving pinhole array. The improved resolution and signal to noise ratio are achieved with improved light intensity in the same exposure time, compared as in a single pinhole system. This new configuration preserves the advantages of pinhole optics while solving the resolution limitation problem and the long exposure time of such systems. The system can also be used as an addition to several existing optical systems with visible and invisible light as well as to x rays systems.

1. Introduction

The advantages of lenses in optics are the main reasons to their popularity in optical imaging systems. Although pinhole optics is probably one of the earlier configurations used in optical imaging systems, over the years, lenses were preferred over pinholes in most optical applications. Low resolution, dim image and long exposure time for photon amount accumulation are the disadvantages in usage of pinholes use. Nevertheless, pinhole optics is still used in many applications today. The disadvantages of using lenses, such as depth of focus problems, aberrations, resolution and limited acceptance angle with micro lenses due to the small numerical aperture (NA), are the main motivation to use pinholes. Pinhole optics is also useful in x-rays and gamma rays imaging or particle streams, where no lens materials are available. On the other hand, the advantages of pinhole optics, beside simplicity, are complete freedom from linear distortion, virtually infinite depth of focus and a very wide angular field that can be made to exceed 90°. In pinhole optics there is also a trade off with large and small pinholes, between resolution and light intensity. A large pinhole can pass large amount of photons in a discrete time exposure but will produce a low resolution image. However, a small pinhole can pass small amount of photons but produces a high resolution image. In pinhole optics, for large pinholes the resolution is characterized by the geometric limit and for very small pinholes by the diffraction limit while the situation between the two limits is more complicated. The optimum pinhole size will be a compromise between the large spot image produced by a large pinhole and the one produced by Fraunhofer diffraction coming from a small pinhole. The optimal image thus occurs roughly where the geometrical optics and Fraunhofer diffraction approximations give the same result. Several methods where suggested to calculate the optimum size for pinhole [1-4].

In order to preserve the advantages of pinhole optics together with solving the problems of light intensity and low signal to noise ratio (SNR), many methods have been proposed over the years. Such methods include the use of high sensitivity detectors and camera films. Another method is based on the concept of using coded aperture imaging (CAI). The basic step in coded aperture imaging systems is the use of multiple pinhole apertures to form numerous images of the object. The resulting recoded image is decoded using a digital or optical method. There are many techniques and applications for multiple pinholes coded aperture systems [5-9]. Over the years, a wide range of pinholes designs have been proposed: random array, mosaic pattern array, cyclic array, uniformly redundant array, some with separate multiple images and some with multiplexed overlapped images [10-23].

In this paper we present a novel technique for lensless imaging system. The system preserves the advantages of pinhole optics together with solving the problems of light intensity and low SNR by using a variable and time multiplexed pinholes array. The system provides the advantages of simplicity, freedom from linear distortion, virtually infinite depth of focus and a very wide angular field. It is done while providing a super resolved image with improved SNR, better light intensity and all in the overall the same exposure time as compared to a single pinhole system or a lenses system.

2. Operation principle

The optical setup is composed of a variable pinholes array consisting from several time varying pinholes arrays (a set of K arrays) set one after the other and a computational reconstruction step (Fig. 1). During the overall capture part, each array in the set is used to capture an image consisting of replications of the object according to the pinholes number and position in the array. The final image is reconstructed after time multiplexing. The time parameter for each image capture is variable and depends on the pinholes arrays design. A single pinhole causes loss of information due to two zero points in its spectral plane. A multi pinholes array also causes loss of information and will badly affect the final reconstruction. The reason for this is that multi pinholes arrays also have zero points in their spectral plane. Therefore, the outcome involves loss of the object's information and the lack of capability to fully reconstruct the image. Using the proposed concept involving a time variable array can solve the above mentioned problem. Although each array in the set has zero points, the overall array maintains all spatial frequencies and it is able to achieve super resolved image (i.e. an image that does not contain partial losses of spectral information). The usage of a set of multi pinholes arrays instead of a single multi pinhole array with the same number of holes is essential in order to cover the spectral information loss that each array by itself has (i.e. in the proposed time multiplexing concept each one of the arrays in the set has zeros in different spectral locations and thus the combination of all arrays avoids the above mentioned spectral loss). The variable array creates the complementary filter with minimum information loss. In order to preserve all the spatial frequencies of the original object image we must control the pinholes arrays design. There are four free parameters in the configuration: the number of pinholes in each array (N_k), the spatial order (location) of the pinholes in each array ($L_{x,y}^{(n,k)}$), the size (diameter) of each pinhole in each array ($D_{n,k}$) and the capture time intervals for each pinholes array (t_k). The use of such kind of variable array does not extend the overall exposure time and enables to perform the reconstruction after a single capture as in a single array.

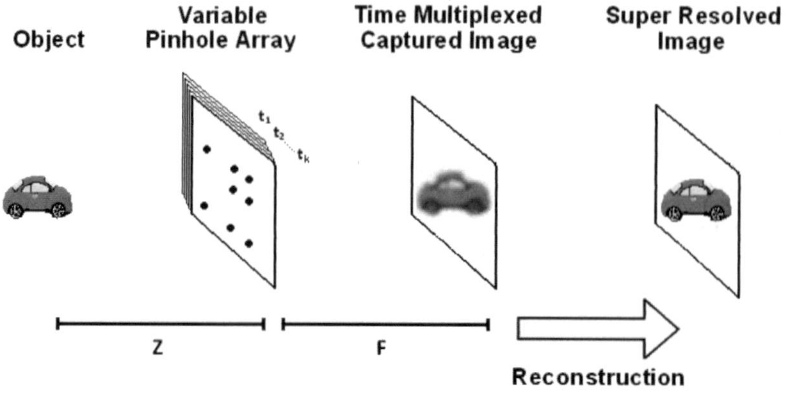

Figure 1. The proposed optical time multiplexed system.

The variable array can be referred as a priori known, unique and changeable filter *F(u,v)* applied in the spatial frequency domain multiplied with the spectrum of the imaged object. This can be clearly seen by computing the spectrum of the spatial distribution of a single time interval capture:

$$F_k(u,v) = \sum_{n=1}^{N} e^{-2\pi i (L_x^{(n,k)} u + L_y^{(n,k)} v)} \tag{1}$$

The spectrum of the spatial distribution is given by the Fourier transform and so the final captured image in the frequency domain is:

$$\Im\{I(x,y)\} = \sum_{k=1}^{K} \Im\{i_k(x,y)\} \cdot t_k = \hat{O}(u,v) \cdot \sum_{k=1}^{K} F_k(u,v) \cdot t_k \tag{2}$$

where I is the captured image, i_k is the captured image distribution from each array position, O is the original object and $\Im\{\ldots\}$ is the Fourier transform operation. Thus, by extracting the spectral distribution, one can reconstruct the original image. The reconstruction is performed in the spatial Fourier plane by, first, applying inversion operation on the filters' sum function, then, multiplying the inverse of the filters' sum function with the final image captured through the variable pinholes array. The final reconstruction is done by applying Wiener filtering that filters out the distortion function, *F(u,v)*:

$$\hat{I}(u,v) \cdot F(u,v)^{-1} = \left[\hat{O}(u,v) \cdot \underbrace{\sum_{k=1}^{K} F_k(u,v) \cdot t_k}_{F(u,v)} \right] \cdot F(u,v)^{-1} = \hat{O}(u,v) \tag{3}$$

In the far field case (*Z*>>*F* in Fig. 1), the difference between the replication distances in the image plane and the pinholes spatial order in the array will be negligible. However in the near field situation the distances between the object's replications in the image plane are not the same as the distances of the pinholes in the array. The ratio between the two set of distances is related to the realized magnification:

$$L'^{(n,k)}_{x,y} = (1+M) \cdot L^{(n,k)}_{x,y} \tag{4}$$

where the magnification factor is $M=F/Z$. As a result, in the near field case, the image reconstruction is performed by using the inverse of function of $F'(u,v)$ that corresponds to the scaled set of distances and ratios as they appear in Eq. 4 (it is the same equation as Eq. 1 for $F(u,v)$ but while using the distances of Eq. 4).

3. Simulations and experimental validation results of the super resolved system

In order to preserve all the spatial frequencies of the original object image we control the variable pinholes array design. We choose the four free parameters (N_k, $L_{x,y}^{(n,k)}$, $D_{n,k}$, t_k) in our configuration such that there will be no frequency loss in the filters' sum $F(u,v)$ as compared to the single pinhole system. Our considerations for choosing the number of pinholes and positions were first to prevent zero points in the filters' sum $F(u,v)$ spectral plane. The second consideration was to make sure that the filters' sum $F(u,v)$ transmission value is higher than the reference single pinhole filter in the region of interest. The last consideration was related to the desired relative light intensity improvement (LII) of the multi pinholes variable array system as compared to a single pinhole system. The relative LII is determined by the number, size and time exposure of each hole in the variable array:

$$Relative\ LII = \sum_{k=1}^{K}\sum_{n=1}^{N}\pi R_{k,n}^{2}\cdot t_{k} \Big/ T\cdot\pi R_{ref}^{2} \tag{5}$$

where R is the radius of the pinhole.

The next $F(u,v)$ simulation's results are related to the pinholes arrays design. We have used three different configurations in order to show the influence of the free parameters on the spatial frequencies of the $F(u,v)$ function. The single pinhole reference system had pinhole size which is referred as 1 radius unit. All three variable arrays had three different positions (K=3). The first two variable arrays had the same hole size (one R unit) for all pinholes in the array ($D_{n,k}$ constant). One variable array had a total holes number of 17 and the other had 38 total pinholes. The third variable array had total holes number of 36 but with different holes size (3, 1, 2/3, 1/2, 1/3, 1/4.5). The relative LII of the three variable arrays were 5.67, 12.67, 13.21 respectively for constant integration time tk in all the arrays. The Three variable arrays are presented in Fig. 2.

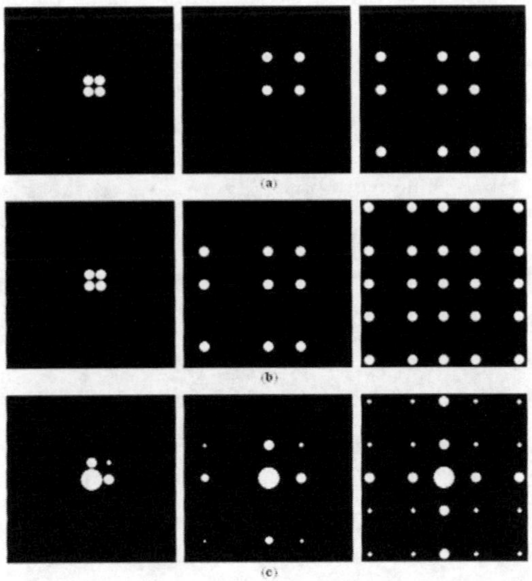

Figure 2. The three variable arrays (K=3): (a). The three positions of the variable array consist of

total 17 pinholes with the same holes size; (b). The three positions of the variable array consist of total 38 pinholes with the same holes size; (c). The three positions of the variable array consist of total 38 pinholes with different holes size (0.5:1:2 ratio).

The number and location of the pinholes in the array and the variable array positions enable to prevent zero points in the spatial frequencies of *F(u,v)*. The difference between the three deigns and the influence on the *F(u,v)* function can be seen in Fig. 3. The time parameter tk for each array position is variable and depends on the pinholes arrays design. In the figure below (Fig. 3), P.T.F.T. stands for P.T.F. x 1 time unit – that is the Pinholes' Transfer Function computed in relative units. One P.T.F.T. unit is the transmission equals the one of a single pinhole, multiplied by the photons integration time t_k. The time parameter helps to achieve the second consideration of the design.

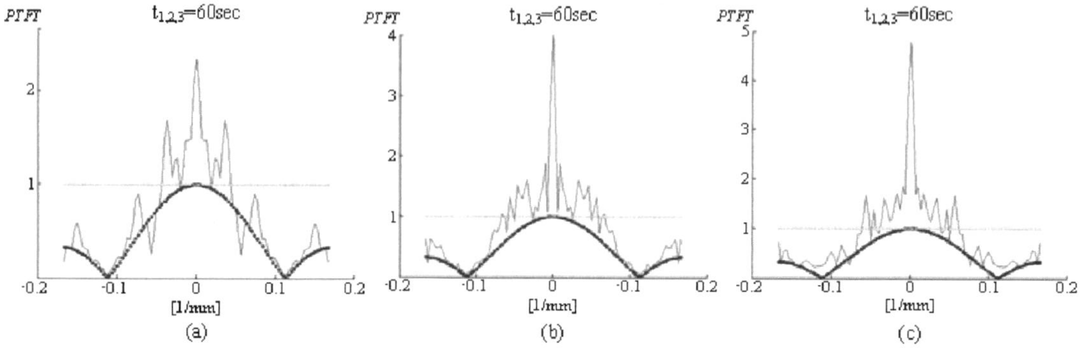

Figure 3. The filters' sum *F(u,v)* comparison (top red line) as compared to one single pinhole (thick blue line) for the three above mention variable arrays (*K*=3). The time interval is equal for each array position (and it equals to 60 sec). The straight green line is the maximum level of the single pinhole reference level (since the designs are equal to the x-y axes for simplicity the simulation was performed in 1D only).

The variable pinholes arrays were designed by simulations done according to the conditions that were mentioned above. The pinholes focal length was 45 mm and the magnification was 1. The light source was a LUXEON Rebel ES neutral white light source with a diffuser and the camera was DCC1545M 1280x1024 pixels, 5.2 µm x 5.2 µm pixel size. The object was a spoke resolution target (Siemens star). Fig. 4 shows the experimental setup that was constructed. The single pinhole reference system was with a hole diameter of 250 µm. To the variable array of total 17 holes (250 µm each) we added another reference system of single pinhole having hole diameter of 350 µm with high light intensity for the same time exposure.

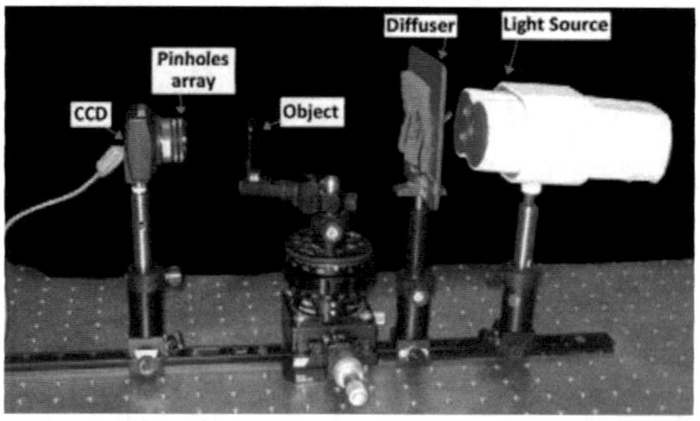

Figure 4. The variable pinholes array experimental setup.

The experimental results are shown in Fig. 5. The results show that in the variable array system compared to the single pinhole reference system with the same hole size (250 μm), we get better resolution with better light intensity (an improvement factor of 5.67) and better SNR. The second reference system with the large hole size (350 μm) achieved high light intensity in the same time exposure but low resolution according to the holes area size ratio.

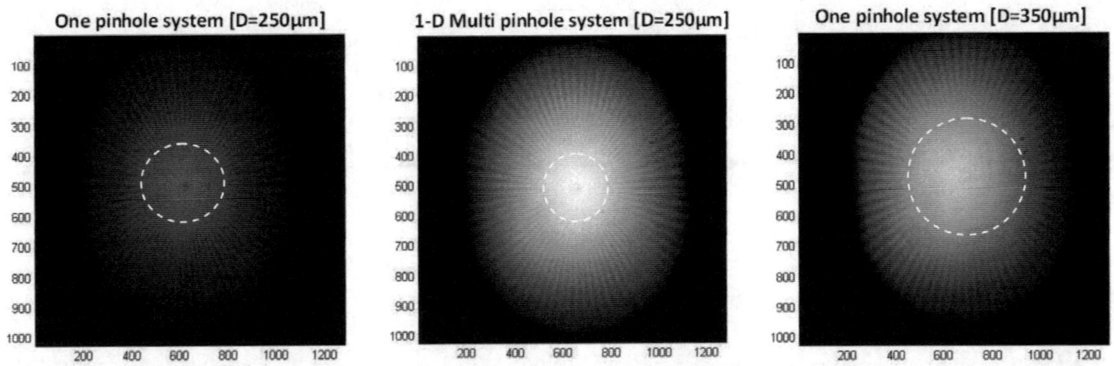

Figure 5. The comparison between systems (the same exposure time): (a). One pinhole reference system (250μm) – low light intensity; (b). Super resolved reconstructed image: the reconstructed image of the variable pinholes array system (250μm) with high light intensity (c). One pinhole reference system (350μm) – high light intensity but with bad resolution.

4. Conclusions

In this paper we have presented a method for super resolved imaging for pinholes optics that can improve light intensity and SNR in the same exposure time as compared to a single pinhole system. These improvements are accomplished by using known designs of variable multiple pinholes array coded aperture. The design considerations and the influence of the various parameters on the obtainable results were presented and discussed. The proposed concept was validated numerically and experimentally. The method can be applied for near and far field systems and for many applications achieving the advantages that pinholes optics can offer.

References

[1] Selwyn E W H 1950 *Photographic J.* **90B** 47
[2] Martin L C and Welford W T 1966 *Technical Optics* Vol. I (London: Pitman)
[3] Hardy A C and Perrin F 1932 *The Principles of Optics* (New York: McGraw-Hill)
[4] Kingslake R 1963 *Lenses in Photography* (New York: A S Barnes and Company)
[5] Mertz L and Young N O 1961 *Proc. Int. Conf. Optical Instruments and Techniques* 305
[6] Dicke R H 1968 *Astrophys. J.* **153** L101
[7] Ables J G 1968 *Proc. Astron. Soc. Aust.* **1** 172
[8] Barrett H H, Wilson D T and Demeester G D 1972 *Opt. Commun.* **5** 398
[9] Barrett H H and Horrigan F A 1973 *Appl. Opt.* **12** 2686
[10] Golay M J E 1971 *JOSA* **61** 272
[11] Klemperer W K 1974 *Astron. Astrophys. Suppl.* **15** 449
[12] Calabro D and Wolf J K 1968 *Inf. Control* **11** 537
[13] MacWilliams F J and Sloane N J A 1976 *Proc. IEEE* **64** 1715
[14] Fenimore E E and Cannon T M 1978 *Appl. Opt.* **17** 337
[15] Gunson J and Polychronopulos B 1976 *Mon. Not. R. astr. Soc.* **177** 485
[16] Skinner G K 1984 *Nucl. Instrum. Methods Phys. Res.* **A 221** 33
[17] Busboom A, Elders-Boll H and Schotten H D 1998 *Exp. Astro.* **8** 97

[18] Gottesman S R and Fenimore E E 1989 *Appl. Opt.* **28** 4344
[19] Rideout M R 1995 *Coded Imaging Systems for X–ray Astronomy*, PhD thesis, University of Birmingham
[20] Accorsi R, Gasparini F and Lanza R C 2001 *Nucl. Instrum. Methods Phys. Res.* **A 474** 273
[21] Aggarwal M and Ahuja N 2004 *Int'l. J. Computer Vision* **58** 7
[22] Georgiev T, Zheng K C, Curless B, Salesin D, Nayar S and Intwala C 2006 *In Proceedings of Eurographics Symposium on Rendering* 263
[23] Mcguire M, Matusik W, Chen B, Hughes J F, Pfister H and Nayar S 2007 *IEEE Computer Graphics and Applications* **27** 32

Journal of Physics: Conference Series **605** (2015) 012010 doi:10.1088/1742-6596/605/1/012010

Image nonlinearity and non-uniformity corrections using Papoulis - Gerchberg algorithm in gamma imaging systems

A Shemer[1], A Schwarz[1], E Gur[2], E Cohen[1], and Z Zalevsky[1]
[1] Faculty of Engineering, Bar Ilan University, Ramat Gan 5290002, Israel
[2] Department of Electronic Engineering, Azrieli - College of Engineering, Jerusalem 9103501, Israel

E-mail: zalevsz@biu.ac.il

Abstract. In this paper, the authors describe a novel technique for image nonlinearity and non-uniformity corrections in imaging systems based on gamma detectors. The limitation of the gamma detector prevents the producing of high quality images due to the radionuclide distribution. This problem causes nonlinearity and non-uniformity distortions in the obtained image. Many techniques have been developed to correct or compensate for these image artifacts using complex calibration processes. The presented method is based on the Papoulis - Gerchberg (PG) iterative algorithm and is obtained without need of detector calibration, tuning process or using any special test phantom.

1. Introduction

Image non-uniformity and nonlinearity corrections exist in many imaging systems at different spectral regions such as visible, infrared optics, X-ray as well as gamma radiation. In the Gamma imaging, one reason for the mentioned problem is due to photomultiplier tubes (PMT) undergoing gain shifts when spatial orientation changes with respect to an external magnetic field [1-2]. These effects can be produced due to several reasons related to a magnetic field e.g. earth's magnetic field. The maximal gain shifts occurs when the PMT rotates through an angular orientation generating a 90° angle with the azimuth and the orthogonal direction which is relative to the external field. It causes a one percent change in the secondary emission ratio per dynode which results in a change of more than 10% in the charge gain of the PMT. Consequently, detector assemblies with inadequate magnetic shielding can experience gain shifts sufficient to cause significant angle-dependent spatial non-uniformities, some appearing in the shape of a half-moon, and some as changes in the observed count rate as a function of angle [3-5]. Another cause is due to thermal gradients in detector housing, which can produce changes in uniformity and other artifacts [6]. Image non-uniformity is very sensitive to the uniformity of the electric field in scintillation cameras. There are two primary reasons for it in flood field images produced from Anger scintillation cameras: one third is caused by photo-peak energy-axis shifts at different points in the field of view [7,8] and the remaining two-thirds are primarily due to spatial nonlinearities (distortion).

Nowadays uniformity requirements are more stringent and correction circuits do not correct the collimator non-uniformities. Spatial non-uniformities in scintillation cameras and collimator defects

Journal of Physics: Conference Series **605** (2015) 012010 doi:10.1088/1742-6596/605/1/012010

may produce count losses. In circular detector orbits cases, these manifest themselves as concentric ring or "bull's eye" artifacts and become even more complicated artifacts for noncircular orbits [9].

In our paper, we refer to the limitation of the gamma detector mentioned above, which prevents the producing of high quality images. We present a novel technique aiming for gamma imaging correction which uses both Gaussian smoothing for low pass filtering and the use of the Papoulis - Gerchberg algorithm for extracting a high resolution images from the non-uniform acquired data. The method is performed on the acquired raw image exploit areas of high resolution to fill up low resolution parts and try to minimize calibration processes both on a daily basis and the less frequent ones.

2. Gamma imaging distortion function

In order to understand the method used, we need to understand the distortion function caused in the gamma detectors as seen in Fig. 1a. We noticed that this distortion function is localized at each PMT position and is similar to a Gaussian smoothing function. It means that at the center of the PMT, it has a high sampling rate and at the edges it has a lower rate. The PG algorithm has two requirements for full reconstruction. One is that it needs sufficient high resolution sampling data from the original, non-distorted image that meets the PG full reconstruction condition, and the second is to know the precise bandwidth, σ, of the original image. In our case both conditions are met and therefore the PG algorithm is found to be very suitable for the PMT distortion case.

3. Mathematical overview using the Papoulis - Gerchberg algorithm

The Papoulis - Gerchberg algorithm method for super-resolution is based on work done by Papoulis [10] and Gerchberg [11]. Gerchberg proposed to perform signal reconstruction given a priori knowledge of the signal diffraction limit and only a portion of its spectrum. In Papoulis work he showed how to extrapolate a band limited signal from part of the original signal, i.e., determining the transform. The Papoulis-Gerchberg algorithm has been used in a modified form by Vanderwalle et al. [12] to super-resolve images when multiple low-resolution (LR) registered images are available and Chatterjee [13] used only a single LR image for the same purpose.

Now, let us express the distortion function as in Eq.1 where we use a distorted image denoted by *g(x)* which is built from high-resolution image, *f(x)*, undergoing a distortion expressed by the function *d(x)*. The distortion function is actually built from a basic Gaussian distortion function which mimics the PMT distortion and distort local areas in the original image according with the PMT hexagonal array arrangement or any other arrangement that exist in Gamma cameras. Hence, we have partial sampling data that causes the non-uniformity and non-linearity.

$$g(x)=f(x)\cdot d(x) \tag{1}$$

The image distortion results in a local clustering with barrel like distortions. The barrel like distortion has both high resolution data on the center and low resolution data on the boundaries. To make a distortion function that mimics the PMT array we use the following function,

$$d(x)=PMT_{Gaussian}(x)*\sum_{i=0}^{i=n}\delta(x-x_i) \tag{2}$$

where * designates a convolution operation. Here we used a Gaussian shape and on the overall result we apply the Papoulis–Gerchberg (PG) algorithm to obtain the reconstructed image.

Note that we assume having sufficient high resolution data to meet the PG algorithm requirements for full reconstruction. The PG algorithm is described briefly in Eq. 3 and Eq. 4. At first, the given image is Fourier transformed (usually after some noise is added to the inaccurate portion of the image to enable fast convergence), then in the Fourier plane one imposes the known low frequency data regions on the result and keeps the rest of the pixels as is. Next, an inverse Fourier transform is generated and, as in Eq. 3, one imposes the known data regions on the result and keeps the rest of the pixels as is. Next, as in Eq. 4, in the Fourier domain – the true low frequency data is imposed and the rest is kept, and so on.

$$g_n(x)=\begin{cases} g_{n-1}(x), & x\in\left[g_0(x)=0\right] \\ g_0(x), & x\in\left[g_0(x)\neq 0\right] \end{cases} \quad \Rightarrow \quad G_n(v)=\int_{-\infty}^{\infty} g_n(x)e^{-2\pi i x v}dx \tag{3}$$

Note that, v is the Fourier transform coordinate and the G_n, G_{n-1} are the Fourier transforms of the nth and (n-1) iterations of the distorted function, $g_0(x)$, and P_σ is the original bandwidth window which is known a priori.

$$G_{n+1}(v)=G_n(v)\cdot \underset{\sigma\text{- Original image Bandwidth}}{P_\sigma(v)} \quad , \quad P_\sigma(v)=\begin{cases} 1, & |v|\leq\sigma \\ 0, & |v|\geq\sigma \end{cases}$$
$$\Rightarrow \quad g_{n+1}(x)=\int_{-\infty}^{\infty} G_n(v)e^{2\pi i x v}dv \tag{4}$$

Now, the reconstructed image is obtained by iterating until the error criterion is reached.
In our simulations we incorporate low pass Gaussian filter that helps smoothing the result prior to the reconstruction GP process as seen in Eq. 5.

$$g_0(x)=LPF_{Gaussian} * g(x) \tag{5}$$

4. Results using the Papoulis - Gerchberg algorithm

The proposed method was simulated with a distortion function that mimics the PMT distortion pattern that exists in the Gamma image. The Gamma imaging methods cause non-uniformity as seen in Fig. 1(a) while the required uniformity should be seen as in Fig. 1(b). Furthermore, local pincushion and barrel like distortions causing the image to have wave-like shapes instead of straight lines are also present. This distortion can be easily seen in Fig. 2(a) and Fig. 2(d) while the required result should be as seen in Fig. 2(b) and Fig. 2(e). When both linear and uniformity image correction are present in the image it should be seen as in Fig. 2(c) and Fig. 2(f).

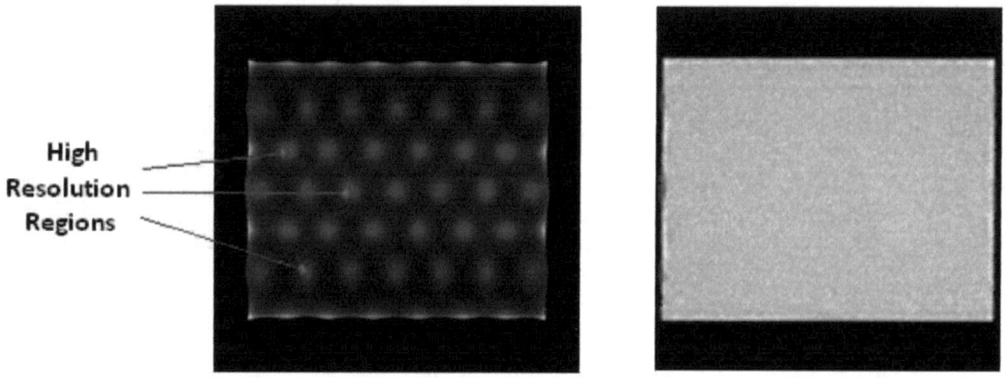

(a) (b)

Figure 1. (a) Gamma distorted image with non-uniformity areas (b) Field flood sample for corrected uniform and linear image

In order to convert the gamma-distorted image to the original image we use the connection between the high-resolution regions of the detector and the linear and uniform one. Furthermore, we enhance the reconstruction result by applying a Gaussian smoothing kernel. The chosen kernel size was relative to the size of the high-resolution regions.

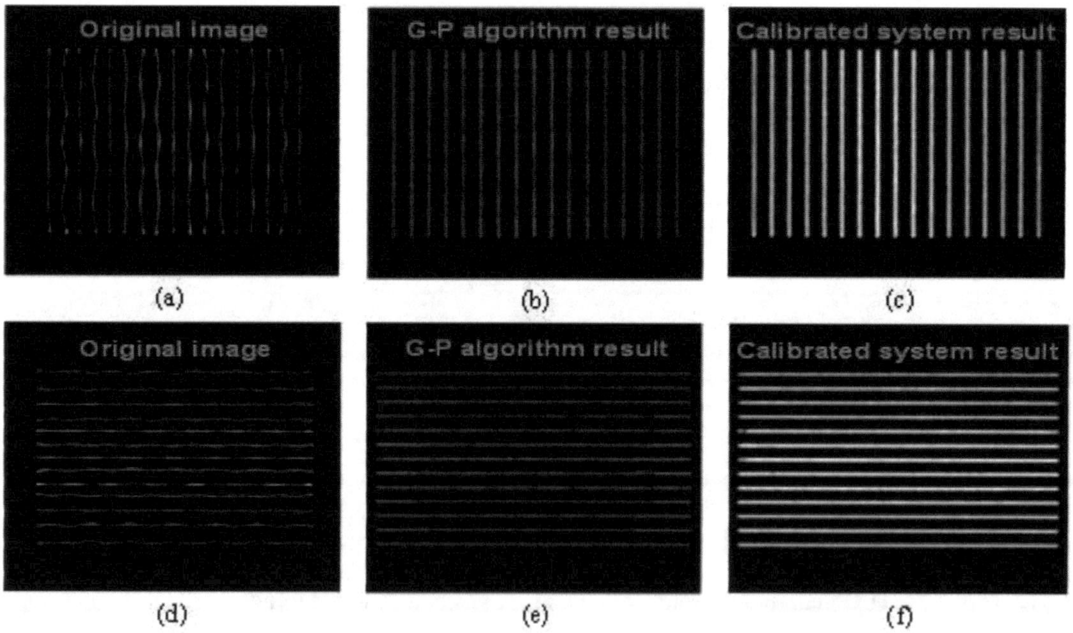

Figure 2. The (a), (d). The original distorted image; (b), (e). The corrected image results from the G-P algorithm technique; (c), (f). The reference system results of the calibrated system.

In gamma imaging, unlike optical imaging, linearity refers to the ability of the system to maintain straight line in lead bar phantom. The measurement parameter is estimated by measuring the amount of deviation from straight line in mm as seen in Fig. 2(a) and Fig. 2(d). The outcome of the PG algorithm as simulated by the authors, are shown is Fig. 2(b) and Fig. 2(e) and the calibrated results for achieving uniformity are shown in Fig. 2(c) and Fig. 2(f). It was found that the correlation between the PG algorithm linearity correction results and the reference calibrated system results was 74.34% for the vertical lines image and 68.87% for the horizontal lines image. Notice that after uniformity correction the correlation should be higher.

To show that the method can correct distortions not just for linear shapes, but also for more complex images, we introduce the Leonardo da Vinci's "Adorazione dei Magi" famous painting as our target image. Fig. 3 shows details from this famous painting by da Vinci. In Fig. 3(a) we see the original image, where regions of interest are circled.

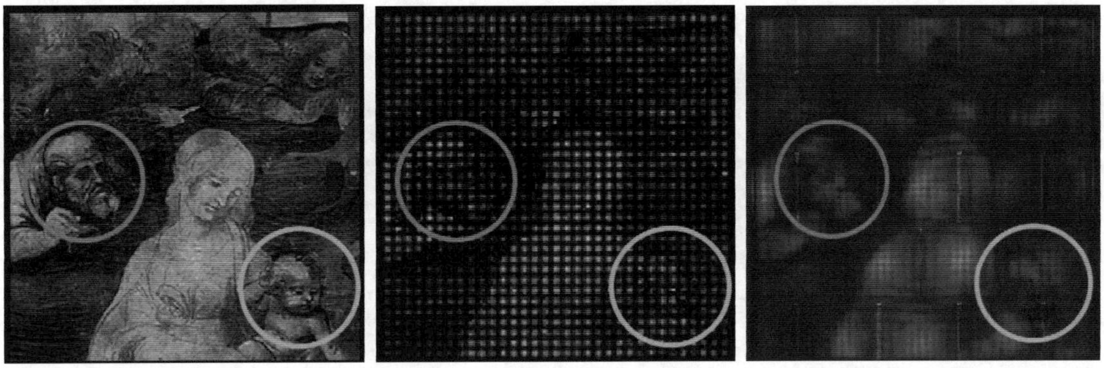

(a) (b) (c)

Figure 3. (a) Part of Leonardo da Vinci's "Adorazione dei Magi" (1481); (b). Simulated gamma camera distorted image; (c). The corrected image results when applying Gaussian smoothing prior the Papoulis - Gerchberg algorithm technique.

In Fig. 3(b) we observe the image obtained by simulating the gamma camera operation. We re sample the image and applied the PMT distortion function and the smoothing function which result in gird like shape. Note that in Fig. 3(c) one can see the corrected image after using the PG algorithm which is used to correct the low resolution regions. However, the linearity correction was only partially corrected and barrel like distortion still partially exist.

5. Conclusions

In this paper, the authors presented a new technique for gamma imaging correction. The method reconstructs the original image by correcting the nonlinearity and non-uniformity in the image. This method aims to reduce the need of complicated calibration process with test phantoms. Further work should be done on real time Gamma image to examine method accuracy. This improvement is accomplished by using only the Papoulis - Gerchberg algorithm and by using the existing connection between the high resolution regions of the detector and a linear and uniform region of the captured image. The proposed concept was validated numerically. It was also shown that the method can be applied not only for gamma imaging but also for optical imaging systems having non-uniformity problems.

Acknowledgments

The authors would like express their appreciation to Dr. O. Amir, Dr. L. Tsukerman and Dr. M. Kogan from the General Electric (GE) Healthcare Company in Haifa, Israel for their valuable help in carrying on this research activity.

References

[1] Coleman C I 1982 *Rev. Sci. Instrum.* **53**, 735
[2] Takasaki F, Ogawa K and Tobimatsu K 1985 *Nucl. Instrum. Meth.* **228**, 369
[3] Rogers W L, Clinthorne H N, Harkness B A, Koral K F and Keyes Jr J W 1982 *J. Nucl. Med.* **23** 162
[4] Larsson S and Israelsson A 1982 *IEEE Trans. Nucl. Sci.* **NS-29** 1331
[5] Bieszk J 1986 *J. Nucl. Med.* **27**. 1902
[6] Johnson T K, Kirch D L, Hasegawa B H, Thompson D and Steel P P 1983 *Emission Computed Tomography: Current Trends*, ed P D Esser (New York: The Society of Nuclear Medicine)
[7] Soussaline F, Todd-Pokropek A E and Raynaud C 1977 *Review of Information Processing of Medical Imaging*, ed A B Brill and R R Price (Oak Ridge, TN: Oak Ridge National Lab) p 375
[8] Todd-Pokropek A E, Erbsmann F and Soussaline E 1977 *Medical Radionuclide Imaging* vol I (Vienna: International Atomic Energy Agency) p 67
[9] Gullberg G T 1987 *Med. Phys.* **14** 105
[10] Papoulis A 1975 *IEEE Trans. on Circuits Syst.* **19** 735
[11] Gerchberg R W 1974 *Optica Acta* **21-9** 709
[12] Vandewalle P, Susstrunk S and Vetterli M 2003 Superresolution images reconstructed from aliased images *SPIE/IS&T Visual Communication and Image Processing Conference*, ed T Ebrahimi, and T Sikora (Lugano, Switzerland, July 2003) p 1398
[13] Chatterjee P, Mukherjeel S, Chaudhuri S and Seetharaman G 2009 *The Computer Journal* **52** 80

Multispectral digital lensless holographic microscopy: from femtosecond laser to white light LED

J Garcia-Sucerquia
Universidad Nacional de Colombia, Sede Medellin, School of Physics, A.A. 3840
Medellin 050034, Colombia

E-mail: jigarcia@unal.edu.co

Abstract. The use of femtosecond laser radiation and super bright white LED in digital lensless holographic microscopy is presented. For the ultrafast laser radiation two different configurations of operation of the microscope are presented and the dissimilar performance of each one analyzed. The microscope operating with a super bright white light LED in combination with optical filters shows very competitive performance as it is compared with more expensive optical sources. The broadband emission of both radiation sources allows the multispectral imaging of biological samples to obtain spectral responses and/or full color images of the microscopic specimens; sections of the head of a drosophila melanogaster fly are imaged in this contribution. The simple, solid, compact, lightweight, and reliable architecture of digital lensless holographic microscopy operating with broadband light sources to image biological specimens exhibiting micrometer-sized details is evaluated in the present contribution.

1. Introduction

The illumination of the samples with broadband light sources allows their spectral analysis and /or their full color imaging. As the sample has microscopic dimensions, multispectral microcopy methods have revealed valuable information of the specimens that no other imaging methodologies can achieve [1]. The achievements of the multispectral microscopy have been also sought with the emerging technology of digital holographic microscopy. The multispectral illumination of the samples has been utilized in digital holographic microscopy to provide full color images with micrometer resolution [2-5] or to enhance the features of the microscope itself [6]. Among the multispectral light sources, ultrafast lasers, with its associated broadband illumination and its intrinsic spatial coherence, have also been utilized in digital holographic microscopy [7-9]; the researches in these works have shown the possibility of doing multispectral digital holographic imaging of specimens with and without internal structure. Despite the good results reported by the use of the ultrafast lasers in digital holographic microscopy, their usually high economical cost and bulky geometry convert that multispectral approach in an exclusively in-lab methodology. With the aim of searching for a multispectral microscope with high degree of portability and ideally low-cost other options of broadband light sources should be explored.

Digital lensless holographic microscopy (DLHM) shows great portability [10] and a set-up that could be potentially of low-cost. DLHM has been developed up to the level that images with micrometer spatial resolution can be achieved with the use of no lenses, at a low computational cost [11] and also basic multispectral applications of the technique have been presented [2, 4]. The simplicity of its architecture has encouraged the researches to utilize diverse light sources, as LED [4, 12, 13], discharge lamp [14], and of course ultrafast light sources [7-9]. In this contribution we want to summarize the main results that our group has achieved with the utilization of broadband light sources. We emphasize the efforts towards the development of a lightweight, low-cost, and portable multispectral microscopy device.

2. Digital lensles holographic microscopy (DLHM)

The possibility of getting rid of lenses to do microscopy brings great advantages: i) there are no optical aberrations associated with the microscope objectives and ii) the microscope could be less expensive, more lightweight and more portable. This possibility is brought into reality if the imaging process is carried out in a two-step methodology as was taught by Gabor [15-17]. The original Gabor´s idea was adopted by Kreuzer et al. [18] who added the power and versatility of the digital world to process the information recorded in a digital screen. This addition allows the overcoming of the flaws that prevented Gabor´s idea to be more attractive; the digital processing powers the elimination of the zero-diffracted order and the reduction of the nuisance introduced by the twin-images [11, 19]. This relatively new technology that merges the bright Gabor´s ideas with the power of the digital processing is named digital lensless holographic microscopy (DLHM).

In DLHM the sample is illuminated by the spherical waves produced by a point source. To produce the point source, the light source is focused down onto a pinhole with a diameter slightly larger than the illuminating wavelength λ; the sample is placed at a distance z from the pinhole. On the surface of a digital screen (CCD or CMOS camera), located at a distance L from the point source, the waves scattered by the sample $U_{scat}\ \mathbf{r}$ are superimposed with the portions of the spherical waves that travel with no perturbation $U_{ref}\ \mathbf{r}$. The intensity recorded on the screen, called in-line hologram, is transferred to a computer for its further processing; see Figure 1. The intensity recorded on the screen with no object present is pixel-wise subtracted from the in-line hologram to produce the modified hologram $\tilde{I}\ \mathbf{r}$ [11] which does not carry the information of the zero-diffracted order:

$$\tilde{I}\ \mathbf{r} = \left|U_{scat}\ \mathbf{r}\right|^2 + \left[U_{scat}\ \mathbf{r}\ U^*_{ref}\ \mathbf{r} + U^*_{scat}\ \mathbf{r}\ U_{ref}\ \mathbf{r}\right]. \quad (1)$$

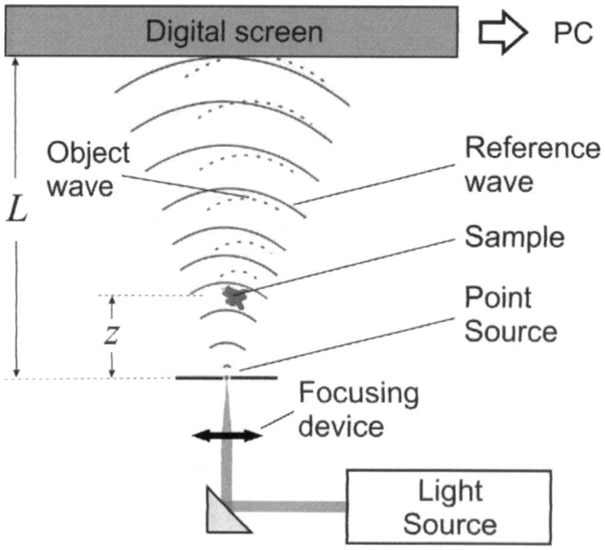

Figure 1. Schematic set-up of DLHM.

Journal of Physics: Conference Series **605** (2015) 012011 doi:10.1088/1742-6596/605/1/012011

Because the preferred samples used in DLHM to work are considered weak scatters, the intensity of the scattered wave $|U_{scat}\ \mathbf{r}\ |^2$ can be neglected [15]. As spherical illumination is utilized the presence of the twin-images, the terms in square brackets in equation (1), does not introduce any nuisance on the reconstructed images. Further discussion about the modified hologram $\tilde{I}\ \mathbf{r}$ and the effects of the twin-images can be read in reference [11]. To recover the information gathered by the waves as they travel through the sample, $\tilde{I}\ \mathbf{r}$ is numerically back-illuminated by a converging spherical wavefront that travels towards the pinhole. This reconstruction process can be correctly described by the Fresnel-Kirchhoff diffraction integral, which written to represent the wavefront propagation in DLHM turns into [11]:

$$U_{scat}\ \mathbf{r}' = \int_{Screen} \tilde{I}\ \mathbf{r}\ \exp\left[i\frac{2\pi}{\lambda}\left(\frac{\mathbf{r}'\bullet\mathbf{r}}{|\mathbf{r}|}\right)\right]d\mathbf{r} \tag{2}$$

with the position vectors $\mathbf{r}=(x,y,L)$, $\mathbf{r}'=(x',y',z)$ representing points at the digital screen and at the reconstruction plane, in that order.

3. Ultrafast laser-DLHM

The needed point source to illuminate the sample is generated in ultrafast laser-DLHM by focusing down onto a pinhole the light from a Ti:Sa laser of 12 fs operating a 80 MHz. The focusing element can be either an achromatic microscope objective (AMO) or a diffractive lens (DL), see panel (a) of Figure 2. As the AMO is utilized for generating the point source, the wavelengths that compose the spectrum of the broadband illumination of the source (panel (b) of Figure 2), generate an in-line hologram that corresponds to the intensity superposition of the individual holograms produced by each of the composing wavelengths $I'\ \mathbf{r},\lambda$ [9]:

$$I\ \mathbf{r} = \int_{\lambda_0-\Delta\lambda/2}^{\lambda_0+\Delta\lambda/2} \rho\ \lambda\ I'\ \mathbf{r},\lambda\ d\lambda \tag{3}$$

with λ_0 the mean wavelength and $\Delta\lambda$ the spectral width of the illuminating source. $\rho\ \lambda$ is the spectral power that weights the individual in-line holograms $I'\ \mathbf{r},\lambda$ which can be written as:

$$I'\ \mathbf{r},\lambda = |U_{scat}\ \mathbf{r},\lambda\ |^2 + |U_{ref}\ \mathbf{r},\lambda\ |^2 + \left[U_{scat}\ \mathbf{r},\lambda\ U^*_{ref}\ \mathbf{r},\lambda + U^*_{scat}\ \mathbf{r},\lambda\ U_{ref}\ \mathbf{r},\lambda\ \right]. \tag{4}$$

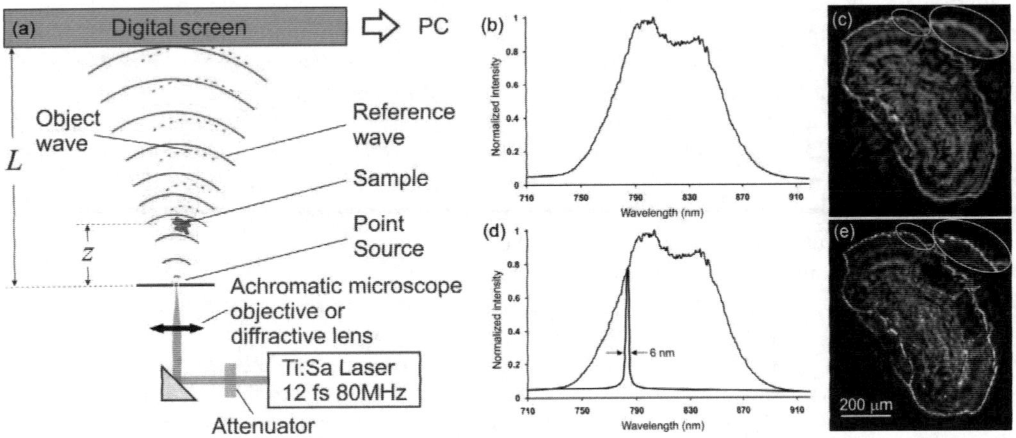

Figure 2. Ultrafast digital laser lensless holographic microscope. (a) Schematic set-up for ultrafast DLHM. (b) Power spectrum of the Ti:Sa laser. (c) Hologram reconstruction obtained as the power spectrum in (b) is utilized. (d) 6 nm tuned power spectrum by using a diffractive lens. (e) Hologram reconstruction obtained as the tuned power spectrum in (e) is utilized.

To eliminate possible intensity inhomogeneities and the zero-diffracted order, the intensity recorded by the digital screen with no sample present is pixel-wise subtracted from the in-line hologram to produce the intensity from the information of the sample is retrieved:

$$\tilde{I}\ \mathbf{r} = \int_{\lambda_0-\Delta\lambda/2}^{\lambda_0+\Delta\lambda/2} \rho\ \lambda \left[I'\ \mathbf{r},\lambda\ - \left| U_{ref}\ \mathbf{r},\lambda \right|^2 \right] d\lambda . \tag{5}$$

This intensity or incoherent superposition given by equation (5), which takes place due to the long integration time of the digital camera in comparison with the pulse duration, introduces the blurring of the reconstructed hologram as shown in panel (c) of Figure 2. No details of the sample are observed rather than the overall shape of it. The other option to produce the point source is the combination of the DL and the pinhole. This arrangement behaves as a tunable spectral filter where the selected wavelength is controlled by the distance from the DL to the pinhole and the spectral width is determined by the pinhole size and the numerical aperture of DL [8]. As the DL is utilized, such that from the spectrum of the femtosecond laser a pulse with 6 nm width is chosen (panel (d) Figure 2).The reconstructed image shows characteristics of the section of the head of the drossophila melanogaster fly that are not visible with the use of the AMO. In panels (c) and (e) of Figure 2, we have 2-time zoomed in similar regions to highlight the differences of the reconstructed images, as the AMO and the DL are utilized to produce the point source. Further details about the underlying phenomena about the coherence of the illuminating light that give rise to these two very different performances of the DLHM can be read elsewhere [8, 9].

4. White light LED-DLHM

The performance of the ultrafast laser-DLHM shows two distinctive performances of this microscopy methodology. While the overall shape of the sample can be reproduced by generating the point source with the AMO, details of the structure of the sample can be recovered by creating the point source with the DL. However, neither of these two options to generate the illuminating broadband source show any promise towards the design of a portable and low-cost multispectral microscope. The use of a super-bright white light LED as illuminating source supplies an alternative path to develop a portable and low-cost multispectral microscope. In the white light LED-DLHM, the needed point source to illuminate the sample is generated by focusing down the light from the LED onto a pinhole, as shown in panel (a) of Figure 3. A regular of these LEDs has a spectral width ranging from 400 nm to 750 nm. This feature allows for using the appropriate bandpass filters to obtain the required illumination for the wavelength of interest. In the case of RGB imaging, one could think a further simplification by using directly a RGB or a tri-LED, however the individual spectral width for each wavelength of at least 30 nm spoils this option [4]. To test the performance of the white light-DLHM, we have initially recorded a set of red, green, and blue holograms to produce a full color image of a section of the head of a drosophila melanogaster fly. In our experiment, the light emitted by the super-bright white-light LED (LXML-PWN2 Luxeon®) was initially collimated. The collimated light was filtered spectrally by means of three bandpass filters from the company Thorlabs®, 632.8 ± 0.2 nm, 532 ± 1 nm and, 405 ± 10nm; the filters were mounted on a motorized wheel. To produce the illuminating point source for each wavelength, the corresponding beam was focused down onto a pinhole of 1 μm in diameter by means of a AMO. The use of this microscope objective minimizes the waist variation of the focused beams for the different wavelengths and therefore smoother spherical waves were produced. The pinhole spatial filters the impinging light on it; the combination of the spectral and the spatial filtering tailors the light of the super-bright white-light LED to have the coherence properties demanded by DLHM to work [11, 13, 20]. The complete experimental set-up is illustrated in panel (a) of Figure 3. The wheel is synchronized with the recording device to record individual hologram for the chosen wavelengths.

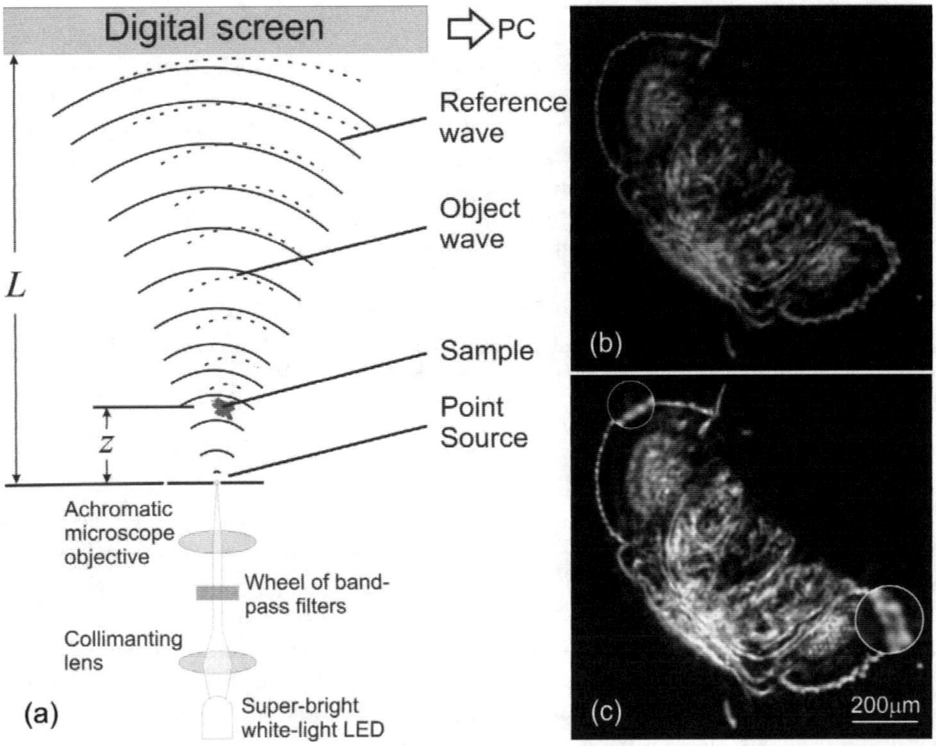

Figure 3. White light LED digital lensless holographic microscope. (a) Schematic set-up for white light LED DLHM. (b) Reconstruction of the green channel of the RGB holograms. (c) Full RGB image of the reconstructed image for a section of the head of a Drosophila Melanogaster fly.

To illustrate the individual reconstructed images, in panel (b) of Figure 3 we show the green channel of the set of images. The fusion of the red, green, and blue reconstructed images is shown in panel (c) of Figure 3. The full color image of the specimen after the correct fusion show details in the range of micrometers. The zoomed-in encircled area in the top shows a hair that measures of the order of 2 μm. At the lower right corner of the same panel the zoomed-in area shows the *omatidia* of the eye of the fly which are in the order of 3 to 2 μm. The quality of the full color reconstructed image in panel (c) validates the feasibility of performing multispectral microscopy with a low-cost, lightweight, and potentially portable device. In similar way as shown in this application for the RGB image, further number of wavelengths can be utilized to have a more complete spectral analysis of the sample while the micrometer-sized resolution with a low-cost, lightweight and potentially portable multispectral microscope is kept.

One key factor to consider in the operation of the white light DLHM is the amount of power available after the spatial filtering process of the LED light. The LED device we have used has a radiant power of the order of 0.5 mW. After the collimation and focalization processes, over the surface of the pinhole we drive an irradiance of about 0.2 MW/cm^2. In contrast, as the femtosecond laser is utilized, that irradiance is of the order of 2.5 TW/cm^2. Fortunately, the versatility provided by the digital recording and the numerical processing, powers this technology with the needed tools to equalize the recorded holograms. The equalization process avoids any perturbations that could be introduced by the dissimilar values of available irradiance upon the sample and hence over the recording device. This claim can be verified in the content of this paper, where the quality of the reconstructed holograms for both types of light sources is fully understood from the coherence properties of the sources rather than irradiance basis analysis would be necessary.

5. Conclusions

The use of multiple wavelengths in microscopy imaging systems allows the study of the spectral response of the observed specimen. That spectral response can be evaluated via digital lensless holographic microscopy (DLHM), one the simplest microscopy methodologies. The simplicity, robustness, and reliability of DLHM ease the utilization of broadband light sources to accomplish the multispectral imaging of microscopic samples with micrometer resolution and without the need of lenses. The use of no lenses to image the sample eliminates any chromatic or optical aberration inherited from the lenses as well as allows the development of lightweight devices.

In this work we have presented the use of a femtosecond-laser and a super bright white LED as light sources in DLHM. With both light sources we have shown that full color images and the spectral response of the sample can be obtained with this simple imaging methodology. Sections of the head of a drosophila melanogaster fly or fruit fly have been imaged with the ultrafast-DLHM and the white light LED-DLHM. The results shown in this contribution are good examples of the versatility of DLHM in the use of light sources which could worth from as large as thousands to as little as cents of dollars. The latter option paves the road to the development of a low-cost, lightweight, low computational cost, and potentially portable multispectral lensless holographic microscope.

Acknowledgments

The author wants to thank the economical support from Universidad Nacional de Colombia grant Hermes code 19384 and the Programa de Internacionalización del Conocimiento. He also wants to acknowledge the priceless cooperation with the group of Institut de Noves Tecnologies de la Imatge (INIT) Universitat Jaume I Castelló, Spain. Special thanks to all those how have contributed to the development of this technology; the inventor and my friend Prof. Hans Jurgen Kreuzer at Dalhousie University, my alumni and my colleagues.

References

[1] M. E. Dickinson, G. Bearman, S. Tille, R. Lansford, and S. E. Fraser, 2001, Multi-spectral imaging and linear unmixing add a whole new dimension to laser scanning fluorescence microscopy, *Biotechniques* **31**.

[2] J. Garcia-Sucerquia, 2012, Color lensless digital holographic microscopy with micrometer resolution, *Opt. Lett.* **37**, 1724-1726.

[3] F. Dubois and C. Yourassowsky, 2012, Full off-axis red-green-blue digital holographic microscope with LED illumination, *Opt. Lett.* **37**, 2190-2192.

[4] J. P. Ryle, S. McDonnell, and J. T. Sheridan, 2011, Lensless multispectral digital in-line holographic microscope, *Journal of Biomedical Optics* **16**, 126004-126004.

[5] S. Seo, T. W. Su, A. Erlinger, and A. Ozcan, 2008, Multi-color LUCAS: Lensfree on-chip cytometry using tunable monochromatic illumination and digital noise reduction, *Cellular and Molecular Bioengineering* **1**, 146-156.

[6] M. K. Kim, 1999, Wavelength-scanning digital interference holography for optical section imaging, *Optics Letters* **24**, 1693-1695.

[7] M. Brunel, H. Shen, S. Coetmellec, D. Lebrun, and K. Ait Ameur, 2012, Femtosecond digital in-line holography with the fractional Fourier transform: application to phase-contrast metrology, *Appl. Phys. B* **106**, 583-591.

[8] O. Mendoza-Yero, E. Tajahuerce, J. Lancis, and J. Garcia-Sucerquia, 2013, Diffractive digital lensless holographic microscopy with fine spectral tuning, *Opt. Lett.* **38**, 2107-2109.

[9] O. Mendoza-Yero, A. Calabuig, E. Tajahuerce, J. Lancis, P. Andrés, and J. Garcia-Sucerquia, 2013, Femtosecond digital lensless holographic microscopy to image biological samples, *Optics Letters* **38**, 3205-3207.

[10] S. K. Jericho, P. Klages, J. Nadeau, E. M. Dumas, M. H. Jericho, and H. J. Kreuzer, 2010, In-line digital holographic microscopy for terrestrial and exobiological research, *Planetary and Space Science* **58**, 701-705.

[11] J. Garcia-Sucerquia, W. Xu, S. K. Jericho, P. Klages, M. H. Jericho, and H. J. Kreuzer, 2006, Digital in-line holographic microscopy, *Appl. Opt.* **45**, 836-850.

[12] P. Petruck, R. Riesenberg, and R. Kowarschik, 2012, Partially coherent light-emitting diode illumination for video-rate in-line holographic microscopy, *Appl. Opt.* **51**, 2333-2340.

[13] J. Garcia-Sucerquia, 2013, Noise reduction in digital lensless holographic microscopy by engineering the light from a light-emitting diode, *Appl. Opt.* **52**, A232-A239.

[14] P. Petruck, R. Riesenberg, and R. Kowarschik, 2012, Optimized coherence parameters for high-resolution holographic microscopy, *Appl. Phys. B* **106**, 339-348.

[15] D. Gabor, 1948, A new microscopic principle, *Nature* **161**, 777–778.

[16] D. Gabor, 1949, Microscopy by reconstructed wave-fronts, *Proc. R. Soc. London A* **197**, 454.

[17] D. Gabor, 1951, Microscopy by reconstructed wave fronts: II, *Proc. Phys. Soc. London B* **64**, 449.

[18] H. J. Kreuzer, H. W. Fink, H. Schmid, and S. Bonev, 1995, Holography of holes, with electrons and photons, *Journal of Microscopy* **178**, 191-197.

[19] J. Garcia-Sucerquia, D. C. Alvarez-Palacio, M. H. Jericho, and H. J. Kreuzer, 2006, Comment on "Reconstruction algorithm for high-numerical-aperture holograms with diffraction-limited resolution", *Optics Letters* **31**, 2845-2847.

[20] D. Alvarez-Palacio and J. Garcia-Sucerquia, 2010, Digital in-line holographic microscopy with partially coherent light: Micrometer resolution, *Revista Mexicana de Fisica* **56**, 445-448.

Journal of Physics: Conference Series **605** (2015) 012012 doi:10.1088/1742-6596/605/1/012012

Optical elements with extended depth of focus and arbitrary distribution of intensity along the focal segment obtained by angular modulation of the optical power

K Kakarenko[1], I Ducin[1], Z Jaroszewicz[2,3], A Kołodziejczyk[1], K Petelczyc[1], A Stompor[1] and M Sypek[1]

[1]Faculty of Physics, Warsaw University of Technology, Koszykowa 75, 00-662 Warsaw, Poland

[2]Institute of Applied Optics, Kamionkowska 18, 03-805 Warsaw, Poland

[3]National Institute of Telecommunications, Szachowa 1, 04-894 Warsaw, Poland

E-mail: mmtzjaroszewicz@post.pl

Abstract. Light Sword Lens (LSL), i.e., an optical element with extended depth of focus (EDOF) characterized by angular modulation of the optical power in its conventional form is characterized by a linear relationship between the optical power and the angular coordinate of the corresponding angular lens sector. This dependence may be manipulated in function of the required design needs. In the present communicate this additional degree of freedom of design is used for elimination of the LSL shape discontinuity.

1. Introduction

One of possible ways to classify the family of EDOF optical elements with explicitly given transmittance can be made according to their mapping function, i.e., the relationship between the part of the element's area, from which light is focused into the corresponding point of the focal segment and this point. The most famous example is axicons, where the mapping function ties an infinitesimal annulus within the aperture of the element with the corresponding focal point [1-5]. In this communicate we will focus our attention on the LSL, where the mapping function joins the infinitesimal sector of the element's surface and the given focal point [6-8].

LSL has been shown recently to have promising properties for presbyopia compensation [9-12], and therefore in the present communicate we would like to address a possibility of the compensation of the shape discontinuity, which is characteristic for the LSL in its hitherto design.

2. Elimination of the LSL's discontinuity of shape

The distribution of the optical path difference due to the LSL is given as follows [9], [10]:

$$\Delta l(r,\ \theta) = -\frac{D_0 r^2}{2} - \frac{\Delta D \theta r^2}{4\pi} \tag{1}$$

where $(D_0,\ D_0 + \Delta D)$ stands for the range of the optical powers created by the element. The phase distribution of the kinoform counterpart of the LSL, the shape of the LSL itself, as well as the geometry of the focal segment formation process is shown in Fig.1.

As one can see, a characteristic feature of the LSL is a discontinuity of its shape (phase discontinuity in the case of the LSL's diffractive counterpart) due to the direct neighborhood of sectors given by $\theta=0$ $(\Delta l=D_0)$ and by $\theta=2\pi$ $(\Delta l= D_0 +\Delta D)$. This discontinuity can be a source of an additional boundary wave forming an undesired interference pattern with the main wave forming the focal pattern or in the incoherent case may be a cause of the noise increase. Additionally, from purely mechanical point of view such sharp step may be reason of inconvenience and may cause additional problems during the manufacture of the LSL.

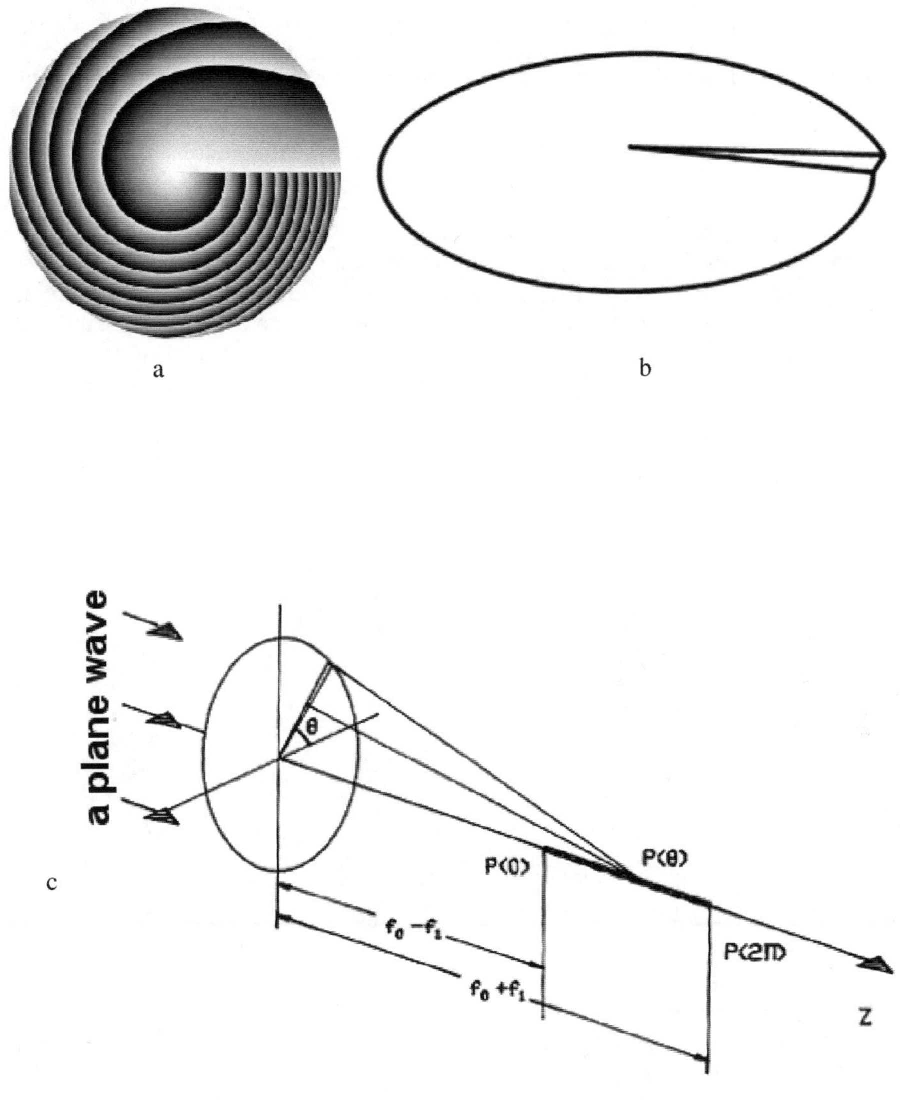

Fig.1. LSL: a) phase distribution of the diffractive kinoform version); b) refractive version; c) geometrical scheme of the design – the infinitesimal angular sector of the LSL with a parameter θ focuses light into a point P(θ) of the focal segment.

Here we would like to draw attention for another feature of LSL's, namely the linear dependence of the optical power on the angular coordinate θ given by Eq.(1) can be changed and consequently the constant value of the axial intensity within the whole focal segment in principle can be substituted by an arbitrary one. The reasons for doing so can be various. One of motives could be a requirement to make some range or ranges of objects location privileged. In our case we are choosing a possibility to remove the phase (or shape in the refractive version) discontinuity present on the radius corresponding to transition from 0 to 2π radians. Let us see that there could be found at least two motives for such proceeding. First one is of technological nature: refractive LSLs are easier to fabricate when such discontinuities of shape do not take place. The second one, in the case of possible ophthalmologic application is of physiological nature: it turns out that a typical value of the jump on the 2π line, e.g., in the case of a presbyopia correcting lens would be of order of several tens of micrometers.
Indeed, assuming the radius of the lens $R=4\ mm$, the focal length $f_0 = 18.5mm$ (which results from typical value of optical power $D_0 = 54D$ and formula $f_0 = 1/D_0$) and $\Delta f=1.5\ mm$, what in turn results from the assumed range of optical power equal to $4D$ we obtain that

$$h \approx \Delta D R^2/2 \text{ or } \Delta f R^2/2f_1^2 \text{ i.e., } 35\mu m. \qquad (2)$$

In order to eliminate this discontinuity we introduce into the LSL surface an angular sector of width equal to Θ radians, where the dioptric power increases gradually from the value D_0 to the value $D_0+\Delta D$ oppositely to the remaining sector occupied by common LSL in such a way that both junctions remain continuous (Fig.2). The optical features of such element for satisfactorily great value of Θ remain almost unchanged, what allows substituting the old structure with the new one.

$\Theta = \frac{\pi}{4}$ (45°)	$\Theta = \frac{\pi}{2}$ (90°)	$\Theta = \frac{3\pi}{4}$ (135°)	$\Theta = \pi$ (180°)

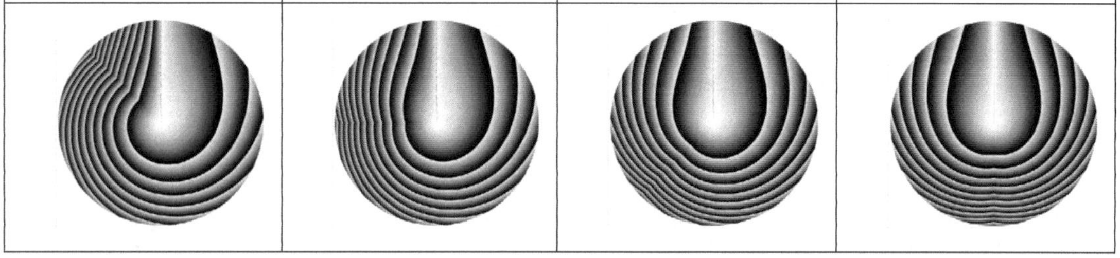

Fig.2. Modified LSL without the phase discontinuity in a kinoform version with an angular sector of oppositely growing dioptric power of width $\Theta = \pi/4$, $\pi/2$, $3\pi/4$, and π, respectively.

We conducted numerical simulations of imaging for all elements shown in Fig.2 as well as for previous version of LSL, marked in the upper row of Fig.3 as LSL, yet with phase discontinuity and without any angular sector by means of software based on the modified convolution method [13, 14]. We assumed monochromatic, spatially incoherent light with a wavelength 555 nm corresponding to the highest sensitivity of the human eye in photopic vision [15, 16].

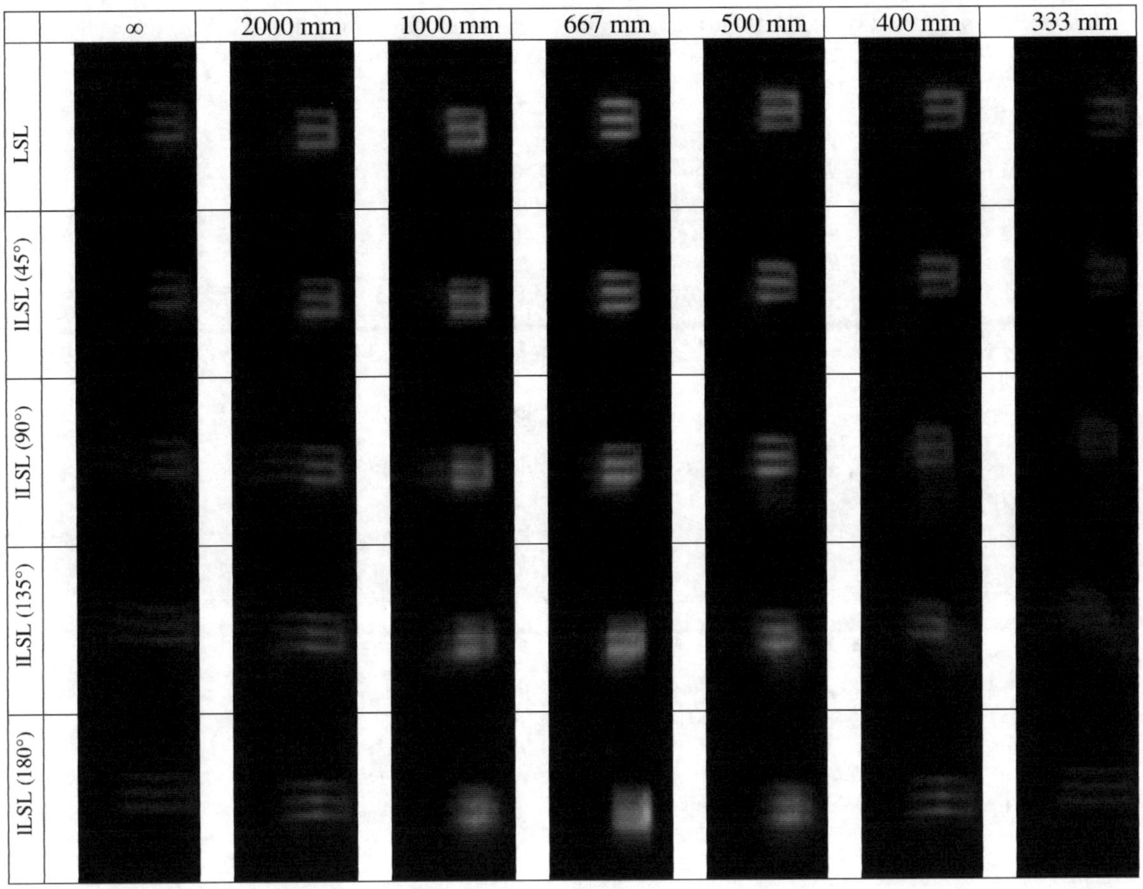

Fig.3. Images of letter E obtained with conventional LSL and modified LSL with an angular sector of oppositely growing dioptric power of width $\Theta = \pi/4$, $\pi/2$, $3\pi/4$, and π, respectively. All elements were assumed in a kinoform version. The distances of the object were equal to 5000 mm, 2000 mm, 1000 mm, 667 mm, 500 mm, 400 mm and 333 mm from left to the right.

It can be seen that the quality of imaging, though continuously deteriorating with the increasing angle of the discontinuity removing sector, still offers satisfactorily broad range of angles, where remains comparable with the conventional LSL.

3. Conclusions

Summing up it can be said that the elements with angular modulation of optical power turn out to be an alternative for the purposes of imaging with extended depth of focus. Such conclusion results from independence of their optical power on aperture [12] and better quality of obtained images in comparison with EDOF optical elements characterized by symmetry of revolution.

Moreover, these elements, as shown in the present communicate, allow additionally for flexible manipulation of their distribution of the optical power and thereby for easier adaptation of the element for the particular needs.

On the other hand it should be bear in mind that the price for these possibilities is their imperfect imaging as well as tiny waving of produced images for different points along the focal segment [7-11].

4. Acknowledgements
This work is supported by the National Centre for Research and Development (NCBiR) under Grant No. PBS/A9/23/2013 entitled: „Model of the intraocular lens for vision with extended depth of view".

5. References
[1] J.H. McLeod J H 1954 *J. Opt. Soc. Am.* **44** 592
[2] Dyson J 1958 *Proc. R. Soc. London Ser. A* **248** 93
[3] Vasara A, Turunen J, and Friberg A.T. 1989 *J. Opt. Soc. Am. A* **6** 1748
[4] Jaroszewicz Z 1997 *Axicons. Design and propagation properties*, in *Research and Development Treatises* vol.5, ed. M Pluta (Warsaw: Polish Chapter of SPIE)
[5] Sochacki J, Kołodziejczyk A, Jaroszewicz Z, and Bará S 1992 *Appl. Opt.* **31**, 5326
[6] Kołodziejczyk A, Bará S, Jaroszewicz Z, and Sypek M 1990 *J. Mod. Opt.* **37**, 1283
[7] Mikuła G, Jaroszewicz Z, Kołodziejczyk A, Petelczyc K, and Sypek M 2007 *Opt. Express* **15**, 9184
[8] Ares García J, Bará S, Gomez García M, Jaroszewicz Z, Kołodziejczyk A, and Petelczyc K 2008 *Opt. Express* **16**, 18371
[9] Petelczyc K, Ares García J, Bará S, Jaroszewicz Z, Kakarenko K, Kołodziejczyk A, and Sypek M 2011 *Opt. Express* **19**, 8693
[10] Petelczyc K, Bará S, Ciro López A, Jaroszewicz Z, Kakarenko K, Kołodziejczyk A, and Sypek M 2011 *Opt. Express* **19**, 25602
[11] Arias Galego A, Bará S, Jaroszewicz Z, and Kołodziejczyk A 2012 *Optom. Vis. Sci.* **89**, 1702
[12] Petelczyc K, Ares Garcia J, Bará S, Jaroszewicz Z, Kołodziejczyk A, Sypek M 2009 *Phot. Lett. Poland* **1**, 55
[13] Sypek M 1995 *Opt. Commun.* **116**, 43
[14] Sypek M, Prokopowicz C, and Gorecki M 2003 *Opt. Eng.* **42**, 3158
[15] Valberg A, 2005 *Light Vision Color* (New York: John Wiley & Sons).
[16] Gross H, Blechinger F, and Achtner B, 2008 *Survey of Optical Instruments* vol.4 *Handbook of Optical Systems,* (Weinheim: Wiley-VCH).

Functional optical coherence tomography (fOCT) biospeckle imaging to investigate response of plant leaves to ultra-short term exposure of Ozone

L K T Srimal[1,2], U M Rajagopalan[3] and H Kadono[1]

[1]Graduate School of Science and Engineering, Saitama University, 255 Shimo-okubo, Sakura-ku, Saitama-shi, Saitama-ken 338-8570, Japan,
[2]Department of Mechanical and Manufacturing Engineering, Faculty of Engineering, University of Ruhuna, Hapugala, Galle 80000, Sri Lanka
[3]Faculty of Food Life Sciences, Toyo University, 1-1-1 Izumino, Itakura-machi, Ora-gun, Gunma-ken 374-0113, Japan

E-mail: thanujasrimal@gmail.com

Abstract. In this study, response of leaves of Chinese chives (*Allium tuberosum*) to ozone stress was investigated using functional optical coherence tomography (fOCT) based on biospeckle. The biospeckles arising out of dynamic motion of organelles can reflect the biological activities of plant. The fOCT biospeckle image was obtained by calculating the standard deviation (SD) of the fOCT temporal signal (biospeckle signal) at each and every point from the successively acquired OCT images. Plant leaves were subjected to treatment under different concentrations of O_3, and imaging data were acquired from back and front surfaces of the leaves. The internal cell structure within the Chinese chives leaves could be clearly visualized in the functional OCT biospeckle image, which was not clearly visible in conventional OCT cross-sectional image. The SDs were found to be increasing significantly, especially in the surface layers of both front and back sides of the leaf with ozone exposure. Thus, the fOCT based on biospeckle is found to be suitable for fast, non-destructive monitoring environmental stresses on plants, which can potentially lead to significant time saving, for which conventional techniques require a few days to a few weeks time.

1. Introduction

Ozone (O_3) is generally conceived as occurring only in the form of beneficial stratospheric O_3, a natural screen from the harmful effects of ultraviolet (UV) radiation. Although stratospheric O_3 is decreasing, concentration of surface O_3 in the troposphere (troposheric O_3) is increasing due to complex photochemical reactions involving volatile organic compounds (VOCs) and nitrogen oxides (NOx) in the presence of UV light [1]. For example, tropospheric O_3 concentrations exceeding 80ppb occur regularly, in China, Japan, and even in rural areas of India [2].

Tropospheric O_3 enters the leaf through stomata, the pores for photosynthetic gas exchange. It is known that O_3 causes biochemical and physiological changes leading to the direct or indirect inhibition of photosynthesis, extra cost of repairing any cellular and metabolic damage, and a consequent decrease in plant growth. Adverse effects on plant photosynthesis were identified as a major factor limiting growth and yields of crop under high O_3 concentrations [3]. Yield loss is caused by the cumulative effect of chronic daily O_3 exposure over the growth period [4].

Biomass, height change, and counting are commonly employed techniques to monitor the effect of O_3 stress on plants [5]. Those require long waiting periods, and the changes in the internal structure of the plant cannot be recognized in real time. So, there is a need of fast and reliable method to assess the structural (i.e. functional) changes in plants against O_3 stress. The majority of methods for investigating internal structure of plants are invasive. They cause damage to the tissue under study, thereby limiting their application for *in-vivo* monitoring. An example of such a destructive approach is conventional optical microscopy which requires a destruction of the sample and a subsequent analysis, thus plants responses to external stress are inferred indirectly. In addition, the analysis cannot be performed in real time.

To overcome these limitations, we proposed to use optical coherence tomography (OCT). OCT [6] is an interferometric technique that detects internal reflected light and can provide non-contact two or three dimensional *in-vivo* tomographic images of the internal tissue structure with very high resolution of a few micrometers. It has generated numerous biological applications, most commonly in the fields of ophthalmology [7] and dermatology [8]. More recently, OCT has also been used in the field of botany to visualize the inner structure of tissue [9] and to study plants suffering from pathogen attack with conventional structural OCT imaging [10]. However, its potential in monitoring functional changes, in other words, biological activities inside tissue has not been well developed. In this study, we proposed to use biospeckle signal to realize a functional OCT (fOCT) [11] to monitor the effect of O_3 on plants.

Processes such as cytoplasmic streaming, organelle movement, cell growth and division, and biochemical reactions are the biological activities occurring inside of plant leaf [12]. Behavior of these activities are changed under the influence of environmental stresses. We developed, and applied, proposed fOCT based on biospeckle to monitor the changes in biological activities of plants. Biospeckle is optical information that has been getting attention during last fifteen years for evaluation of properties of biological materials to monitor biological activities of plants [13].

The speckle pattern is generated by interference created by the backscattering of a random distribution of scatters at some distance from a scattering object under the illumination of a coherent light such as a laser. If there is no movement in scatters and no change in the structure of the object, the speckle pattern is stable in time. However, in the case of biological samples, the speckle pattern shows a dynamic property in time due to the moving organelles and particles within the sample. This dynamic speckle pattern is characteristic for biological tissues and has been called a biospeckle [14]. Therefore, the dynamic characteristics of the biospeckle can reflect the biological activities of a living substance [15].

To realize the biospeckles fOCT, OCT cross-sectional images were acquired successively. Magnitude of the biospeckle signal, in other words, standard deviation (SD) of the signal was calculated along time axis of each spatial position from the array of OCT cross-sectional images. Then a fOCT biospeckle image was constructed from the SDs of biospeckle signals. Magnitude (SD) of the time varying biospeckle signal reveal different biological information and it was proposed as a measure to evaluate the changes in biological activities against O_3 stress on plants.

In the experiments, *Allium tuberosum,* commonly known as Chinese chives was exposed to different concentrations of O_3, and the effects to both back and front side of the leaf were observed to validate the developed fOCT based on biospeckle.

2. Optical System and Method

2.1. Optical system

SD-OCT system was constructed in this study to achieve high resolution OCT imaging. The schematic diagram of the SD-OCT system is shown in Figure 1. The light source used in the system is a super luminescent diode (SLD) with a central wavelength 836.1nm and a band width of 55.2nm. A 1×2 fiber coupler splits the incident light into sample (99%) and the reference (1%)beams. Reference arm consist of collimating lens L1, objective lens L4, and mirror M1. The backscattered light from the sample and the light reflected from the reference mirror are recombined by the coupler. By means of a circulator, those lights are guided into a spectrometer consisting of a collimator, lens L5, and grating. The collimated light enters grating, and the spectral interference signal is obtained by the line-scan camera through lens L6. We used a line scan CCD camera (L104k-2k, BASLER, Germany) with 2048 pixels and acquisition rate of 25,000 lines per second. The axial resolution (depth resolution) is given by [16],

$$\Delta z = \frac{2\ln \lambda_o^2}{\pi n \Delta\lambda}, \quad (1)$$

where, λ_o , Δ□are central wavelength, and band width of the light source, respectively. n is the refractive index of the tissue. The depth resolution in free space was estimated to be 6μm in and the lateral resolution was estimated to be 22μm (NA=0.024).

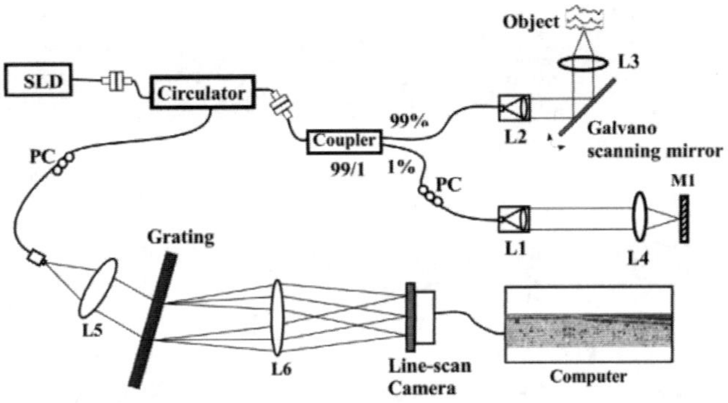

Figure 1. Experimental setup of spectral domain optical coherence tomography system. SLD:super luminescent diode, L1~L6:lenses, M1:mirror, PC:polarization controller, DAQ:data acquisition device, IMAQ:image grabber.

Galvano-mirrorscanners in the sample arm permits a lateral scanning in x and y directions. The incident power delivered to the leaf sample was approximately 5mW, that gave an irradiance lower than the damage threshold for plant leaves [17].

2.2. Standard deviation of biospeckle signal

Acquired spectral interference signal from the line scan camera was first transformed into wave number, k, space from wavelength, λ, space that is called as a rescaling. Next, the signal was Fourier transformed in order to obtain the depth resolved reflectivity profile of the sample at the focal spot of the sample beam for a fixed lateral position (A-scanning). multiple A-scans were performed while the galvano-mirror swept the probing beam across the sample, and the sequentially acquired line image (B-scanning) were stored in the computer into a two-dimensional cross-sectional image of the sample. One image of size 1024×2500 pixels was acquired with an acquisition rate of 10 frames per second.

A certain target region of the leaf image of size 65×2500 pixels (223×4000 μm^2) acquired at time t_i was chosen from the whole cross-sectional image $I_{OCT}(x,y;t_i)$. Then, a fOCT biospeckle image $I_b(x,y;t_i)$ was obtained by calculating the standard deviation (SD) of the fOCT biospeckle signal at each and every point in the selected portion of the OCT cross-sectional image along the time axis (from the successively acquired OCT cross-sectional images) as given by,

$$I_b(x,y,t_i)=\left[\frac{1}{N}\sum_{j=1}^{N}\left\{I_{OCT}(x,y,t_{i+j})-\langle I_{OCT}(x,y,t_i)\rangle\right\}^2\right]^{\frac{1}{2}} \tag{2}$$

Where,

$$\langle I_{OCT}(x,y,t_i)\rangle=\frac{1}{N}\sum_{j=1}^{N}I_{OCT}(x,y,t_{i+j}) \tag{3}$$

and N is the number of averaging frames.

2.3. *Plant and ozone exposure system*

The O_3 exposure system consists of air pump, charcoal filters (CF), an O_3 generator (OES 10A, Dylec Inc., Japan), an O_3 monitor (Model 1150, Dylec Inc., Japan), main O_3 chamber (33.5×33.5×58 cm^3). A specially designed leaf O_3 chamber (10×10×2 cm^3), made up of plastics, was connected to the main O_3 chamber to the leaf of the plant under O_3 exposure. Generated O_3 was fed into the main O_3 chamber, and its concentration was precisely controlled by measuring the O_3 concentrations with the O_3 monitor and feeding the signal back to the generator. The O_3 generation system approximately took 9 minutes, 10 minutes, and 14 minutes to reach the target O_3 concentrations of 80 ppb, 120 ppb, and 240 ppb, respectively. The O_3 feedback control system could precisely stabilize the concentration of O_3 within the leaf and main O_3 chambers at an accuracy of around 0.5% during the experiments.

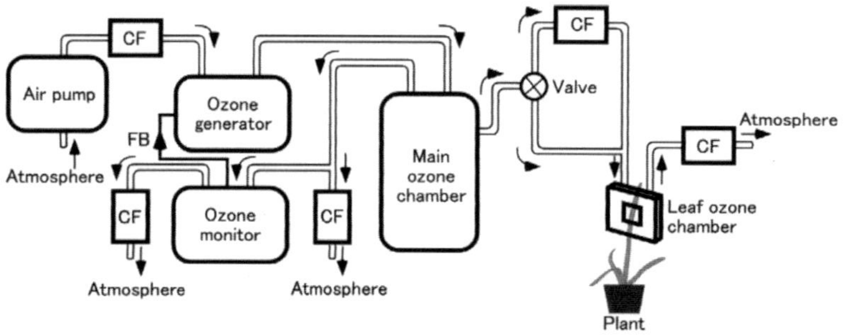

Figure 2. Schematic of O_3 exposure system used in this study. CF:charcoal filter, FB:feed back signal.

Chinese chives plants were grown in a growth chamber (Conviron, Controlled Environmental Ltd, Canada) for 30 days. We maintained the relative humidity at 55%-65%, and respective day/night cycle of 12/12 hours at a light irradiance of 260-350 $\mu molm^{-2}s^{-1}$/ 0 $\mu molm^{-2}s^{-1}$ and an air temperature of 25°C/18 °C. The plants were kept regularly watered. In the experiment, healthy young leaves of the plants, one to two weeks old, were used and mounted in the leaf O_3 chamber. Three halogen lamps with the fiber optic light guides (PHL-150, MEJIRO PRECISION, Japan) were used to illuminate the plant leaf with a light irradiance of 500 $\mu molm^{-2}s^{-1}$ during experiments.

2.4. *Timing protocol and data acquisition*

The experiments were carried out with Chinese chives under O_3 concentrations of 0, 80, 120, and 240 ppb to investigate the influence of O_3. The entire experimental duration was 8.5 hours (Figure 3). As an initial condition, a continuous air flow (10 liter/min.) after passing the CF was fed into the leaf O_3 chamber (Figure 2), where the leaf was mounted, for two hours to acclimate to the environmental condition prior to the experiment. After acclimation, for each O_3 exposure condition, next half an hour

Journal of Physics: Conference Series **605** (2015) 012013 doi:10.1088/1742-6596/605/1/012013

was allowed for settling of the desired O_3 concentration levels (0, 80, 120, and 240 ppb) in the main O_3 chamber. At the end of this half an hour period, the air containing O_3 with the desired concentration will reach leaf without passing through the CF by switching off the valve as shown in Figure 2. The O_3 exposure was carried out for three hours followed by another three hours of CF air. The O_3 concentrations of 120 ppb and 240 ppb were chosen based on the criterion used in Japan, where a warning and a serious warning are issued for 120 ppb and 240 ppb of O_3 concentrations, respectively, in Japan. Meantime, 0 ppb was chosen for control experiments. Both the temperature and the humidity were kept constant during a single session of experiment with respective variations, ±2% and ±1% respectively. Measurements were performed at the position 5 cm away from the apex of the leaf.

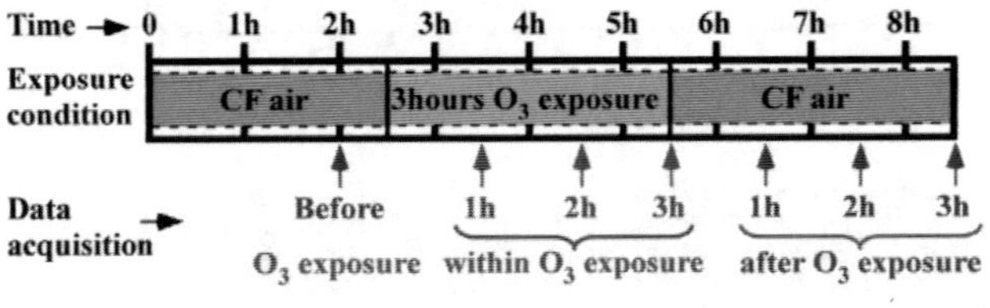

Figure 3. Timing protocol of O_3 exposure and data acquisition.

OCT imaging data were acquired (Figure 3) sequentially, before, and at 1 hour, 2 hours, and 3 hours after starting O_3 exposure (*within*), and at 1hour, 2hours, and 3hours after stopping of O_3 exposure (*after*). Same procedure was applied to the measurements for the front and back surfaces of leaves identically. O_3 concentrations of 0, 80, 120, and 240 ppb were used. Six replicate experiments were performed under each O_3 concentration.

3. Results and discussion

3.1. Anatomy

Micro-structural cross-sectional observation was done for a Chinese chives leaf (Figure 4). Epidermal cells (10 μm thick layer), oval shape mesophyll cells (around 100 μm in height), vascular bundle, stomata were clearly visible. Here, epidermal layer is the outer most cell layer of the plant leaf, which protects leaf against water loss, and regulate gas exchange. It is a single layered structure. Mesophyll cell is the primary location of photosynthesis in the leaf. And these cells govern most of the leaf interior. Vascular tissues help for the transport system for fluids and nutrient internally. They are arranged in long, discrete strands called vascular bundle [18].

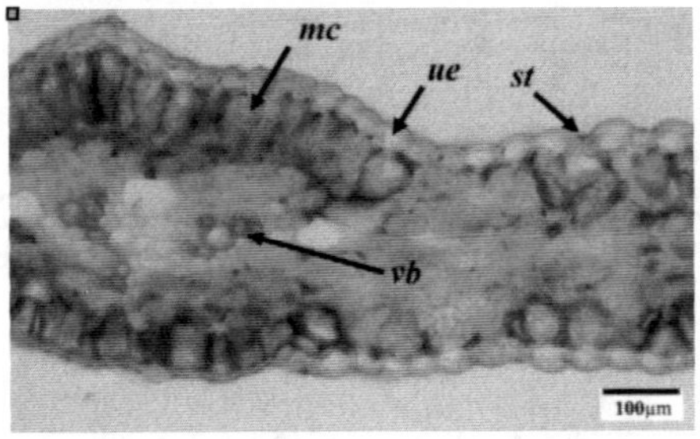

Figure 4. Optical microscopy analysis of cross-section of a leaf of Chinese chives. ue:upper epidemis, mc:mesophyll cell, st:stomata, vb:vascular bundle.

3.2. OCT observation-structural and biospeckle imaging

Conventional OCT cross-sectional imaging can visualize internal structure of biological tissues. Figure 5(a) shows conventional OCT cross-sectional image of a Chinese chives leaf. Epidermal cell layer could be clearly seen. However, layer such as mesophyll cells was not clearly visible. Figure 5(c) shows corresponding fOCT biospeckle image. The depth structure became clear. Comparing with the microscopic image in Figure 4, upper epidermal layer and the distinctive more oval shaped mesophyll cells were clearly seen in the fOCT biospeckle image.

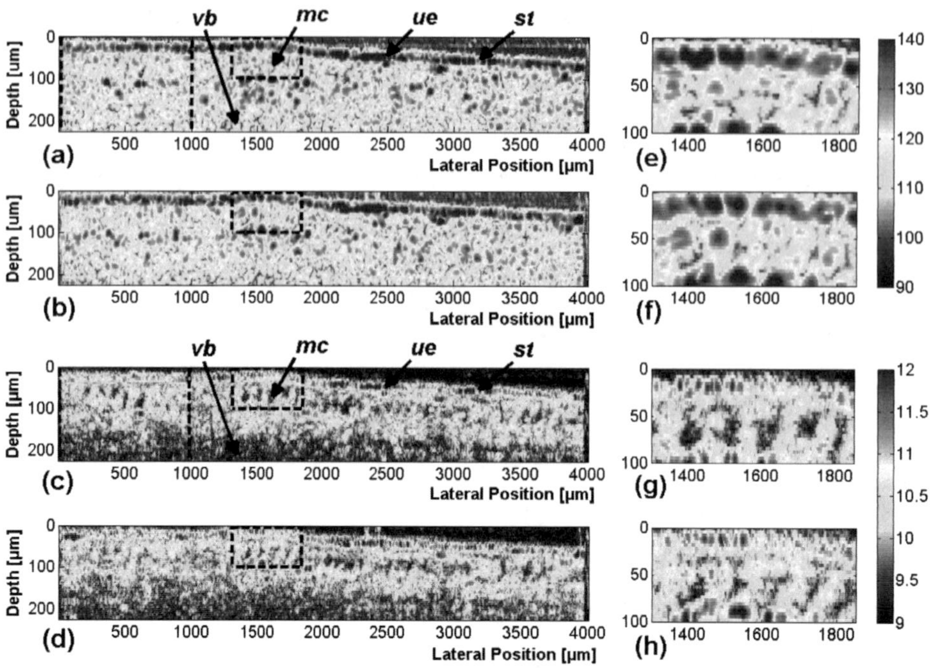

Figure 5. Conventional structural OCT cross-sectional images (a) before exposure of O_3, (b) 3 hours after starting O_3 exposure. Corresponding fOCT biospeckle images (c) before exposure of O_3, (d) 3 hours after starting O_3 exposure. Magnified views of the mesophyll cell layer encased in black dotted squares shown in (a), (b), (c), and (d) are shown, respectively in (e), (f), (g), and (h). ue:upper epidemis, mc:mesophyll cell, ve:vesicular bundle, st:stomata.

3.3. OCT observation-O_3 effect on leaf

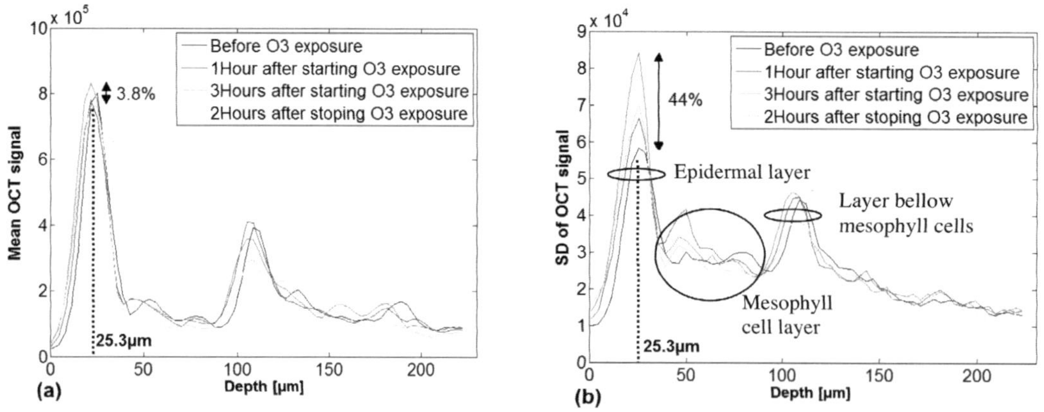

Figure 6. (a) Depth profiles of selected area of conventional structural OCT cross-sectional images, (b) Corresponding depth profiles of fOCT biospeckle images.

The experiments of O_3 exposure were conducted according to the timing protocol described in section 2.4. Figures 5(a), and (b) show conventional structural OCT image of Chinese chives leaf (back surface) acquired before O_3 exposure, and 3 hours after starting O_3 exposure, respectively. Figures 5(c), and (d) represent the corresponding fOCT biospeckle (SD) images before O_3 exposure, and 3 hours after starting O_3 exposure. It was difficult to find a clear difference in conventional OCT images with O_3 exposure (Figures 5(a), (b), (e), and (f)). However, a variation due to O_3 exposure could be seen in the fOCT biospeckle images (Figures 5(c), (d), (g), and (h)). Oval shaped periodic structure that corresponds to mesophyll cells became unclear with even 3 hours short term O_3 exposure in biospeckle images. Earlier studies confirmed that of cells belonging to mesophyll layer as the main target of O_3 injury [19]. So, biospeckle signals coming from different depth layers were analyzed for their temporal characteristics (Unpublished).

Laterally averaged depth profile of conventional structural OCT images so as to improve S/N ratio were obtained under conditions of before, 1, and 3 hours after starting as well as 2 hours after stopping O_3 exposure and are shown in Figure 6(a). Figure 6(b) shows corresponding depth profiles of the fOCT biospeckle images. Here, laterally averaged depth profiles, were obtained from the lateral position of 0 to 1000 μm as indicated by the dark dotted line in the OCT structural image and the corresponding biospeckle image (Figures 5(a), and (c)).

Under all the different O_3 exposure conditions, maximum values were obtained at a depth position of 25.3 μm in the conventional structural OCT images, and it corresponded to the upper epidermal layer of the leaf. The variation of the values between before and after 3 hours of O_3 exposure was 3.8%. On the other hand, the corresponding value of fOCT biospeckle images was significantly increased to 44%. This variation with O_3 exposure in the depth profile of fOCT biospeckle image was larger for epidermal layer in comparison the other deeper layers. Here, the effect of O_3 on mesophyll cell layer seen in Figures 5(g) and (h) was degraded in the depth profile of fOCT biospeckle images due to the lateral average. Therefore, the maxima of averaged depth profiles corresponding to the epidermal layer of fOCT biospeckle images at each time interval such as before, after starting and after stopping of O_3 exposure were used for further analysis in following discussion.

In order to evaluate the effect of exposure of O_3, a normalized parameter was introduced. SDs of biospeckle signals obtained from the epidermal layer corresponding to a depth position of 25.3 μm under O_3 exposure conditions were normalized by the SDs obtained before O_3 exposure, SD_b. Furthermore, averaged normalized standard deviation (ANSD) was defined as,

$$ANSD = \left\langle \frac{SD}{SD_b} \right\rangle, \tag{5}$$

where, $\langle ... \rangle$ denotes average over 18 measurements. The normalized SDs were calculated across each hour of 1 hour, 2 hour and 3 hours and averaged over six different samples thus totaling to 18 measurements. The term '*within*' was used to refer to the average of SDs after starting O_3 exposure while term '*after*' referred to the average after stopping O_3.

The measurements for back and front sides of the leaf were performed independently as a different series of experiments. The experimental results for the back and front surfaces of the leaf are shown in Figures 7(a), and (b), respectively for O_3 concentrations of 0, 80, 120, and 240 ppb.

It can be seen from the Figures 7(a), and (b) that SD becomes larger for both back and front surfaces as the concentration of O_3 becomes larger. For the back surface of the leaf, the differences in ANSDs between before and within O_3 exposure to 80, 120, and 240ppb were significant (t-test, $P < 0.01$, Figure 7(a)). In addition, the differences in ANSDs between before and after O_3 exposure for 80, 120, and 240ppb were also significant (t-test, $P < 0.01$, Figure 7(a)).

Comparing Figures 7(a), and (b), the ANSDs of the front surface were smaller than those on back surface. For the front surface of the leaf, the differences in ANSDs between before and within, and also before and after O_3 exposure of 240 ppb concentration were significant (t-test, $P < 0.01$, Figure 7(b)). However, the differences in ANSDs between before and within, and also before and after O_3 exposure to 80, and 120 ppb concentrations were not significant.

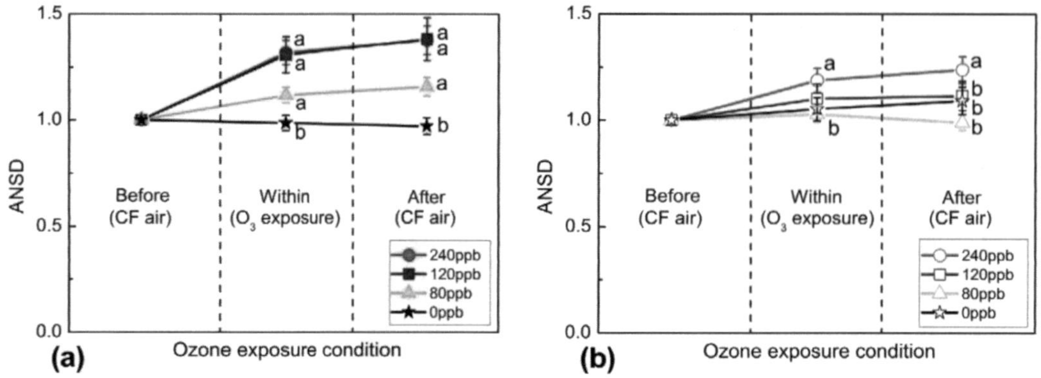

Figure 7. ANSDs of biospeckle signals of Chinese chives leaves under O_3 exposures of 0, 80, 120, and 240ppb. Each average value is plotted with standard error for 18 measurements, for (a) back surface, and (b) front surface. The ANSD within O_3 exposure and after exposure were compared with that before exposure using statistical two samples t-test. (a, $P < 0.01$; b, not significant).

The differences in ANSD between the front and back surfaces for concentrations of 120 ppb and 80 ppb O_3 exposure were 26.7% and 13%, respectively. On other hand, the corresponding values for 240 ppb O_3 exposure was reduced to13%. This may due to the saturation of the reaction to O_3 exposure of the back surface of the leaf. For the different responses of the front and back sides of the leaf for O_3 exposures, the stomata may play a significant role since O_3 is taken up through the stomata.

3.4. Stomatal influence on ANSD under O_3 exposure

O_3 taken up through stomata reacts with water and intercellular components to generate highly reactive oxygen species (ROS), that include peroxides and free radicals. The free radicals are probably believed to be the real cause of the negative effect of O_3. Plants have a number of mechanisms that chemically and enzymatically remove these toxic compounds avoiding cellular damage [20]. It may be the reason for increase of fluctuation of biospeckle signal coming from outermost epidermal layer. O_3 stress affects the photosynthetic activity while altering leaf and root respiration rates [21]. Therefore, stomata densities on both sides of the leaves were examined. Figure 8 shows a microstructural observation of epidermal layer removed from a Chinese chives leaf. Number of stomata was counted within a circular area of radius 940 μm, and the stomata densities were calculated. The stomata densities on the front and back sides of the leaves were 11.1±1.7 and 12.3±1.5, respectively. Therefore, a significant difference could not be found.

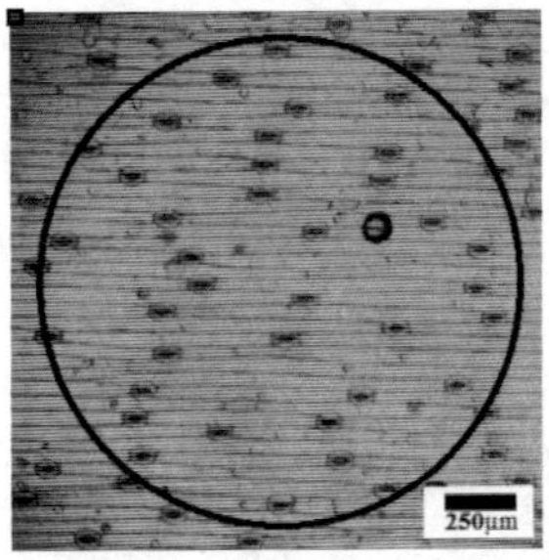

Figure 8. Optical microscopy analysis for stomata on leaf back surface.

Moreover, from the cross-sectional image of the leaf shown in Figure 4. it was difficult to find a clear structural difference between front and back sides. However, roughness of back surface of the leaf is larger and hence has a larger surface area than front surface. This may result in the difference in the total number of stomata on the front and back surfaces. In addition, stomatal conductances [22] may be different for the front and back surfaces. However, they remain the focus for our future research.

4. Summary

We proposed, developed, and applied the fOCT biospeckle imaging technique for monitoring O_3 stress on plants. The internal cell structure, specially mesophyll cell layer, within the Chinese chives leaf can clearly been seen in the fOCT biospeckle image, which is not clearly visible in conventional OCT cross-sectional image. Furthermore, it was experimentally demostrated that the influence of O_3 on each plant structure, e.g., epidermal and mesophyll cell layers, could be individually monitored even for short-term O_3 exposures.

In this study, we found that, epidermal layers of both surfaces of the leaf showed an increased fluctuation in biospeckle signal under the O_3 stress. Further, the study also found a larger fluctuation against O_3 exposure for back side than those for front surface. These results indicate that the epidermal layer of back side of Chinese chives leaf is more sensitive to O_3 stress than the front surface.

The presented study suggests possible applications of developed fOCT biospeckle imaging and signal analysis technique, for investigating the immediate impact of environmental factors. Especially even for O_3 exposure, for a very short span of time such as a number of seconds or a few minutes the proposed fOCT biospeckle was found to be effective which is not possible with conventional experimental methods and even with conventional OCT imaging.

Acknowledgements

This work was partly supported by the Grant-in-Aid for Science Research in JSPS (No. 24651023) of the Ministry of Education, Culture, Sports, Science and Technology in Japan.

References

[1] Oltmans S J *et al* 1998 *Geophysical Research Letters* **25** 139–42
Ashmore M R 2005 *Plant, Cell and Environment* **28** 949–64
Rathnayake A P, Kadono H, Toyooka S and Miwa M 2007 *Journal of Forest Research* **12** 393–402
Ainsworth E A 2008 *Global Change Biology* **14** 1642–50
[2] Thilakarathne B L S, Rajagopalan U M, Kadono H and Yonekura T 2014 *SpringerPlus* **3** 89
[3] Kobayashi K and Okada M 1995 *Agriculture, Ecosystems and Environment* **53** 1–12
[4] Lefohn A S 1992 *Surface-level Ozone Exposure and Their Effects on Vegetation* (Lewis Publishers) pp157–270
[5] Díaz-de-Quijano M, Schaub M, Bassin S, Volk M and Peñuelas J 2012 *Environmental Pollution* **169** 250–7
[6] Huang D *et al* 1991 *Science* **254** 1178–81
Fercher A F 1996 *Journal of Biomedical Optics* **1** 157–73

[7] Drexler W, Morgner U, Ghanta R K, Kärtner F X, Schuman J S and Fujimoto J G 2001 *Nature Medicine* **7** 502–7

Wojtkowski M, Leitgeb R, Kowalczyk A, Bajraszewski T and Fercher A F 2002 *Journal of Biomedical Optics* **7** 457–63

[8] Gambichler T, Moussa G, Sand M, Sand D, Altmeyer P and Hoffmann K 2005 *Journal of Dermatological Science* **40** 85–94

[9] Hettinger J W, Mattozzi M D L P, Myers W R, Williams M E, Reeves A, Parsons R L, Haskell R C, Petersen D C, Wang R and Medford J I 2000 *Plant Physiology* **123** 3–15

Reeves A, Parsons R L, Hettinger J W and Medford J I 2002 *Journal of Microscopy* **208** 177–89

[10] Boccara M, Schwartz W, Guiot E, Vidal G, Paepe R D, Dubois A and Boccara A C 2007 *The Plant Journal* **50** 338–46

[11] Maheswari R U, Takaoka H, Homma R, Kadono H and Tanifuji M 2002 *Optics Communications* **202**, 47–54

Maheswari R U, Takaoka H, Kadono H, Homma R and Tanifuji M 2003 *Journal of Neuroscience Methods* **124** 83–92

Chen Y, Aguirre A D, Ruvinskaya L, Devor A, Boas D A and Fujimoto J G 2009 *Journal Neuroscience Methods* **178** 162–73

[12] Braga R A, Dupuy L, Pasqual M and Cardoso R R 2009 *European Biophysics Journal* **38** 679–86

[13] Kurenda A, Adamiak A and Zdunek A 2012 *Postharvest Biology and Technology* **67** 118–23

[14] Aizu Y and Asakura T 1996 *Trends in Optics* (San Diego: Academic Press) pp27–49

[15] Ansari M D Z and Nirala A K 2013 *Optik* **124** 2180–86

[16] Fujimoto J G and Drexler W 2008 *Optical Coherence Tomography: Technology and Applications* (Springer)

[17] Chow T H, Tan K M, Ng B K, Razul S G and Tay C M, 2009 *Journal of Biomedical Optics* **14** 014006

[18] Taiz L and Zeiger E 202 *Plant Physiology* (Sunderland: Sinauer Associates)

[19] Vollenweider P, Ottiger M and Goerg M S G 2003 *Environmental Pollution* **124** 101–18

Faoro F and Iriti M 2009 *Environmental Pollution* **157** 1470–77

[20] Kangasjärvi J, Talvinen J, Utriainen M and Karjalainen R 1994 *Plant, Cell and Environment* **17** 783–94

Sharma Y K and Davis K R 1997 *Free Radcal Biology & Medicine* **23** 480–88

Sandermann H, Ernst D, Heller W and Langebartels C 1998 *Trends in Plant Science* **3** 47–50

Dizengremel P 2001 *Plant Physiology and Biochemistry* **39** 729–42

Calatayud A, Iglesias D J, Talón M and Barreno E 2003 *Plant Physiology and Biochemistry* **41** 839–45

[21] Walter A and Schurr U 2005 *Annals of Botany* **95** 891–900

[22] Kinose Y, Azuchi F, Uehara Y, Kanomata T, Kobayashi A, Yamaguchi M and Izuta T 2014 *Environmental Pollution* **194** 235–45

Usual and unusual CIELAB color parameters for the study of peat organic matter properties: Tremoal do Pedrido bog (NW Spain)

P Sanmartín[1*], N Silva-Sánchez [2], A Martínez-Cortizas [2] and B Prieto[1]

[1] Departamento de Edafoloxía e Química Agrícola, Facultade de Farmacia, Universidade de Santiago de Compostela. 15782-Santiago de Compostela, Spain
[2] Departamento de Edafoloxía e Química Agrícola, Facultade de Bioloxía, Universidade de Santiago de Compostela. 15782-Santiago de Compostela, Spain

E-mail: patricia.sanmartin@usc.es, psanmartin@seas.harvard.edu

Abstract. We have tested the practical application of color measurements in the study of organic matter properties (C and N content, C/N ratios, degree of peat humification –DPH) of a 335 cm long peat core sampled at Tremoal do Pedrido bog. Usual and unusual CIELAB color parameters were measured on samples that were sectioned at high resolution (slices of 1 cm in thickness). The objective of the study is twofold: (*i*) describe a rapid, cost-effective and non-destructive method of assessing peat properties without the need of extractions and chemical methods and (*ii*) contribute to further research on applied colorimetry using the well-known CIELAB coordinates: L*, a*, b*, C^*_{ab} and h_{ab} ('usual CIELAB color parameters') and the less well-known CIELAB parameters: [a* x b*], [a*/b*], [(a*/b*) x 1000], [1000 x a*/(L*+ b*)], [2000 x a*/(L* x b*)] and $R_{Lab}= [a^*(a^{*2}+b^{*2})^{1/2}\ 10^{10}]/(b^* \times L^{*6})$ ('unusual CIELAB color parameters'). Our findings show that L* and h_{ab} coordinates as well as [(a*/b*) x 1000], [2000 x a*/(L* x b*)] and R_{Lab} parameters give the best bivariate Spearman's correlations. Linear regression equations were calculated to predict peat properties from all CIELAB parameters under study and a notable fit (R^2: 0.65-0.79) was obtained. The evaluation presented here indicates that the determination of usual and unusual CIELAB parameters offers potential for the study of peat organic matter properties and encourages the routine application of this methodology on other peat cores and organic soils.

1. Introduction

There is a significant interest in the development and implementation of new methodologies for color measurement. They can be applied, for example, in the study of environmental and microbial ecology, microbial bioremediation, soil science and materials conservation. In this sense, the study of peatlands using such methods deserves further attention.

Peatlands (Figure 1) are wet ecosystems where reducing and acidic conditions limit the decomposition of plant material, favouring the accumulation of peat: a dark to brown organic substrate

---

[1] To whom correspondence should be addressed.

* Current address: Laboratory of Applied Microbiology, School of Engineering and Applied Sciences, Harvard University, 58 Oxford St., Cambridge, MA 02138, USA.

with high contents of organic carbon. Peatlands are among the most important carbon reservoirs on Earth [1, 2]. They also provide long-term perspectives on past environmental variability, especially on the Holocene (last~12,000 years). Peat accumulation dynamics on peatlands allows the preservation of many biotic and abiotic proxies that can be effectively used to reconstruct past changes on vegetation, soil erosion, climate or human impact [3,4]. Although the use of peat records as a tool in the reconstruction of past environmental conditions started more than a century ago, the most noticeable advances were made in the last few decades [5]: first in the European peatlands, particularly those from the UK and NW Europe, and then in the Southern Hemisphere and tropical regions. In these studies, several authors [e.g. 6- 8] reported variations in the color of peat deposits related to changes in the degree of peat degradation and vegetation remains, which were climatically controlled. In this sense, broadly highly decomposed peat is usually dark brown-blackish whereas less decomposed peat is light brown-yellowish in color. A chemical method to determine the degree of peat humification (DPH) was originally developed by Overbeck [9], and later modified by other investigators [10, 11] becoming the standard method. It is based on the colorimetric determination of an alkaline extract of the peat. The absorbance/transmittance is related to the amount of dissolved humic substances and therefore to the degree of humification and the concomitant decomposition of the peat. Since decomposition largely depends on surface moisture on the peatland, a record of humidity variations can be obtained by relating higher transmittance values to increased moisture (i.e. effective precipitation), and low transmittance values to drier conditions (i.e. lower effective precipitation). However, this is a destructive, time consuming method. For this reason, identifying innovative tools for determining changes in peat humification or decomposition is an important goal in the field of paleoenvironmental reconstruction. Techniques based on optical methods, as the instrumental color measurement with a chromameter/colorimeter or spectrophotometer for solids, could be an interesting option.

Figure 1. Color variations in a (left) peat cut and in a (right) fresh peat core.

Used since 1990s, these contact-type color measuring devices provide a low cost technology to quantify soil color related to soil taxonomy, soil quality, and fertility, through quantitative color space models like the widespread accepted in scientific and industrial fields CIELAB color space [12]. Previous research has provided evidence that CIELAB color parameters may be of application in the development of predictive models for different types of soils. Thus, for example, using a hand-held chromameter Lilies et al. [13] demonstrated the suitability of L* (associated to lightness), a* (associated to redness) and b* (associated to yellowness) CIELAB coordinates to predict forest soil C concentrations from the major conifer forest ecosystems in California (USA). Guimarães et al. [14]

worked with the same color coordinates recorded with a spectrophotometer in palaeoecological studies of Amazon wetlands. Similarly, Moritsuka et al. [15] used L* and b* coordinates for statistically estimating total carbon (C), total nitrogen (N) and active iron (Fe) contents in Japanese agricultural soils. In a 2010 publication, Sanmartin et al. [16] described the usefulness of h_{ab} (which refers to the dominant wavelength and represents the major color perception attribute) in a preliminary approach for the study of the relationship between color changes in a granitic Galician soil (*viz.* Umbric Regosol) and the temperature reached as a consequence of forest fire. More recently, Cancelo-González et al. [17] followed up on this research with other types of granitic Galician soils, adding Leptic Umbrisol and Humic Cambisol to the aforementioned Umbric Regosol, and different processed burnt samples (undisturbed, homogenized and free of organic matter) and concluded that the severity of forest fire can be estimated by determining variations in soil color and organic matter content. Using a dataset derived from the National Soil Inventory of Scotland-NSIS database (from Scottish soils of 721 locations throughout Scotland, which includes also peatlands soils), Aitkenhead et al. [14] reported successful results for ash, clay, sand, Ca, Mo, Ti, N, C:N ratio, organic matter content, pH and texture, and unsuccessful ones for P and K, all predicted by color measurements represented in the RGB and CIELAB systems. In this study, CIELAB system was found to be better overall predictor of soil characteristics than RGB (Red, Green, and Blue digital values) system. Although the authors point that this is balanced against the ease of use of RGB digital values obtained with a digital camera instead a color-measuring device (for further information, see the reference [18]).

Because most studies on soil color using the CIELAB system (including the above-mentioned works) comprised only the five CIELAB coordinates, *viz.* L*, a*, b*, C^*_{ab} and h_{ab}, the applicability of CIELAB parameters to soils is unclear. Despite many other CIELAB parameters, such as R_{Lab} (equation 6), have been used successfully in the study of soils and related environments, their application is still limited. The present study was conducted to shed light in this regard, while for the first time color measurements were employed for the study of a Histosol, a type of soil with cherished information for paleoenvironmental reconstruction and climate studies. We performed colorimetric measurements and recorded CIELAB color parameters, in a 335 cm long core (c. 6600 years) sampled at the Tremoal do Pedrido bog (Xistral Mountains, NW Spain). We hypothesize that CIELAB color parameters may be helpful to assess the organic matter properties (content of C and N, and DPH) of the peat. This study is a first approximation towards developing an understanding of the array of color parameters that can be used to infer Holocene environmental changes using peat cores.

2. Methodology

2.1. Sampling and physico-chemical peat properties analyzed

The peat core used for this study was sampled in Tremoal do Pedrido, a raised bog located in the Xistral Mountains (Figure 2) at an elevation of 695 m a.s.l., and 29 km south of the northern coast of Galicia, NW Spain (for further information, see the reference [19]).

Sampling was done with a Waardenarpeat corer to a depth of 100 cm and a Russian corer below 1m. The peat sections were wrapped in plastic film, protected in PVC hemi-tubes and taken to the laboratory, where they were cut into 1 cm slices, in the upper meter, and 2 cm below. Samples were placed in polyethylene bags and preserved at 4 °C in a freezer.

Before chemical analysis, samples were dried at 35 °C, milled and homogenized. Carbon and N were determined using an elemental analyzer LECO TruSpec CHN. The degree of peat humification (DPH) was determined following the conventional method by extracting humic acids in 8% NaOH and measuring its absorbance at 540 nm in a spectrophotometer [20]. C/N ratios were also calculated, since some authors consider their variations in peat cores are also related to peat decomposition [21].

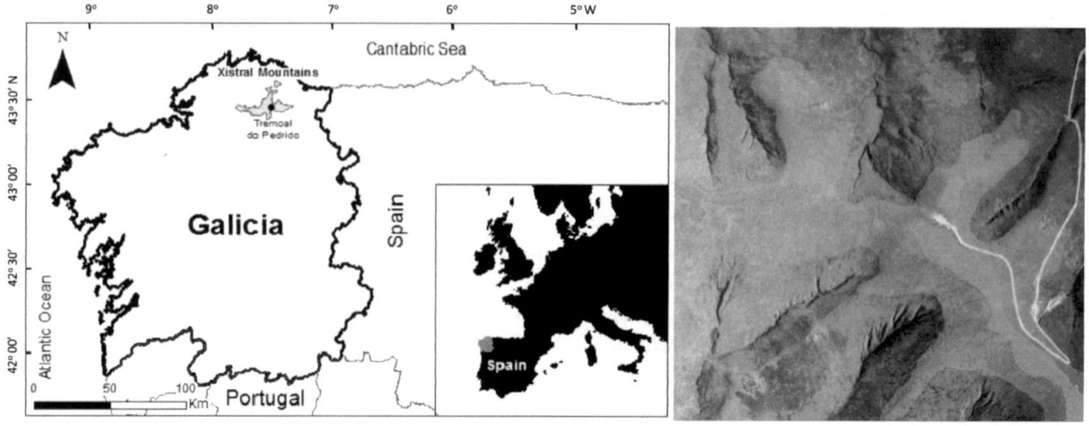

Figure 2. Left) Location map of the Xistral Mountains and the Tremoal do Pedrido mire (after Pérez-Rodríguez and Matínez-Cortizas [19]). Right) Aerial image of the peatlands in the Xistral Mountains (Northwestern Spain) (after Martínez-Cortizas [22]).

2.2. Usual and unusual CIELAB color parameters

In most studies (see e.g. [23], [24]) the characterization of soil color is only assessed by examination with the naked eye. Examination based on soil color charts from Munsell color system [25] is considered a standard method, although variation between existing charts can prevent accurate assignment of color to a sample [26]. Viscarra Rossel et al. [27] also demonstrated that color systems based on a Cartesian approach (like the CIELAB system) are more appropriate that the Munsell system for predictive analysis. Furthermore, color is a subjective property that depends on an individual's perception, and its quantification and representation in a color space is, therefore, a better option.

We have chosen for this study the CIELAB space for several reasons, including: its widespread acceptance in scientific and industrial applications, the opposing character of the coordinates L*a*b*, the association of another set of coordinates, $L^*C^*_{ab}h_{ab}$, with the three color attributes (lightness, chroma and hue), and its analogy relative to the Munsell color system, with which many people are familiar, specially researchers in Soil Science and therefore researchers in peatlands. Furthermore, as already stated, one of the aims of the study was to test a larger number of CIELAB parameters, not just limited to the five coordinates. Thus, the well-known CIELAB coordinates: L^*, a^*, b^*, C^*_{ab} and h_{ab}, namely 'usual' CIELAB color parameters, and the less well-known CIELAB parameters:

$$[a^* \times b^*] \quad (1)$$
$$[a^*/b^*] \quad (2)$$
$$[(a^*/b^*) \times 1000] \quad (3)$$
$$[1000 \times a^*/(L^*+ b^*)] \quad (4)$$
$$[2000 \times a^*/(L^* \times b^*)] \quad (5)$$
$$R_{Lab}= [a^*(a^{*2}+b^{*2})^{1/2}\ 10^{10}]/(b^* \times L^{*6}) \quad (6)$$

namely 'unusual' CIELAB color parameters, were recorded with a GretagMacbeth (now XRite) portable spectrophotometer (CE-XTH) for solids, performed on the dried, milled and homogenized samples (Figure 3).

Figure 3. Example of color variations in consecutive samples of a peat core.

3. Results and discussion

Various modeling and statistical methods have been used to relate soil color characteristics to soil properties and environmental factors. In this study, the color data were used to correlate to the degree of peat humification (DPH) and compositional data of the peat, to determine to which extent peat color reflects changes in peat properties and the related environmental conditions. The relationships between the data were assessed by two tailed bivariate Spearman's correlations and stepwise linear multiple regression models. Statistical analyses were performed with SPSS (SPSS v21.0 for Windows).

In general terms usual CIELAB parameters (Table 1) result in lower bivariate Spearman's correlations with peat properties than unusual CIELAB parameters (Table 2). Among the usual CIELAB color parameters, the highest correlations were found for L* (color lightness) and h_{ab} (the hue angle), while the lowest correlation was found for a*. This suggests that polar coordinates ($L^*C^*_{ab}h_{ab}$), rather than Cartesian coordinates (L*a*b*), better reflect peat organic matter properties. Importantly, the former lead to quantification of a color's similarity to its perceived color, i.e. by taking into account the three attributes of visual assessment [28]. For the unusual parameters, the correlation was higher for (a*/b*) x 1000, 2000 x a*/(L* x b*) and R_{Lab}. The first two parameters, (a*/b*) x 1000 and 2000 x a*/(L* x b*), have been mainly used to study the pigment content of several food types. R_{Lab} has been used recently in Quaternary paleosols and sediments on the Balearic Islands (Spain) to determine climatic preconditions for soil reddening [29].

Table 1. Bivariate Spearman's correlation matrix among geochemical data and usual CIELAB color parameters

CIELAB color parameter	C	N	C/N	DPH
L*	**-0.73****	**0.77****	**-0.78****	-0.68**
a*	-0.26**	0.25**	-0.24**	-0.31**
b*	-0.64**	0.66**	-0.67**	-0.65**
C^*_{ab}	-0.60**	0.60**	-0.61**	-0.61**
h_{ab}	**-0.75****	**0.87****	**-0.88****	-0.66**

n = 227, in bold correlation coefficients equal or higher than 0.7
significance level ** $p < 0.01$
C and N: carbon and nitrogen concentrations
C/N: carbon/nitrogen ratio
DPH: degree of peat humification, i.e. absorbance of the sodium hydroxide extract

Table 2. Bivariate Spearman's correlation matrix among geochemical data and unusual CIELAB color parameters.

CIELAB color parameter	C	N	C/N	DPH
a* x b*	-0.51**	0.51**	-0.51**	-0.54**
a*/b*	0.68**	**-0.76****	**0.78****	0.55**
(a*/b*) x 1000	**0.75****	**-0.86****	**0.88****	0.66**
1000 x a*/(L*+ b*)	0.63**	**-0.77****	**0.77****	0.50**
2000 x a*/(L* x b*)	**0.77****	**-0.85****	**0.86****	**0.71****
R_{Lab}	**0.74****	**-0.80****	**0.82****	0.68**

n = 227, in bold correlation coefficients equal or higher than 0.7
significance level ** $p < 0.01$; * $p < 0.05$.
Acronyms are defined in Table 1.

Regardless of the group of CIELAB parameters considered (usual or unusual), the highest correlation was found for C/N ratio, followed by N and, to a lesser extent, C content. The DPH always showed the lowest correlation. The sign of N correlation was opposite to the sign of C/N, C and DPH correlations. Strong relationships between soil color and the C/N ratio have been found in previous research (e.g. [14]), and Qian et al. [30] found that color was a strong predictor of soil N content in mineral forest soils.

Stepwise multiple linear regression was applied to obtain simple expressions for estimating peat properties using the CIELAB color parameters considered in this study (Table 3). The results indicate that color parameters are of relatively high predictive value (R^2: 0.65-0.79), particularly for the C/N ratio, followed by C and N; the lowest predictability was always obtained for DPH. No large differences were found regardless of the group of color parameters used. Nevertheless, polar coordinates showed a slightly better prediction for C and DPH and the unusual parameters for N and C/N ratios. Lilies et al. [13] found that the best predictive model for C included L* and a*.

The C/N ratio in peatlands usually increases with depth/age, due to an increase in C and a decrease in N concentrations, and it has been associated to long-term peat decomposition [31]. Thus, the correlations and the regression models imply that decomposition is accompanied by a reduction in lightness, red and yellow components, hue and chroma of the peat. The lower predictive value for the DPH may be due to the fact that it is not a property of the bulk peat or, as recently suggested [21], that both C/N ratios and DPH are related to different aspects of peat decomposition.

Table 3. Stepwise multiple linear regression equations for the prediction of the concentrations in carbon (C) and nitrogen (N), carbon/nitrogen ratio (C/N) and degree of peat humification (i.e. absorbance of sodium hydroxide extract, DPH).

Parameter Estimated	Predictive Equation	Adjusted R^2
C[a]	$C = 88.24 - 0.49h_{ab} - 0.22L^*$	0.69
N[a]	$N = -5.92 + 0.11h_{ab} + 0.04C^*_{ab}$	0.68
C/N[a]	$C/N = 300.24 - 4.12h_{ab} - 1.29C^*_{ab} + 0.60L^*$	0.77
DPH[a]	$DPH = 2.80 - 0.01L^* - 0.03h_{ab} - 0.01C^*_{ab}$ $Abs = 3.13 - 0.03h_{ab} - 0.02C^*_{ab}$	0.65
C[b]	$C = 55.47 + 0.02L^* + 2.31a^* - 1.46b^*$ $C = 55.60 + 2.27a^* - 1.42b^*$	0.73
N[b]	$N = 1.22 - 0.03L^* - 0.41a^* + 0.28b^*$	0.67
C/N[b]	$C/N = 41.27 + 0.62L^* + 12.28a^* - 7.72b^*$	0.72
DPH[b]	$DPH = 0.92 + 0.12a^* - 0.09b^*$ $DPH = 0.95 + 0.11a^* - 0.08b^*$	0.69
C[c]	$C = 35.27 + 0.04[(a^*/b^*) \times 1000] - 0.04[a^* \times b^*]$ $C = 36.70 + 0.02[(a^*/b^*) \times 1000] - 0.02[a^* \times b^*] + 0.15[2000 \times a^*/(L^* \times b^*)]$	0.70
N[c]	$N = 3.34 + 0.01[a^* \times b^*] - 0.01[1000 \times a^*/(L^* + b^*)] + 0.02[2000 \times a^*/(L^* \times b^*)]$	0.71
C/N[c]	$C/N = -36.63 + 0.12[(a^*/b^*) \times 1000] + 0.32[2000 \times a^*/(L^* \times b^*)]$	0.79
DPH[c]	$DPH = -0.28 + 0.02[2000 \times a^*/(L^* \times b^*)] - 0.49[a^*/b^*]$	0.66

[a] Usual color parameters (polar coordinates): L^*, C^*_{ab} and h_{ab}

[b] Usual color parameters (Cartesian coordinates): L^*, a^* and b^*

[c] Unusual color parameters

4. Conclusions and prospects

These results demonstrate the utility of quantifying the color of the peat to obtain predictive relationships of soil properties and, at the same time, propose new proxies in the field of paleoenvironmental reconstruction and paleoclimate studies. The open question is how to quantify this relationship and provide accurate predictive power for several properties of peatlands. In this regard, it will be worthwhile to further investigate other type of mires and organic soils, (*i*) increasing the number of samples analyzed, (*ii*) considering the effect of moisture content on the color (since soil moisture generally increases soil darkness reducing the ability to generate accurate predictive models), (*iii*) analyzing more in detail short-term changes, (*iv*) comparing the results with records of other validated proxies and (*v*) introducing other, more sophisticated, statistical tools as PCR and PLS modeling. In any case, results are encouraging and further investigation to determine the value of color as proxy for Holocene environmental change is warranted.

Acknowledgments
This research was supported by the projects CGL2010-20672 (Spanish Ministry of Science and Innovation) and 10PXIB200182PR (Xunta de Galicia). We thank Eva Peiteado for assistance in sample preparation. Noemí Silva-Sánchez is supported by a FPU predoctoral grant (AP2010-3264) funded by the Spanish Government. Dr. Patricia Sanmartín is supported by a postdoctoral contract within the framework of the 2011-2015 Galician Plan for Research, Innovation and Growth (Plan I2C) for the year 2012.

References

[1] Clymo R S, Turunen J and Tolonen K 1998 *Oikos* **81** 368-388
[2] Yu, Z C 2012 *Biogeosciences* **9** 4071–4085
[3] Aaby B 1976 *Nature* **263** 281–284.
[4] Martínez-Cortizas A, Pontevedra-Pombal X, García-Rodeja E, Nóvoa-Muñoz J C and Shotyk W1999 *Science* **284** 939-942
[5] Bindler R 2006 *Global Planet Change* **53** 209-221
[6] Crum H A 1988 *A Focus on Peatlands and Peat Mosses* (Ann Arbor: Univ. of Michigan Press)
[7] Franzén L G 2006 *Mires Peat* **1**1–16
[8] van Geel B and Mauquoy D 2010 *PAGES News* **18** 11–14
[9] Overbeck F 1947*Planta* **35** 1–56 (in German)
[10] BahnsonH 1968 *Meddelande Dansk Geologisk Forening* **18** 55–63 (in Danish)
[11] Blackford J J and Chambers F M 1993 *Int Peat J* **5** 7–24
[12] CIE Publication 15-2. Colorimetry 1986 CIE Central Bureau, Vienna
[13] Liles G C, Beaudette D E, O'Geen A T and Horwath W R 2013 *Soil Sci Soc Am J* **77** 2173–2181
[14] Aitkenhead M J, Coull M, Towers W, Hudson G and Black H I J 2013 *Geoderma* **200** 99–107
[15] Moritsuka N, Matsuoka K, Katsura K, Sano S and Yanai J 2014 *Soil Sci Plant Nutr* **60** (4) 475-485
[16] Sanmartín P, Cancelo-González J, Rial M E, Silva B, Díaz-Fierros F and Prieto B *Opt Pur Apl* **43** 167–172 (in Spanish)
[17] Cancelo-González J, Cachaldora C, Díaz-Fierros F, Prieto B 2014 *Ecol Indic* **46** 92–100
[18] Sanmartín P, Chorro E, Vázquez-Nion D, Martínez-Verdú F M, Prieto B 2014 *Measurement* **56** 194–202
[19] Pérez-Rodríguez M and Matínez-Cortizas A 2014 *Spanish J Soil Science* **4** (2) 158-178
[20] Chambers F M, Beilman D W, Yu Z 2011 *Mires Peat* **7** (7) 1-10
[21] Biester H, Knorr K-H, Schellekens J, Basler A and Hermanns Y-M 2014 *Biogeosci* **11** 2691-2707
[22] Martínez-Cortizas A 2001 Las turberas: terminología, tipos y clasificaciones. Turberas de Montaña de Galicia (A. Martínez Cortizas, E. García-Rodeja Gayoso, eds), 99-123. Consellería de Medio Ambiente, Xunta de Galicia
[23] Pereira P, Úbeda X, Mataix-Solera J, Oliva M, Novara A 2014 *Solid Earth* **5**(1) 209-22
[24] Sánchez-Marañón M,García P A, Huertas R, Hernández-Andrés J, Melgosa M 2011 *Soil Sci Soc Am J* **75**(3) 984-993
[25] Munsell, A H 1905 Una notación del color, Boston: G. H. Ellis Co.
[26] Sánchez-Marañón M, Huertas R, Melgosa M 2005 *Aust J Soil Res* **43**(7) 827–837
[27] Viscarra Rossel R A, Minasny B, Roudier P, McBratney A B 2006 *Geoderma* **133**(3–4) 320–337
[28] Wyszecki G and Stiles W S 1982 *Color Science. Concepts and Methods, Quantitative Data and Formulae* (New York: Wiley–Interscience)
[29] Wagner S, Eckmeier E, Skowronek A and Gu̇nster N 2014 *Catena* **112** 112–124
[30] Qian H, Klinka K, Lavkulich L M 1993*Can J Soil Sci* **73** 61–72
[31] Khury P and De Vitt D H 1996 *Ecology* **77** 216-218

Journal of Physics: Conference Series **605** (2015) 012015 doi:10.1088/1742-6596/605/1/012015

Null Ronchi Gratings as a function of period

D Aguirre-Aguirre, F S Granados-Agustín, R Izazaga-Pérez, B Villalobos-Mendoza, and A Cornejo-Rodríguez.
Optics Departament, Instituto Nacional de Astrofísica, Óptica y Electrónica, INAOE, Apdo. Postal 51 and 216, C. P. 72000, Puebla, Pue., México.

E-mail: daguirre@inaoep.mx

Abstract. In this work is presented in a simple manner, a method to generate null Ronchi gratings. This method consider the period of the Ronchi grating as a function of the x and y positions over the grating, compared to the currently known methods which utilize ray tracing or spot diagrams for the null Ronchi gratings calculus. We compare null Ronchi gratings reported in the literature against gratings calculated with our method. We have found that our method can calculate the Ronchi gratings for any conical surface in a satisfactory way.

1. Introduction

It is well known that a pattern of curved fringes is observed when the Ronchi test is used to test aspherical surfaces [1] (see fig. 1). Thus, for the technicians in an optical workshop is more difficult to observe the deviation of the pattern produced by aspherical surfaces compared with the pattern generated by spherical surfaces, where the fringe pattern is straight and equally spaced.

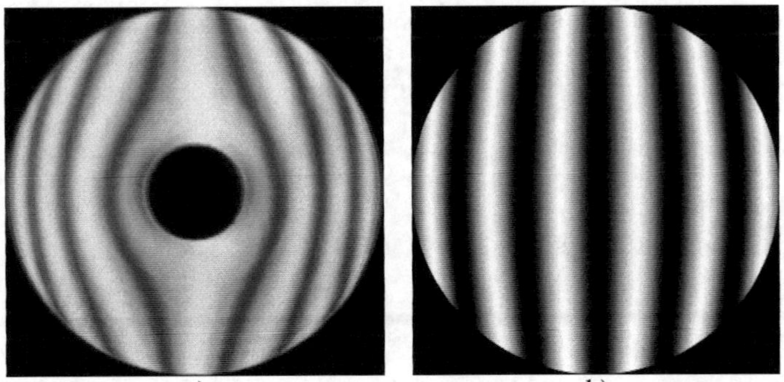

a) b)

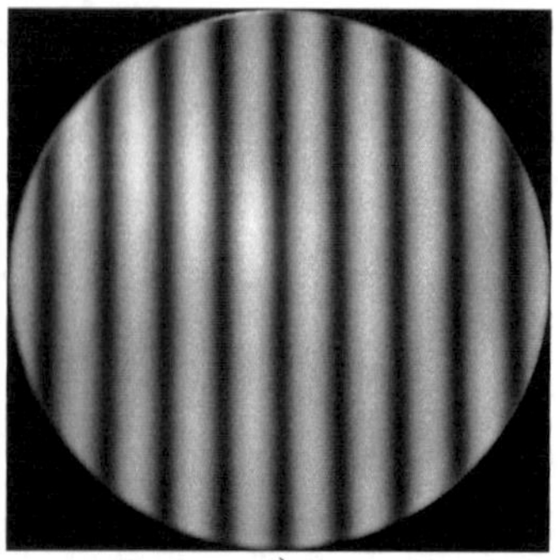

c)

Figure 1. Ronchigrams, a) hyperbolic, b) parabolic and c) spherical surface.

The technique that is used to eliminate the curved fringes of the aspherical surfaces pattern is to use a special kind of Ronchi grating, which has curved fringes unequally spaced (Fig. 2), these kind of gratings are also called null Ronchi gratings [1].

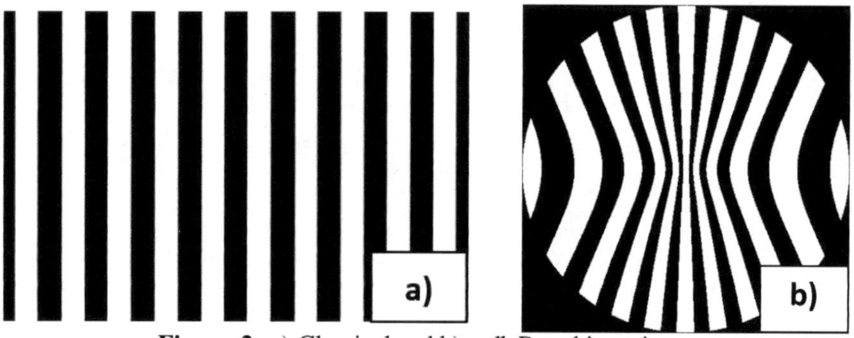

Figure 2. a) Classical and b) null, Ronchi grating.

This kind of gratings acts as an optical compensator, which compensates the asphericity of the surface, causing that the observed fringe pattern will be only influenced by the aberrations of the system; if the system has no aberrations, the observed pattern will have straight and equally spaced fringes [1,2].

Ray tracing and solving a system of linear equations are the most common used method to calculate these kind of gratings [2]. A simpler way for this calculation was proposed by Hopkins and Shagam [3], which use a spot diagram to design null Ronchi gratings.

In this work, a new technique for generating null Ronchi gratings is proposed. This technique is based in the lateral share interferometry formalism. The technique considers the period of the grating as a function of the x and y positions, and not as constant, as it is traditionally done.

2. Null Ronchi gratings calculus

The wavefront aberrations are defined in the exit pupil of the system under test, by using the Rayces equations given in 1964 [1], as

$$\frac{\partial W(\mathrm{x},\mathrm{y})}{\partial x} = -\frac{TA_x}{r}; \quad \frac{\partial W(\mathrm{x},\mathrm{y})}{\partial y} = -\frac{TA_y}{r}. \tag{1}$$

Assuming a Ronchi grating with a spacing *d* between the slits from a point (x, y) in the m-th fringe, we may write, in general, as

$$\frac{\partial W(x,y)}{\partial x} = -\frac{md}{r}, \tag{2}$$

where *W(x,y)* represents the original wavefront which can be expressed by a polynomial function (i.e. Zernike, Seidel, Kingslake, etc.), or by the sagitta differences between the conical surface under test ($z(x,y)$) and the osculating sphere ($z_o(x,y)$) [1, 4], in our case the wavefront of an optical system, which only presents primary aberrations can be described by the polynomial aberration given by Kinglake in 1925 [5], as

$$W(x,y) = A(x^2+y^2)^2 + By(x^2+y^2) + C(x^2+3y^2) + D(x^2+y^2), \tag{3}$$

where *A*, *B*, *C* are the spherical aberration coefficients, coma and astigmatism, respectively. The last coefficient *D* is the defocusing term, given by the distance *l* from the Ronchi grating to the paraxial center of curvature, as

$$D = \frac{l}{2r^2}. \tag{4}$$

Due to the conical surfaces only present spherical and focus aberration, the Eq. 1 is reduced to the following manner

$$\frac{\partial\left[A(x^2+y^2)^2 + D(x^2+y^2)\right]}{\partial x} = \frac{-md}{r}. \tag{5}$$

Solving Eq. (5) for the grating period, we obtain

$$\frac{-r}{m}\left[\frac{\partial\left(A\left(x^2+y^2\right)^2 + D\left(x^2+y^2\right)\right)}{\partial x}\right] = d. \tag{6}$$

Notice that spherical aberration in a conical surfaces only depends to the conic constant and the paraxial radius of curvature, which can be described as

$$A = \frac{|k|}{8r^3}. \tag{7}$$

Substituting Eq. (4) and Eq. (7) in Eq. (6), we obtain the equation for calculate the period of the grating point to point on the null Ronchi grating, as

$$\frac{-r}{m}\left[\frac{\partial\left(\frac{|k|}{8r^3}\left(x^2+y^2\right)^2 + \frac{l}{2r^2}\left(x^2+y^2\right)\right)}{\partial x}\right] = d. \tag{8}$$

3. Experimental Results

In the last section, we have defined the set of equations needed to simulate the null Ronchi gratings for any conical surface. To validate the above equations, it was performed a simulation of a hyperbolic surface of 35 cm diameter, 110 cm paraxial radius of curvature and a conic constant value of -3.5. The position of the ruling was 109.15 cm from the vertex of the mirror. The null Ronchi grating was calculated by using ray tracing program and also with our method. We found that by using our method the null Ronchi grating was a similar to the ray tracing but not identical (Fig. 3).

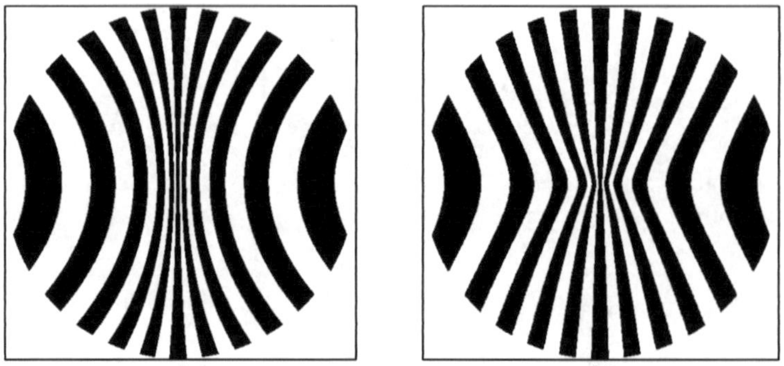

a) b)

Figure 3. Calculated grating with a) our method and b) ray tracing.

This error is due to the form of calculating the null grating because our method only calculate the period of the grating in the x direction, but when the slits are bent produces diffraction in both x and y directions. A solution to this problem was to make a cubic fitting to the data set. By applying this fitting to the null grating computed with our method, we obtain the same result that the calculated grating calculated with ray tracing (Fig. 4).

a) b)

Figure 4. Calculated grating with a) our corrected method and b) ray tracing.

As an example, we used the analysis described by D. Malacara and A. Cornejo [2] to calculate the null Ronchi grating corresponding to a parabolic surface with 30 *cm* in diameter and 302 *cm* of curvature radius; where they used ray tracing and the solution of a linear equation system.

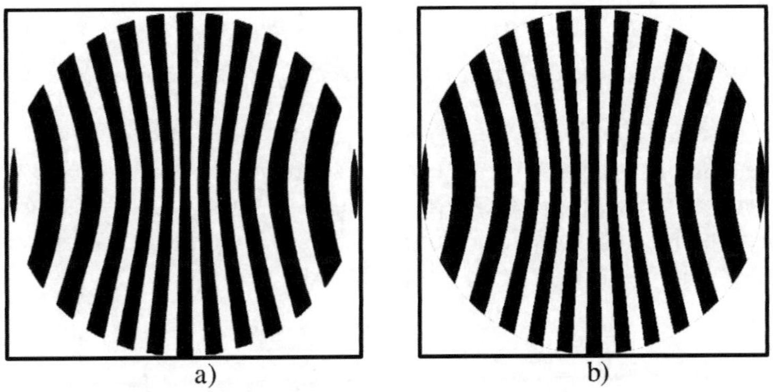

Figure 5. Grating calculated by a) D. Malacara and A. Cornejo (From Appl. Opt. 13, 1778, 1974) and b) our proposal.

We can notice that the grating generated by our corrected method (Fig. 5b) is identical to the grating calculated by D. Malacara and A. Cornejo (Fig. 5a).

A simpler method that can be found in the literature is the analysis described by G. Hopkins and R. Shagam [3] to calculate the null grating by using a spot diagram, corresponding to a parabolic surface with 20 cm in diameter and an $f/3$.

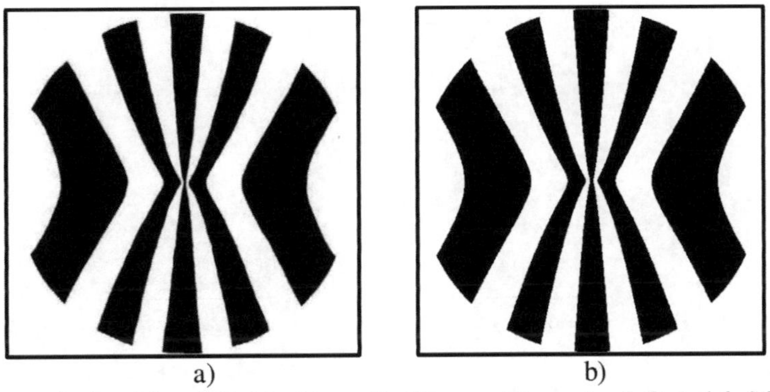

Figure 6. Grating calculated by a) G. Hopkins y R. Shagam (From Appl. Opt. **16,** 2602, 1977) and b) our proposal.

We can notice that our corrected method can also reproduce the results obtained by G. Hopkins and R. Shagam in 1977 (Fig. 6).

4. Conclusions

It has been shown that by considering the period of the Ronchi grating as a function of the x and y positions, it can be calculated in a simple manner the null Ronchi grating associated to different conical surfaces, this is because our proposal is able to reproduce in a satisfactory manner the results presented by D. Malacara and A. Cornejo in 1974 and the results presented by G. Hopkins and R. Shagam in 1977.

Acknowledgments

Villalobos-Mendoza and Izazaga-Pérez thank the CONACyT (México) the graduate studies scholarships granted, CVU: 272864 and 270587, respectively.

References

[1] Cornejo A 2007 Ronchi Test *Optical Shop Testing*, 3th. Ed., ed Malacara D chapter 9 pp 317-360
[2] Malacara D and Cornejo A 1974 Appl. Opt. vol. **13**(8), 1778-1780
[3] Hopkins G and Shagam R 1977 Appl. Opt. vol. **16**(10), 2602-2603
[4] Aguirre-Aguirre D, Izazaga-Pérez R, Granados-Agustín F, Villalobos-Mendoza B, percino-Zacarías M and Cornejo-Rodríguez A 2013 Opt. Eng. **52**(5) 053606
[5] Kingslake R 1925 Trans. Opt. Soc. **27**, 95-105

Obtaining the curve "Phase shift vs gray level" of a spatial light modulator Holoeye LC2012

B Villalobos-Mendoza, F S Granados-Agustín, D Aguirre-Aguirre and A Cornejo-Rodríguez

Instituto Nacional de Astrofísica Óptica y Electrónica Calle Luis Enrique Erro No. 1, Sta. Ma. Tonantzintla, Puebla, Méx. C.P.72840

E-mail: bvillalobosmendoza@gmail.com

Abstract. In this work, the process to obtain the curve "Gray level vs phase shift" of a transmissive spatial light modulator (SLM) Holoeye LC2012 is described. This work arises from the need that exists for having a new optical surface testing method at INAOE´s Optical Workshop. The SLM was placed in one arm of a Twyman-Green interferometer. The fringe shifts in the interference patterns were produced by displaying the different gray levels in the SLM. The gray level images displayed in the SLM were divided in two equal parts, the upper part was varying the different gray levels from 0 to 255 and the lower part stayed fixed with a gray level of 0 as reference. We show the different phase shifts and the experimental interferograms. From this analysis it was found out that a noticeable phase shift can be obtained from the 50 to the 190 gray levels.

1. Introduction

In recent years, the use of spatial light modulators has been applied in several optical applications [1-5]. In the field of optical testing for example such SLM device has been used in the Ronchi test. The purpose in such Ronchi's method is modifying the structure of the Ronchi grating, either in its classical structure or for producing the so called substructured Ronchi gratings [6, 7].

The used techniques for measuring the phase that is introduced by these SLM are varied, one of the methods for phase "extraction" use the correlation matrix of intensities to determine the phase [3]. For performing the calibration of the SLM, one of the most used techniques is interferometry [8, 9]. In particular in this paper a Twyman-Green interferometer was utilized.

In this work, the process to obtain the curve "Phase shift vs gray level"of a SLM Holoeye LC2012 by using a Twyman-Green interferometer is reported. The main aim was to measure the interference fringes shifts which were produced by modifying the gray level displayed in this SLM. For this, an absolute reference which consists in displaying a gray level image dived in two equal parts was used. The upper part of this gray level image was varying the gray level from 0 to 255 meanwhile the other lower part remained fixed in cero as reference.

2. Experimental methodology

We used for the experimental setup a diode laser of $532\,nm$ as source light; a spatial filter of 40x-$10\,\mu m$; a CMOS detector Pixelink PL-B776 with a resolution of 2048x1536 pixels; a beam splitter 50/50 and three flat mirrors with an optical quality of $\lambda/4$. The Mirror M2 was placed in a linear stage with a tilt mounting and M1 was placed in a tilt mounting. A polarized to compensate the visibility in both arms was used.

Before to introduce the SLM in the experimental setup of Fig. 1, the interferometer must be adjusted to have nearly the same optical path for its two arms. In the second step, the SLM is placed into one of the interferometer arms and the optical path produced by its thickness has to be compensated by moving the mirror M2 (in or out) in the other arm.

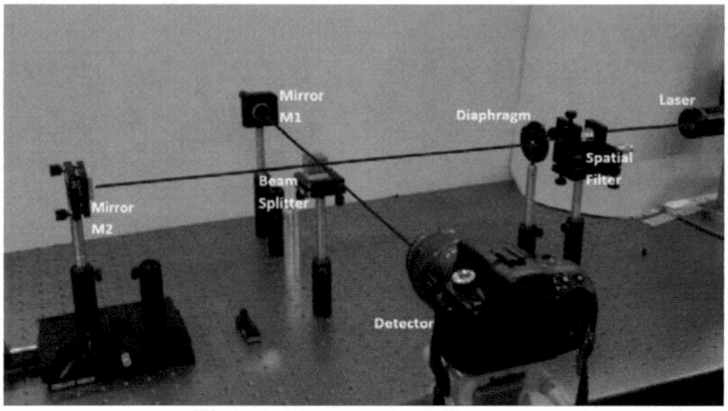

Figure 1 Experimental set up.

The first step for matching the optical paths in the interferometer for both mirrors from the beam splitter was to measure the distances with a Vernier as a first approximation. The next step starts by observing the interference patterns. In Fig. 2 are shown three typical interference patterns that can be observed depending of the adjustments done on mirror M2, and keeping mirror M1 fixed. The closer position of both mirrors with respect to the beam splitter is reached when the pattern of Fig. 2b has the minimum number of interference fringes, similar to the pattern of Fig. 3. This process to get the best matching distances can be repeated many times as needed, in order to achieve a better approach to have the same optical path in both interferometer arms.

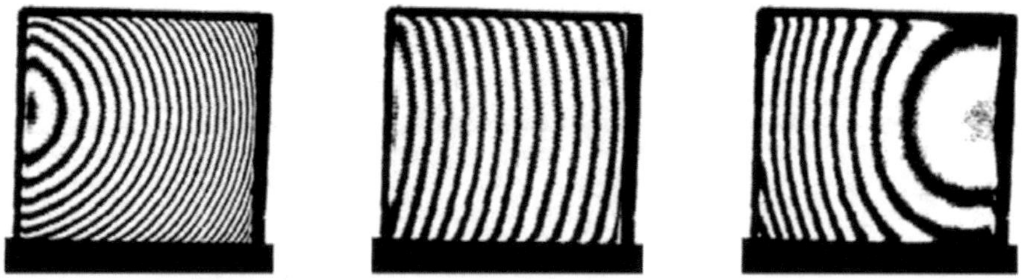

(a) (b) (c)

Figure 2 Experimental interferograms. (a) Right displacement of M2 from the beam splitter; (b) Tilt introduced to M1; (c) Left displacement of M2 from the beam splitter.

Figure 3 Interference fringes with practically the same optical path difference.

Because it is necessary to take into account the thickness of the SLM, when is introduced in the interferometric arrangement of Fig. 1, it is compulsory to have in the produced interferograms a reference fringe. Such reference fringe can only be observed if a white light source is used. Therefore, in our experiment the laser was substituted for a white light source as is shown in Fig. 4. The adjustment of mirror M1 must be repeated with this new light source because given its low temporal coherence, implies more fine applied movements. However, with the previous positions with the laser, the task with the white light source is less difficult.

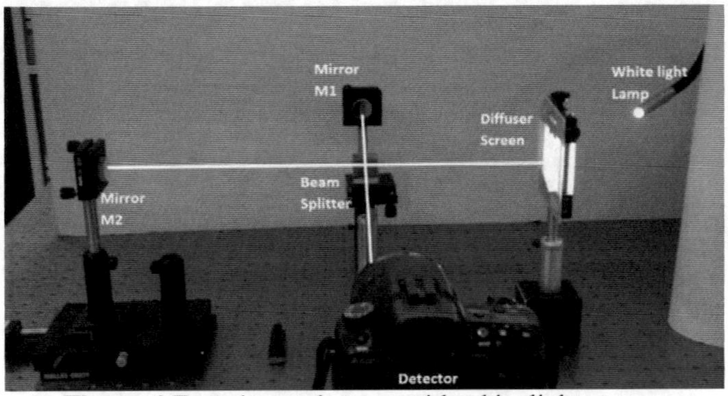

Figure 4 Experimental set up with white light source.

With the white light source in the setup of Fig. 4, a similar procedure of the previous alignment must be done to get the minimum interference fringes in the observed pattern; when such pattern is obtained a dark fringe is observed, as the one seen in Fig. 5. This dark fringe will be the reference interference fringe. When the SLM is introduced in the experimental setup of Fig. 4, the pattern of Fig. 5 changes because the thickness of the SLM. After new adjustments of mirror M2 (displacements in/out), the interference pattern with a dark fringe is recovered, which means that the matching distances of mirrors M1 and M2 from beam splitter has been properly adjusted.

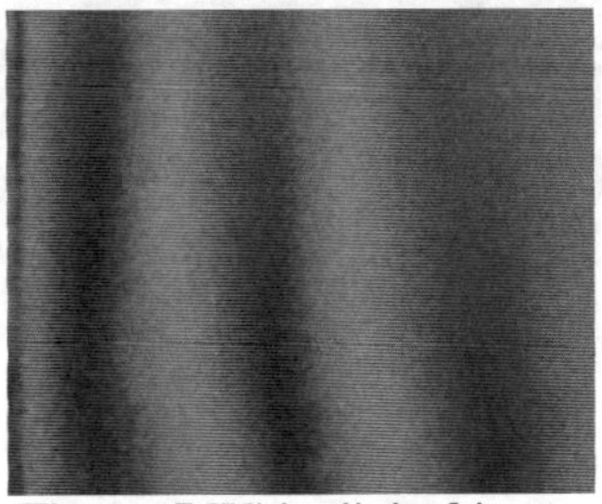

Figure 5 White light fringes.

Journal of Physics: Conference Series **605** (2015) 012016 doi:10.1088/1742-6596/605/1/012016

When the SLM is introduced in the interferometer, it produces a reduction in the contrast of the interference pattern that can be compensated by using a beam splitter with a different percent of intensities in each arm, or can be used also some filter. In our case, a polarized was used, Fig. 6.

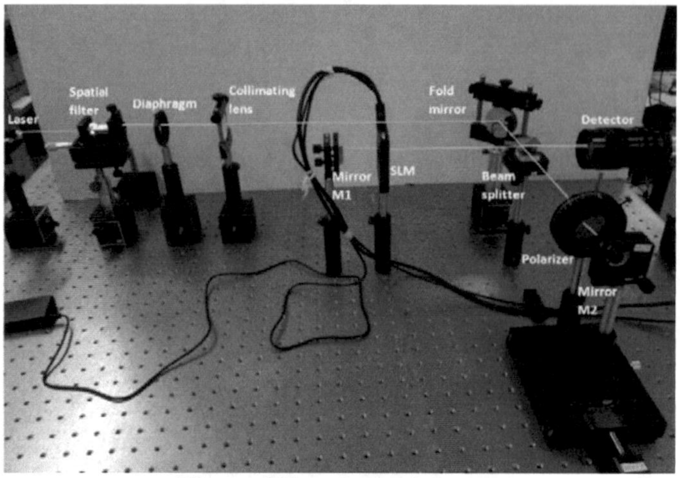

Figure 6 Experimental set up.

An algorithm to display the different grey levels in the SLM was developed. This algorithm generates a matrix with the same resolution of the SLM (1024x768pixels); this matrix is divided in two identical parts, the upper part is varying the grey levels from 0 to 255 and the lower part remains fixed with 0 as reference, Fig. 7. This change in the grey level produces a phase shift in the wavefront that can be seen in a fringe shift in the interference pattern which is captured in the detector, Fig. 8.

The next step to obtain the curve "Gray level vs phase shift" of the SLM Holoeye LC2012 was to capture the 256 interferograms with phase shift. For this, the 256 images with the different grey levels were displayed in the SLM and at the same time its interferogram was captured, Fig. 8.

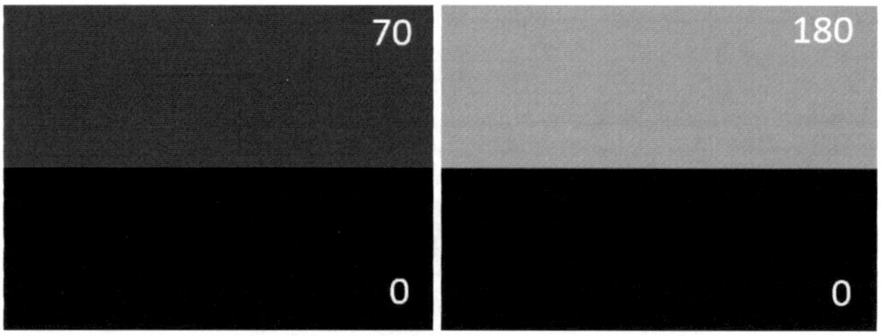

Figure 7 Different gray levels displayed in the SLM.

3. Experimental Results

In Fig. 8 the experimental interferograms captured with the CMOS detector are shown. A fringe shift in each interferogram that is more remarkable each time that the gray level is increased can be seen.

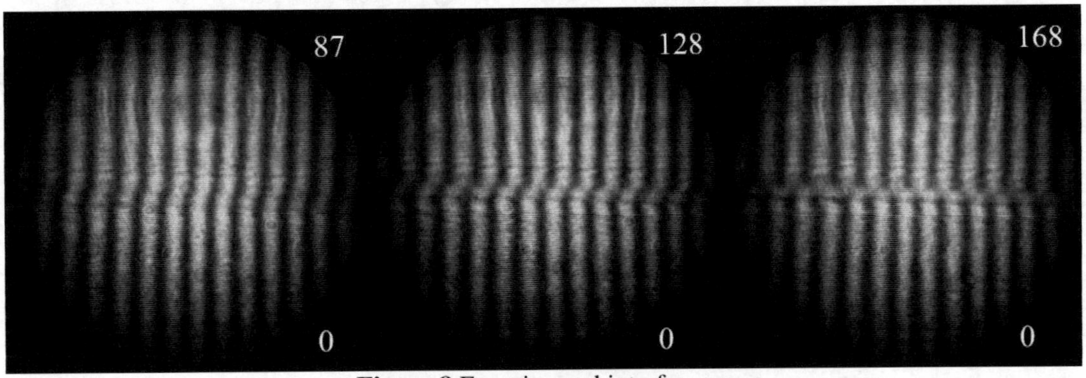

Figure 8 Experimental interferograms.

A computational algorithm for measuring the fringe shift in each interferogram was developed. This algorithm finds the value and position of each maximum in the experimental interferograms. After this, the algorithm takes two rows in the interferogram, one corresponding to the part which the gray level was changed and the second row corresponding to the part where the gray level remained fixed. The algorithm makes the subtraction of this two positions giving the introduced shifting by each gray level, Fig. 9.

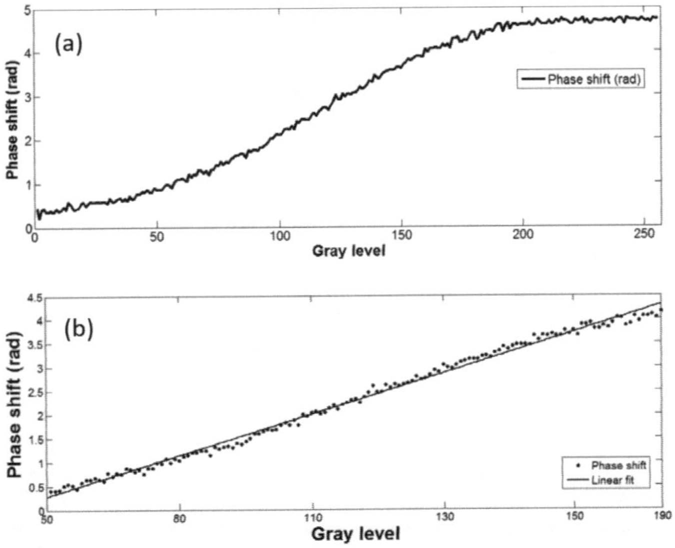

Figure 9 Phase shifts. (a) All the gray levels from 0 to 255 and (b) linear fit from 50 to 190 gray levels.

4. Conclusions

It has been shown that varying the gray levels in the SLM Holoeye LC2012 different fringe shifts can be obtained. The best rank of work was found between the 50 to the 190 gray levels. A linear fit between the 50 to 190 gray levels with an error of rmse=0.0980 was found. The fringe shift associated with the different gray levels displayed in the SLM by using a reference in the interference patterns, was calculated. From this analysis it was found out that a noticeable linear phase shift can be obtained with the Holoeye LC2012 between the 50 to 190 gray levels, Fig. 9. From these initial results we are in the position to do a more complete phase shift analysis of our experimental interferograms.

References

[1] Mahmud M S *et al* 2008 *J. Opt. A: Pure Appl. Opt.* **10** 085007
[2] Hu L *et al* 2004 *Opt. Express* **12** 26 6403-6409
[3] Yue H *et al* 2006 *IEEE Conference on Optoelectronic and Microelectronic Materials and Devices* 105-108
[4] Zhang Z, You Z & Chu D 2014 *Light: Science & Applications* **3** 10 e213
[5] Aguirre-Aguirre D, Granados-Agustín F S, Campos-García M and Cornejo-Rodriguez A 2011 *J. Phys.: Conf. Ser.* **1** 274
[6] Castro-Ramos J, Vázquez-Montiel S, and Padilla-Vivanco A *5th Iberoam. Meet. on Opt. and 8th Latin Ame. Meet. on Opt., Lasers, and Their Appl., A.*
[7] Bergeron A, Gauvin J, Gagnon F, Gingras D, Arsenault H H, & Doucet, M 1995 *Appl. Opt.* **34** 23 5133-9
[8] Cai L Z, Liu Q, and Yang X L 2003 *Opt. Lett.* **28** 19 1808-1810

Evaluation of the performance of orthodontic devices using FBG sensors

L Carvalho[1,3] , P Roriz[1], O Frazão[1,2] and M B Marques[1,2]
[1] INESCTEC, Campus da FEUP, Rua Dr Roberto Frias, 4200-465 Porto, Portugal
[2] Faculty of Sciences of the University of Porto, Rua do Campo Alegre, 687, 4150-179 Porto, Portugal

E-mail: lmrc@inesctec.pt

Abstract. Cross-bite, as a malocclusion effect, is defined as a transversal changing of the upper dental arch, in relation to the lower arch, and may be classified as skeletal, dental or functional. As a consequence, the expansion of maxilla is an effective clinical treatment used to correct transversal maxillary discrepancy. The maxillary expansion is an ancient method used in orthodontics, for the correction of the maxillary athresia with posterior crossbite, through the opening of the midpalatal suture (disjunction), using orthodontic- orthopaedic devices. Same controversial discussion arises among the clinicians, about the effects of each orthodontic devices as also about the technique to be employed. The objective of this study was to compare the strain field induced by two different orthodontic devices, named disjunctor with and without a connecting bar, in an acrylic model jaw, using fiber Bragg grating sensors to measure the strain patterns. The orthodontic device disjunctor with the bar, in general, transmits higher forces and strain to teeth and maxillae, than with the disjunctor without bar. It was verified that the strain patterns were not symmetric between the left and the right sides as also between the posterior and anterior regions of the maxillae. For the two devices is also found that in addition a displacement in the horizontal plane, particularly in posterior teeth, also occurs a rotation corresponding to a vestibularization of the posterior teeth and their alveolar processes.

1. Introduction

Cross-bite, as a malocclusion effect, is defined as a transversal changing of the upper dental arch, in relation to the lower arch, and may be classified as skeletal, dental or functional. This type of malocclusion is frequent in some populations, appearing in early stage development of teenagers and does not present self-correction. As a consequence, the expansion of maxilla is an effective clinical treatment used to correct transversal maxillary discrepancy, in children and teenagers that present maxillary atrophy, until the final phase of puberty (end of the active growth). The maxillary expansion is an ancient method used in orthodontics, for the correction of the maxillary athresia with posterior crossbite, through the opening of the midpalatal suture (disjunction), using orthodontic- orthopaedic devices [1-3]. The expansion is achieved when a force is applied to the dental-alveolar structures, through a removable or fixed device, by exceeding the necessary limits for the expansion, anticipating the cellular reaction of the periodontal ligament and favouring the dissipation of the forces through the

[3] To whom any correspondence should be addressed.

maxilla sutures[4-6]. According to de Assis et al [7] several studies have confirmed that the activation of the tooth-borne (Hyrax) or tooth-tissue-borne (Haas) expanders after osteotomies dissipates tensions responsible for the lateral movement of the maxilla and all adjacent structures, such as the teeth and the bones of the face and skull, which affects the nasal cavity, nasal septum, lateral walls and floor of the nose and nasal area, as well as the upper lip, alar base, gingiva and facial soft tissues.
Due to anatomic characteristics of the jaw and its resistance of several cranial-facial sutures (pterigo-palatal, to fronto-maxillary, to naso-maxillary and to zigomatic-maxillary), the expansion occurs in the horizontal plane with a pyramidal form, where the base is directed to the anterior inter-incisive region (anterior nasal spine), and the vertex located posterior, in direction to the posterior nasal socket (posterior nasal spine) (Fig. 1). In the frontal plane the disjunction of the jaw obeys to the same geometric configuration, with the vertex located close to the fronto-nasal suture and with the base at the oclusal plane (figure 1).

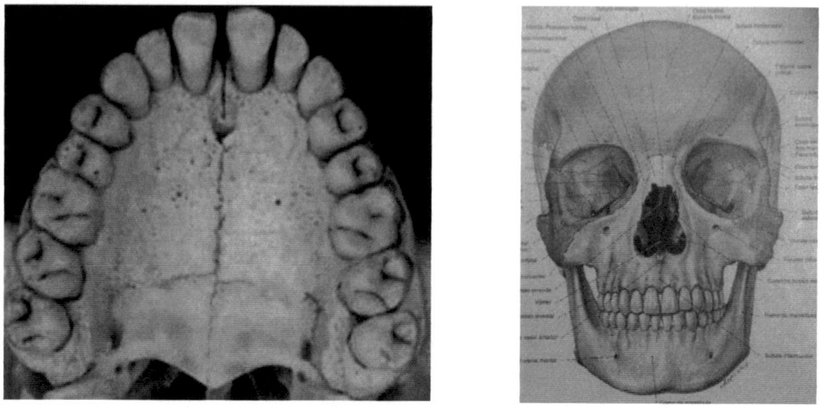

Figure 1. The maxillary expansion in the frontal and horizontal planes.

In clinical practice, after a few weeks of using the orthodontic-orthopaedic devices, the expansion in the horizontal plane induces the appearance of a maxillary midline diastema (figure 2).

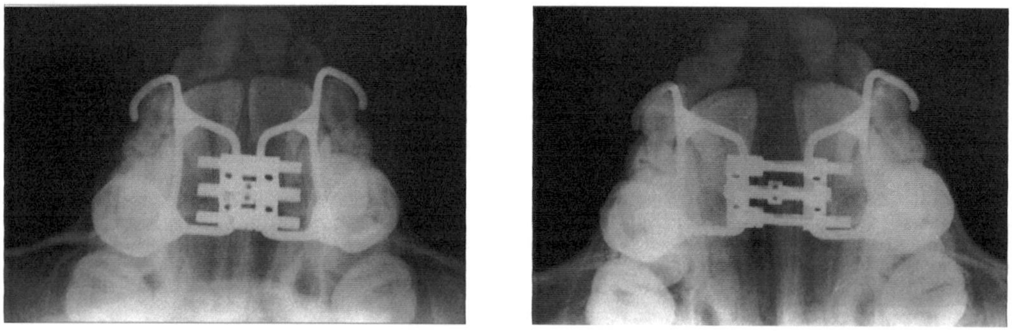

Figure 2. Radiographs showing clinical appearance of a maxillary midline diastema, in the horizontal plane.

Same controversial discussion arises among the clinicians, about the effects of each orthodontic devices as also about the technique to be employed, in order to achieve their goals of treatment, minimizing the negative effects for the patient [8-11]. Some authors [12,13] have reported that short-term rapid maxillary expansion (RME) with conventional appliances (Haas and Hyrax models) promotes the anterior and inferior dislocation of the maxilla, inclination of the alveolar process,

Journal of Physics: Conference Series **605** (2015) 012017 doi:10.1088/1742-6596/605/1/012017

extrusion and buccal inclination of the posterior teeth, with a consequent posterior-inferior rotation of the mandible. Awareness of these effects motivated researchers to develop other types of appliances.
The objective of the present study was to compare the strain field induced by two different orthodontic devices, named disjunctor, in an acrylic model jaw.

2. Materials and methods

For this study it was used an acrylic model of a human jaw. The maxillae suture was simulated using a hard silicone at the interface of the two midpalatal parts. Periodontal ligament, who maintains teeth attached into alveolar socket, was simulated with a fine membrane of polyethylene, also to permit natural tooth mobility. The combination of mechanical properties of materials used within the study were taken into account the relationship between naturals ones. The two devices have been manufactured with a Hyrax screw of 10mm from Forestedente®. Two kinds of disjunctor were made, one with a bar connecting the metallic bands around teeth of the same side, and the other one without that bar (figure 3). The metallic bands were putted in pre-molar and molar teeth. Teeth were manufactured in metal alloy, to be rigid as natural teeth are and also to avoid teeth deformation during the experiment. Were used two optical fibers, each one with four Bragg grating sensors, one in the anterior region and the other in the posterior region of the maxilla (figure 3).

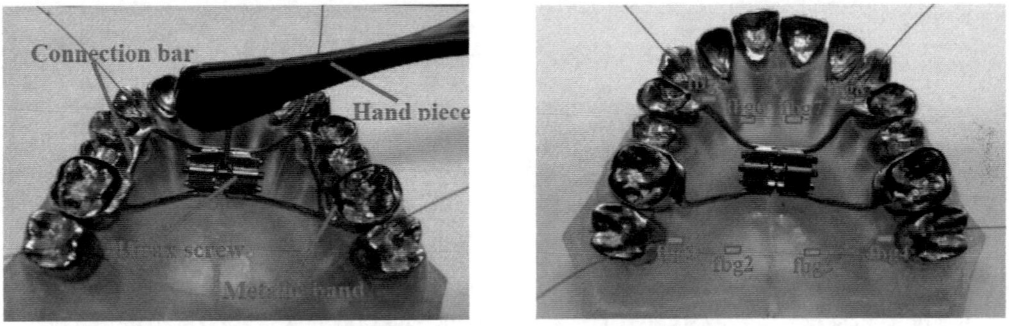

Figure 3. Overall view of the two devices, with the location of 8 FBG sensors.

The applied load was made twirling the screw Hirax one quarter turn, each 5 minutes, with a hand piece, in a total of twelve times. Data acquisition was made at 1sample/2seconds, with BraggMeterTM from FiberSensing.

3. Results and discussion

To make a better evaluation and comparison of the strain pattern, they were grouped for different locations, for the two devices. The term _b in each graph legend, corresponds to the device with the metallic bar connecting teeth in the same side. In figure 4 the strains located near the alveolar sockets, both in left and right sides, are displayed.
The maximum values of strain were not similar in both sides and this is due to the fact that the device was handmade and the technician has to adapt the device to model anatomy/geometry, forcing the arms of hyrax screw to adapt and consequently, sometimes they do not have the same length. For the region near the maxillae suture, there was obtained the same pattern in both sides, as it can be seen in the graphs displayed in figure 5. For that location, it was also observed that the strain difference between the two devices maintains constant during the entire applied load.

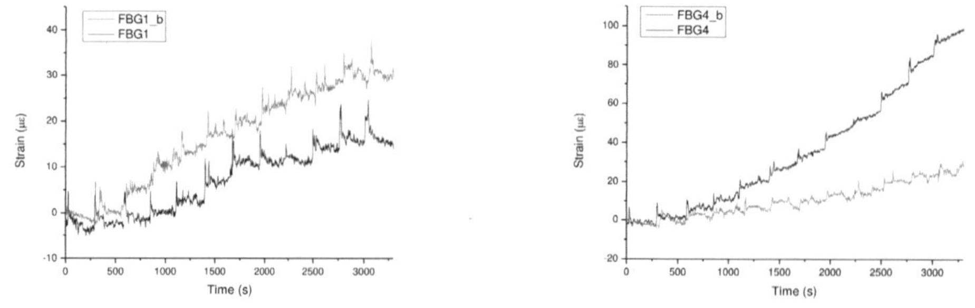

Left side Right side

Figure 4. Comparison of the strains versus time in the posterior region, for the two orthodontic devices.

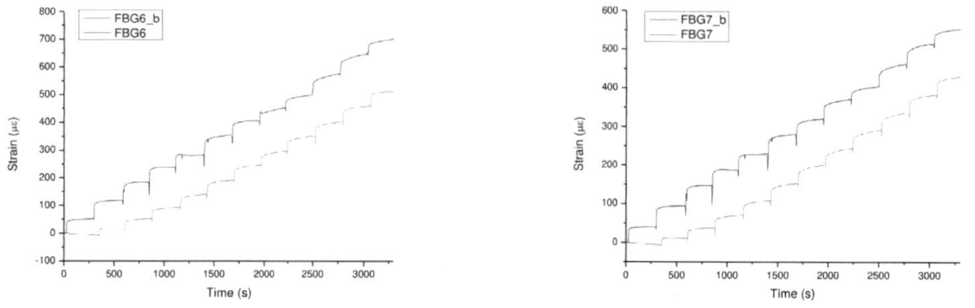

Left side Right side

Figure 5. Comparison of the strains versus time in the anterior region, for the two orthodontic devices.

The patterns observed for all measurements reflect the viscoelastic behaviour of the model, as it happens with natural bone. This behaviour is shown by the pic of strain when the screw was twirled, followed by a strain relaxation. The great amount of force produced and transmitted from the devices into the bone were located at the anterior region, what corroborates the appearance of a middle diastema, seen in clinical practice. The observed strains for the device disjunctor with the bar are, in general, greater than with the disjunctor without bar. The observed strains are greater in the middle of the maxilla then near the teeth. The observed strains are greater in the anterior region of the maxilla. At the posterior teeth, in some cases it was observed initially a compression and after an expansion. This was due to the fact that the location of the sensors in that region was near vertical, proving that these devices provoke a rotation for vestibular position of posterior teeth and alveolar process. These results were also observed in others studies [7, 12]. For the two devices it has also been determined that, besides the displacement in the horizontal plane, especially on the posterior teeth, there is also a labial movement of the posterior teeth and alveolar processes, and a back rotation of the anterior section of the jaw, as already was verified by some authors [7, 11-12].

For each device, it was also performed the comparison between anterior and posterior region, for each side of the maxillae in both devices. In figure 6 are displayed the results in the posterior and anterior regions, for the two orthodontic devices, in the left side.

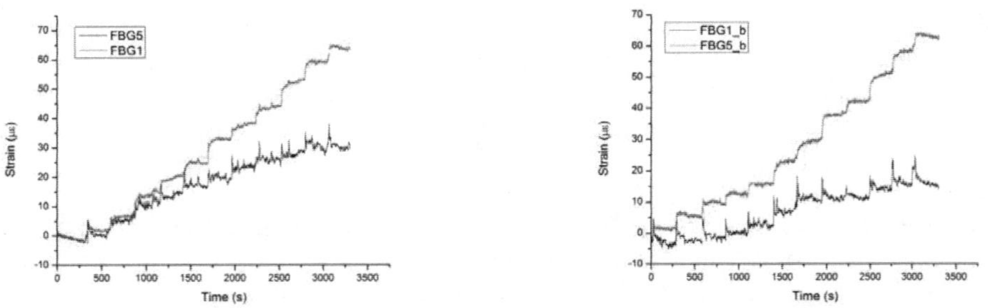

Disjunctor without bar Disjunctor with bar

Figure 6. Comparison of the strains versus time in the posterior and anterior regions, for the two orthodontic devices, in the left side

In figure 7 are displayed the results for the right side of the maxillae, in anterior and posterior regions. As it can be seen, only in the case of the disjunctor with bar, the strain difference maintains constant, during the applied load. For the other cases, there is a greater increase in strain at the posterior region than at the anterior region.

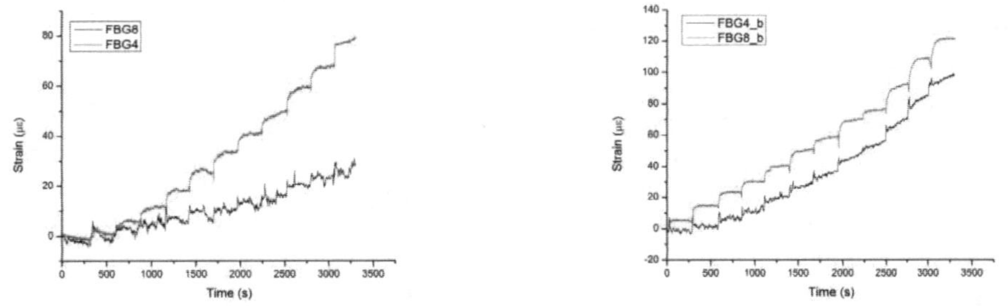

Disjunctor without bar Disjunctor with bar

Figure 7. Comparison of the strains versus time in the posterior and anterior regions, for the two orthodontic devices, in the right side

In order to compare quantitatively the induced strains by the two orthodontic devices, the results were organized in two tables and represent the strain measured at the maximum amount of applied load, corresponding to twelve turns of the Hyrax screw. In table 1 are displayed the results for the disjunctor with the bar and in table 2 are displayed for the disjunctor without the bar.

Table 1. Strains in the anterior region, for the dijunctor with bar.

	Left	Centre-Left	Centre-Right	Right
Anterior region	63.4 (με)	701.2 (με)	553.0 (με)	121.6 (με)
Posterior region	15.0 (με)	902.0 (με)	1423.3 (με)	98.5 (με)

Table 2. Strains in the anterior region, for the dijunctor without bar.

	Left	Centre-Left	Centre-Right	Right
Anterior region	63.8 (με)	512.8 (με)	430.0 (με)	78.9 (με)
Posterior region	29.4 (με)	785.4 (με)	1276.7 (με)	29.2 (με)

For both types of disjunctors, the highest values are measured in posterior region, in the middle of maxillae, in the right side of the suture. Comparing the two devices at the same location, higher strains are obtained with the disjunctor with bar varying between 112% and 154%, except in left posterior region, around 51%, and in the anterior region was almost the same value. The greatest difference was achieved at the right posterior region, reaching almost 334%.

4. Conclusions

FBG sensors were able to measure the strain field pattern in an acrylic model of a maxillae and it was possible to compare the performance of two orthodontic devices. The orthodontic device disjunctor with the bar, in general, transmits higher forces and strain to teeth and maxillae, than with the disjunctor without bar. It was verified that the strain patterns were not symmetric between the left and the right sides as also between the posterior and anterior regions of the maxillae. For the two devices is also found that in addition a displacement in the horizontal plane, particularly in posterior teeth, also occurs a rotation corresponding to a vestibularization of the posterior teeth and their alveolar processes.

Acknowledgments

This work was co-financed by the North Portugal Regional Operational Programme (ON.2 – O Novo Norte), under the National Strategic Reference Framework (NSRF), through the European Regional Development Fund (ERDF).

References

[1] Hass A J 1965 *Angle Orthodontist* **35** 200
[2] Linder-Aronson S and Lindgren J 1965 *Br. J. Orthod.* **6** 25
[3] McNamara J A 2006 *Am. J. of Orthod. and Dentofacial Orthop.* **129** S71
[4] Isaacson R J and Ingram A H 1964 *Angle Orthodontist* **34** 261
[5] Isaacson R J, Word J L and Ingram A H 1964 *Angle Orthodontist* **34** 256
[6] Andrew B, Marțin G, Ansgar H, Arthur W, Christian S, Franz M S and Franz G S 2008 *J Biomechanics* **41** 376
[7] de Assis D S F R, Xavier T A, Noritomi P Y, Gonçales A G B, Ferreira Jr O, de Carvalho P C P and Gonçales E S 2013 *Int. J. Oral Maxillofac. Surg.* **42** 1093
[8] Martinelli F L, Couto P S and Ruellas A C O 2006 *Angle Orthodontist* **76** 1047
[9] Sari Z, Uysal T, Usumez S, Basciftci F A 2003 *Angle Orthodontist* **73** 654
[10] Sada J G V 2005 *Ortod. Esp.* **45** 64
[11] Memikoglu U T 1999 *Angle Orthodontist* **69** 251
[12] De Rossi M, De Rossi A and Abrão J 2011 *Braz. Dent. J.* **22** 334
[13] Oneţ M, Grivu O, Podariu A C, Popovici R A, Sava Roşianu R and Oancea R 2011 *Medicine in Evolution* **17** 495

Uncertainty evaluation and electronic improvements of a wavemeter to measure the wavelength of an external cavity diode laser

I Outumuro[1], J L Valencia[1], J Diz-Bugarin[2], J Blanco[2], B V Dorrío[2]

[1]R&D Department, Laboratorio Oficial de Metroloxía de Galicia (LOMG), Parque Tecnolóxico de Galicia, 32901 Ourense, Spain
[2]Applied Physics Department, University of Vigo, Campus Universitario, 36310 Vigo, Spain

E-mail: ismael.outumuro.gonzalez@xunta.es

Abstract. A Michelson wavemeter was developed to test the accuracy and give traceability to the wavelength of external cavity diode lasers. These lasers were stabilized using a Littrow configuration and an iodine gas cell as frequency reference, and they were used as light sources in the assembly of a new interferometric system for the gauge block calibration. Previously, a microcontroller counting device with a Vernier logic and the uncertainty evaluation of the Michelson wavemeter had to be made.

1. Introduction

The development of an interferometric system for gauge block (GB) calibration, based on frequency stabilized ECDL at 633 nm, involves measuring and giving traceability to the diode wavelength [1]. In order to carry out these measurements, a Michelson wavemeter with a Vernier counter was developed and calibrated to ensure the traceability and evaluate the uncertainty of the wavelength of external cavity diode lasers (ECDL).

2. Design

The wavemeter is based on a Michelson interferometer [2-4] which measures the number of fringes for two counter propagating laser beams, λ_R reference laser and λ_U unknown laser, produced due to the movement of a motor (DS) in the arms of the interferometer. The relationship between the wavelengths of the two lasers and their number of fringes is:

$$\lambda_U = \left(\frac{N_R}{N_U}\right)\lambda_R \frac{n_U}{n_R} \quad (1)$$

Where, N_R and N_U are the number of fringes counted for the reference laser and the unknown one, n_R and n_U are the refraction indices in air for these lasers.

The experimental setup employed for the wavemeter comprises a He-Ne (REO 32734) reference laser, a 220 mm linear motor (DS) (Thorlabs DDS220/M), an optic mount and an electronic system in order to count the interference fringes using the Vernier method to improve system accuracy, figure 1.

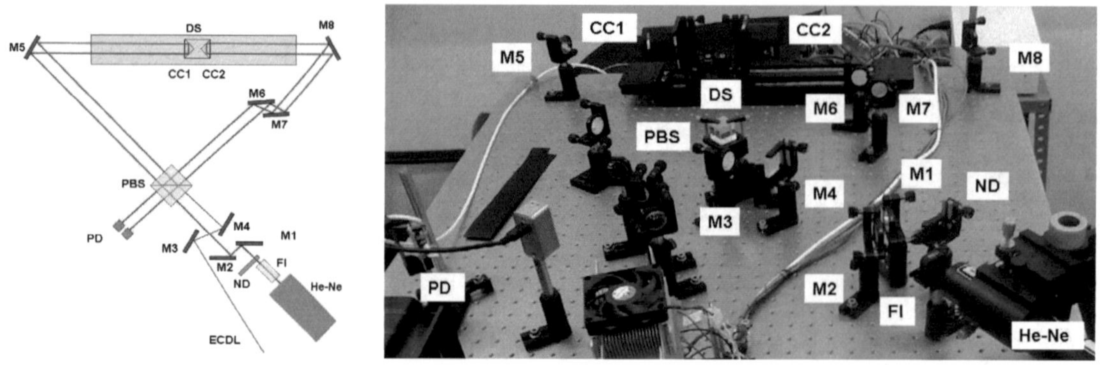

Figure 1. Wavemeter setup: CC corner cubes, M mirrors, FI Faraday isolator, PD photodetector, ND neutral filter, and PBS polarized beam splitter.

3. Electronic. Vernier counter

The Vernier counter is based on a microcontroller to measure and track the count of the fringes, and it is divided into three modules: an analog, figure 2, a digital coincidence detector, figure 3a, and a microcontroller, figure 3b.

The analog module is composed of two identical sections with a transimpedance amplifier, voltage preamplifier with filter, variable gain amplifier and Schmitt-trigger comparator. A positive 5V power supply with a LDO regulator, (LP2950-3.3), was used to provide the reverse polarization for the photodiodes and reference for the preamplifiers. This configuration allows the use of a single supply for all the electronics circuits.

The operational amplifier, (Texas Instruments TLE2074), was used in this module because of its good performance characteristics (low noise, 10 MHz bandwidth, 45 V/µs slew rate) and wide supply range. Noise immunity is achieved with the Schmitt-trigger comparator and its hysteresis level of ±200 mV. As it is critical for the accuracy of the system due to a small signal-to-noise ratio of the fringes, many tests were made to verify the reliability of the module.

The digital coincidence module provides the digital input for the microcontroller and it is responsible for the coincidence detection required to start and stop the fringe measurements in the Vernier method.

So far, the counter logic has been made with a microcontroller, instead of the fixed counters found in literature [2,3], and a coincidence detector. This coincidence detector generates short pulses at the positive edges of the input signals, which are combined in a high speed 74HC00 logic gate that activates a RS flip-flop. The output of this circuit triggers an external interruption on the microcontroller which starts and stops pulse counter. Microcontroller resets the flip-flop when the count is finished to allow the detection of next coincidences. For the first prototype, pulse duration was set to 100 ns to avoid performance problems on the system when input frequency is slower than 100 KHz, a resolution of 10^{-2} per fringe can be achieved.

Moreover the microcontroller module performs the user input, the visualization system and the counter logic, and it improves the flexibility of the system to introduce changes in hardware and software, in contrast to older designs based in wired logic [2, 3]. It was developed with MCS-51 family of microcontrollers,(Atmel AT89S52), a 24 MHz quartz crystal, a 2x16 monochrome LCD display and four pushbuttons. This limits the frequency counter of the microcontroller to a maximum of 500 KHz, enough for the incoming of the signal.

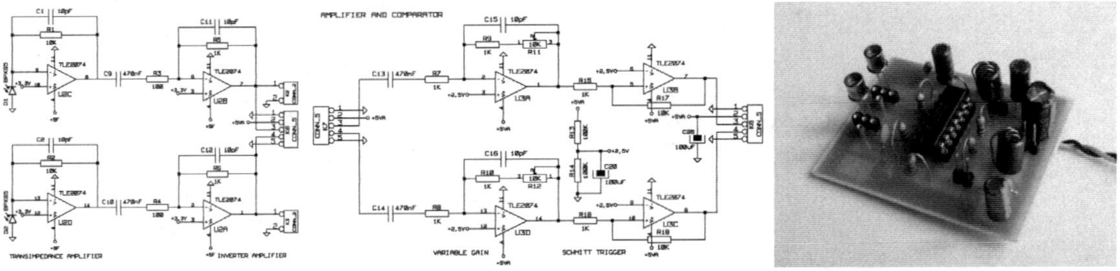

Figure 2. Schematics and photography of the analog module.

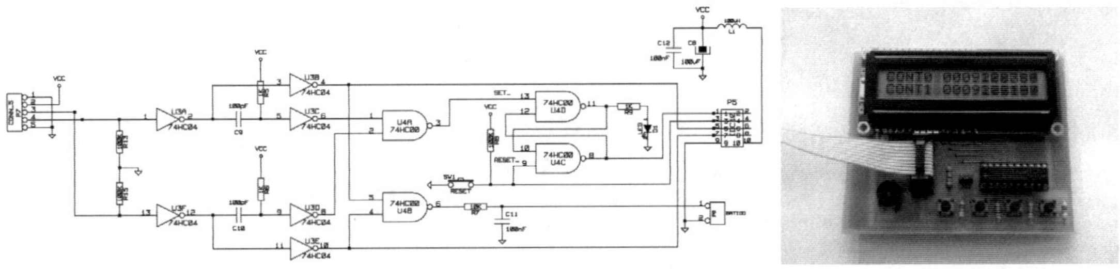

Figure 3. a) Vernier detector schematic b) Photography of the microcontroller board.

4. Uncertainty evaluation

Equation (1) can be simplified to $\lambda_U = \lambda_R\ (N_R/N_U)$ due to the difference between the refractive index of air of both laser beams is less than 0.01ppm, in our laboratory environmental conditions through Ciddor equation [6]. This occur because of the similarity between the wavelength of reference and unknown lasers (<0,005 nm).

The uncertainty of the model function, equation (2), was developed from the simplified equation (1) taking into account all contribution errors that affect the measurement found in bibliography [3,4], and following the recommendation for expressing uncertainty given by the BIPM [5]. Accordingly, the uncertainty components were calculated through table 1.

$$U(\lambda_U) = \sqrt{U^2(N_R) + U^2(N_U) + U^2(\lambda_R) + U^2(\lambda_{WCR}) + U^2(\lambda_{WCU}) + U^2(\lambda_{ALIGR}) + U^2(\lambda_{ALIGU}) + U^2(\lambda_S) + U^2(\lambda_{CIDDOR}) + U^2(\lambda_{TEMP}) + U^2(\lambda_{VIB})} \qquad (2)$$

The uncertainty function was applied to two lasers, the stabilized ECDL and an HP 5519A which was used to check the accuracy of the system and it has a wavelength of 632.991 368 7 nm with an uncertainty of 0.000 013 nm. The results obtained from the experimental data of tables 2 and 3 were a wavelength of 632.991 41 nm with uncertainty (at 95% with k=2) of 0.000 16 nm (0.26 ppm) for the HP 5519A, which agrees with the results of the calibrated laser given by the reference laboratory; and a wavelength of 632.992 2 nm with uncertainty of 0.0001 3 nm (2.0 ppm) for the ECDL, a value which has a high uncertainty that it does not allow to identify the $^{127}I_2$ stabilized transition [7].

Table 1. Contributions to the uncertainty of the wavemeter.

U(λ_U)	Description	Method of calculus
$U^2(N_R)$	Reference laser fringe count	$(\lambda_R / N_U)^2 \Delta N_R^2$
$U^2(N_U)$	Unknown laser fringe count	$(-\lambda_R / N_U^2)^2 \Delta N_U^2$
$U^2(\lambda_R)$	Reference laser wavelength	$(N_R / N_U)^2[(U(\lambda_R)/2)^2 + (D(\lambda_R)/\sqrt{12})^2]$
$U^2(\lambda_{WCR})$	Wavefront curvature of the reference laser	$((\lambda_R \Delta\delta_R^2)/4)^2$
$U^2(\lambda_{WCU})$	Wavefront curvature of the unknown laser	$((\lambda_U \Delta\delta_U^2)/4)^2$
$U^2(\lambda_{ALIGR})$	Reference laser alignment	$\{(\lambda_R/2)[(\Delta x/L)^2 + \Delta\theta_{DS}^2 + \Delta\theta_{CC}^2]\}^2$
$U^2(\lambda_{ALIGU})$	Unknown laser alignment	$\{(\lambda_U/2)[(\Delta x/L)^2 + \Delta\theta_{DS}^2 + \Delta\theta_{CC}^2]\}^2$
$U^2(\lambda_S)$	Repeatability of the measurements	$(\sigma(\lambda_S))^2$
$U^2(\lambda_{TEMP})$	Thermal dilation of the optics system	$[(\lambda_U + \lambda_R)/2(\Delta T d\alpha / d)^2]^2$
$U^2(\lambda_{CIDDOR})$	Refraction index of air [6]	$(\lambda_U \Delta n_{CIDOR})^2$
$U^2(\lambda_{VIB})$	Vibrations in the optics system	$[(\lambda_U + \lambda_R)/2(\Delta A / d)^2]^2$

Table 2. Experimental values of the wavemeter for the HP 5519.

Quantity	Symbol	Value
Reference laser wavelength	λ_R	632,9910325 nm
Reference laser wavelength uncertainty	$U(\lambda_R)$	0.0000031 nm
Reference laser deviation per year	$D(\lambda_R)$	0.000007 nm
Reference laser beam divergence	$\Delta\delta_R$	0.6 mrad
Unknown laser beam divergence	$\Delta\delta_U$	0.2 mrad
Reference and unknown wavelength fringe error	ΔN_R *and* ΔN_U	0,1 fringe
Reference wavelength fringe count	N_R	1332045,4 fringes
Unknown wavelength fringe count	N_U	1332044,6 fringes
Reference and unknown position misalignment	Δx_R *and* Δx_U	0.5 mm
Alignment distance	L	2000 mm
Direct drive stage misalignment	$\Delta\theta_{DS}$	0,25 mrad
Corner cubes misalignment	$\Delta\theta_{CC}$	0,015 mrad
Standard deviation of the measured wavelength	$\sigma(\lambda_U)$,	0,000031 nm
Air diffraction index uncertainty	Δn	0,01 ppm
Corner cube base lineal expansion coefficient	α	22 μm/Km
Temperature gradient	ΔT	0,5 K
Corner cube distance	d	62 mm
Vibration amplitude	ΔA	50 nm

Table 3. Contributions to the uncertainty of the wavemeter.

Quantity	Symbol	Value
Unknown laser beam divergence	$\Delta\delta_U$	2.0 mrad
Reference wavelength fringe count	N_R	1332023,2 fringes
Unknown wavelength fringe count	N_U	1332020,8 fringes
Standard deviation of the measured wavelength	$\sigma(\lambda_U)$,	0,000043 nm

5. Conclusions

According to the results obtained with the HP laser calibration, the wavelength measuring system works properly but it is not yet able to reach the required precision for GB calibration. As a result, it is suitable for making a first approximation of the measurement of the ECDL wavelength and it provides a solution for those applications where an uncertainty on de order of 1 ppm in wavelength is required.

In future, a set of improvements will be made to allow wavemeter to reduce the uncertainty and enhance the precision. Furthermore, the uncertainty equation will be extended to other wavelengths, taking into account the changes produced in the refraction index of air with Ciddor's equation [6], and the traceability of the system will be verified with lasers away from 633 nm. In addition, an evaluation of the stability of the laser diode will be made through an Allan variance [8].

References

[1] Diz-Bugarín J, Outumuro I, Dorrío B V, Blanco, Miranda J M and Valencia J L 2013 Design of an interferometric system for gauge block calibration *Opt. Eng.* **52**(4), 045601-7

[2] Kahane A, O'Sullivan M S, Sanford N M and Stoicheff B P 1983 Vernier fringe counting device for laser wavelength measurements *Rev. Sci. Instrum.* **54**, 1138-1142

[3] Fox P J, Scholten R E, Walkiewicz M R and Drullinger R E 1999 A reliable, compact, and low-cost Michelson wavemeter for laser wavelength measurement *Am. J. Phys.* **67**(7), 624-630

[4] Monchalin J P, Kelly M J, Thomas J E, Kurnit N A, Szdke A, Zernike F, Lee P H and Javan A 1981 Accurate laser wavelength measurement with a precision two-beam scanning Michelson interferometer *Applied Optics* **20**(5), 736-757

[5] BIPM 2008 Evaluation of measurement data – Guide to the expression of uncertainty in measurement JCGM 100:2008

[6] Ciddor P E 1996 Refractive index of air: new equations for the visible and near infrared *Appl. Optics* **35**, 1566-1573

[7] BIPM 2003 MEP 2003, IODINE ($\lambda\approx$633 nm) Absorbing molecule $^{127}I_2$, a^{16} or f component, R(127) 11-5 transition http://www.bipm.org/en/home/mep.html

[8] Allan D W 1987 Time and frequency (time domain) characterization, estimation and prediction of precision clocks and oscillators *IEEE trans. UFFC*, vol **UFFC-34**, n° 6, Nov

Journal of Physics: Conference Series **605** (2015) 012019 doi:10.1088/1742-6596/605/1/012019

Influence of magnification on extraction efficiency in laser resonators with non-overlapping beams

M G González [1,3], A L Peuriot [2], M T Garea [1] and G D Santiago [1]

[1] Grupo de Láser, Óptica de Materiales y Aplicaciones Electromagnéticas (GLOMAE), Facultad de Ingeniería, Universidad de Buenos Aires, Paseo Colón 850 (1063), Buenos Aires, Argentina.

[2] Centro de Investigaciones en Láseres y Aplicaciones, CITEDEF, Juan Bautista de Lasalle 4393 (1603), Buenos Aires, Argentina.

[3] Consejo Nacional de Investigaciones Científicas y Técnicas (CONICET).

e-mail: gsantia@fi.uba.ar

Abstract. The magnification and the Fresnel number determine the mode profile and losses in a bare unstable resonator. Upon inclusion of gain, both the beam pattern and the reflectivity are changed, more than in a stable cavity, because the counter-propagation intensities differ spatially and saturate the amplifier in a way that alters the mode profile, the reflectivity and the conditions of optimal operation. In this paper we present a numerical study of two types of cavities and compute the mode profile and losses in presence of an amplifier that saturates homogeneously. We compare these results with experimental data obtained on a TEA CO_2 laser.

1. Introduction

Unstable resonators have been extensively studied since the seminal paper of Siegman [1]. These resonators have some desirable properties: large volume mode, single transverse mode operation, lower sensitivity to misalignment and good beam quality. Their main drawback is the rather high losses, which hinders the use in low gain amplifiers. Early studies addressed computations on mode structure and losses in bare unstable resonators. In 1974 and 1975 Siegman and Siklas published the first studies on a loaded unstable resonator [2, 3]. The results showed an unstable resonator with gain has an effective reflectivity lower than bare cavity value and the figure depends on the excess gain over threshold. The behavior, although difficult to compute, can be easily understood on simple physical grounds. In an unstable resonator the two counter-propagating waves occupy different volumes, leading to a strong position-dependent gain that prevents us from employing a model similar to Rigrod's in stable resonators [4]. The outer region of the output beam is amplified more than the central zone; thus reducing the effective reflectivity.

In large Fresnel-number unstable resonators the counter propagating beams still resemble spherical waves; thus an approximate analytic analysis can be carried out. A similar situation arises when a Gaussian output mirror is employed. On the other hand, hard-edge cavities with low-to-medium

Fresnel number demand numerical simulations. The results depend on the optical configuration and are less general than those obtained in stable cavities.
In this paper we analyze, numerically, the influence of an active medium on mode pattern and effective reflectivity in two types of cavities we have previously studied neglecting the presence of gain [5,6]. These simulations attempt, among other goals, to determine the conditions of optimum coupling and compare the performance of a given unstable resonator with a stable one.
We consider two cases, the self-filtering unstable resonator [7] (SFUR) and the dot mirror cavity [8]. Even though the latter is a stable one, the energy is coupled out through diffraction in a way similar to the unstable case. These resonators have a medium-to-low Fresnel number and produce very clean, near-gaussian, lowest order mode which leads to better beam quality. In addition, lower losses per round trip (compared to other designs) can be attained. We present computed mode patterns in loaded resonators as well as measured effective reflectivity in a TEA CO_2 laser.

2. Experimental differences between bare and loaded cavities.

In previous papers, we reported on the extension of the SFUR resonator to a ring cavity [5] and the characteristics of the dot mirror when the Fresnel number is outside the range originally conceived by Pax and Weston [6]. The SFUR cavity is a confocal unstable one that belongs to the negative branch. Two concave mirrors of focal lengths f_l and f_s ($f_l>f_s$) define a geometrical magnification $|m|=f_l/f_s$. An aperture, placed at the common focal point, provides spatial filtering and outcoupling. The dot mirror cavity is a semiconfocal stable one with a concave mirror of focal length f and a flat one placed at the focal point. The plane mirror's radius a is smaller than the spot size of the associated Gaussian beam w_0. Due to diffraction effects some energy is coupled out around the plane mirror, in a way similar to unstable cavities. This way, a pseudo magnification can be defined in terms of the Fresnel number $N_F=a^2/f\lambda$ in the following way: $m=0.61/N_F$ [6].

Figure 1 shows a schematic of both resonators. Following the traditional approach, we consider two counter-propagating waves whose intensities are I^+ and I^-. The difference I^+-I^-, evaluated at the filtering aperture (SFUR) or the dot mirror, gives the output intensity I_{out}. The dashed box represents an amplifying medium of length L_a.

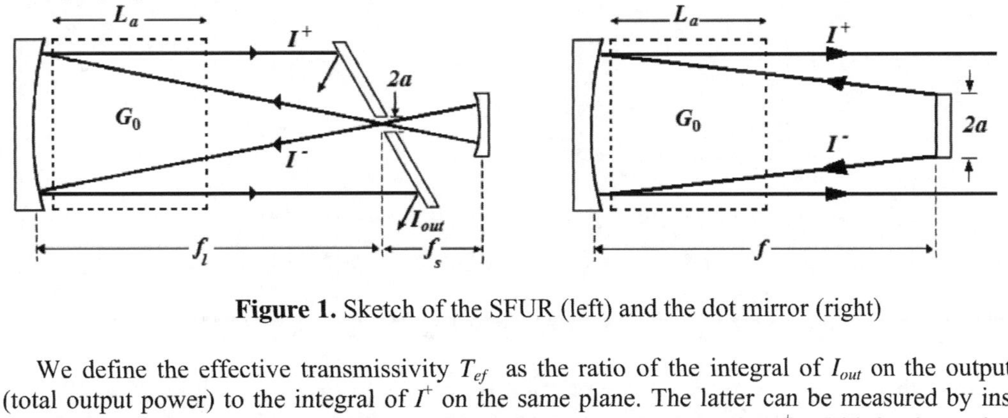

Figure 1. Sketch of the SFUR (left) and the dot mirror (right)

We define the effective transmissivity T_{ef} as the ratio of the integral of I_{out} on the output plane (total output power) to the integral of I^+ on the same plane. The latter can be measured by inserting, inside the resonator, a low-reflectivity (1%) beam splitter that samples I^+. With lossless mirrors the effective reflectivity is computed as R_{ef}=1-T_{ef}. This procedure, defined for a continuous-wave laser, can be extended to a pulsed one if the respective powers are replaced with energies.

In both cases, the values of R_{ef} we obtained were always below the theoretical predictions, and the difference increased as the gain rose. However, when we operated the laser near the threshold we registered a quantity close to the expected result. Figures 2 and 3 present some experimental results that follow the same tendency: the higher the gain, the lower the measured reflectivity.

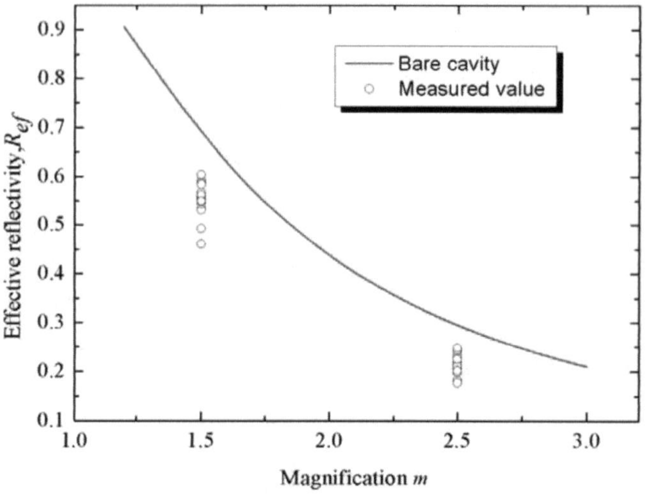

Figure 2. Computed (line) and measured (dots) reflectivity for a SFUR resonator, b) Dot mirror.

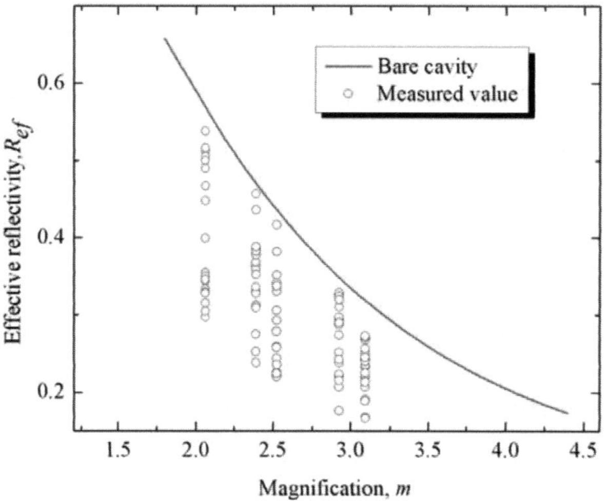

Figure 3. Computed (line) and measured (dots) reflectivity for a dot mirror resonator

To quantify the dependence of reflectivity on gain we resorted to the Fox and Li method [7] but, instead of propagating the beam in a round trip, we applied successive steps of free propagation and lumped saturated amplification. From a formal point of view, the Fox and Li method is only valid in a continuous wave laser; nevertheless, a TEA CO_2 laser provides a pulse that may last as long as 1 μs (including the Nitrogen "tail"). Therefore, a laser pulse includes many round trips and the Fox and Li method provides fair results. We considered a homogenously broadened medium of small signal intensity gain α_0, saturation intensity I_s and length L_a. We did not take into account interference effects between the counter propagating beams I^+ and I^-. Therefore, the medium saturates by the simple addition of intensities; this fact determines a saturated gain: $\alpha=\alpha_0/[1+(I^++I^-)/I_s]$. We divided the amplifier in N_s sections, propagated the beams a distance $\Delta z = L_a/N_s$ by means of the spectral method and adjusted the amplitudes using a saturated gain computed with the intensities of the previous round trip. The initial amplitudes were set to a value well below saturation and the spatial pattern to that of the bare cavity. We increased N_s until further increments did not bring a noticeable change in beam pattern and reflectivity. For most cases $N_s \approx 10$ proved to be large enough. The repetitive procedure stopped when the output power fractional change was less than 10^{-3} in a round trip.

This problem has several independent parameters: small signal gain, position and length of the amplifier, etc. Hence, we will restrict the analysis to a set of values close to the ones we had in our experiments. Nevertheless, the qualitative characteristics are similar and general tendencies can be derived.

Figures 4 and 5 depict the beam profile in a bare ($G_0=\alpha_0 L_a=0$) and loaded ($G_0 > 0$) resonator and the magnification as a parameter. The position has been normalized to the aperture (SFUR) or radius (Dot mirror) a. The presence of the amplifying medium determines an increased energy content of the outer lobes; this accounts for the smaller effective reflectivity for more energy is coupled out. In addition, it can be seen the loaded cavity mode approaches the bare condition either when the total gain G_0 is low, or the counter propagating beams highly overlap (low magnification m), or the losses are high. At intermediate situations a noticeable distortion appears; this may lead to confusion when computing the optical efficiency. These simulations assumed the amplifying medium has no transverse limits; however, the gain zone has some finite extent.

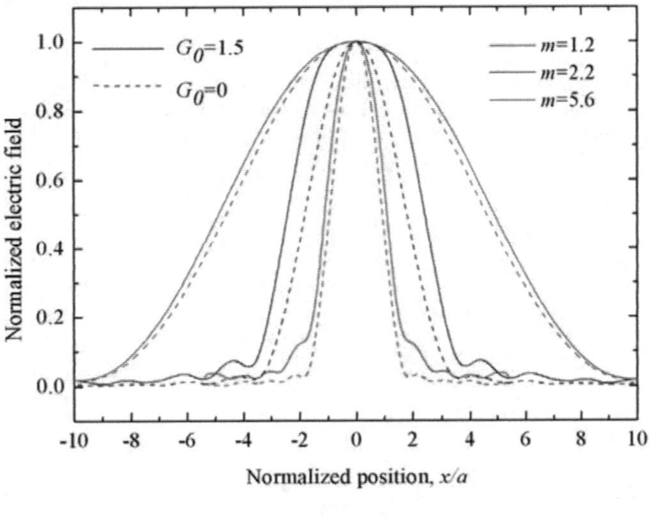

Figure 4. Electric field pattern versus normalized position for a SFUR resonator.

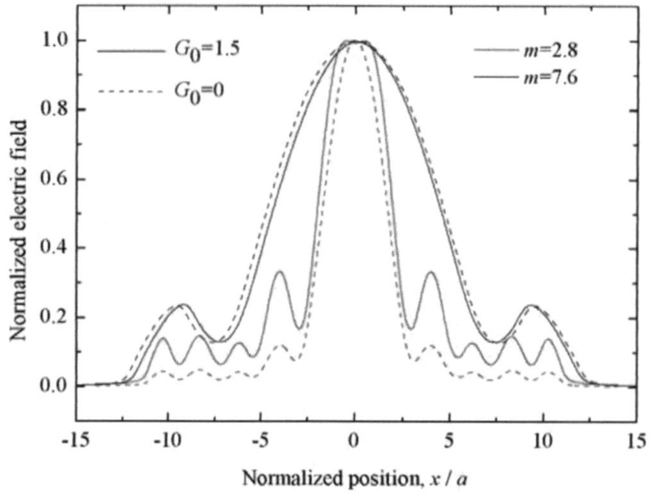

Figure 5. Electric field pattern versus normalized position for a dot mirror resonator.

We repeated the procedure but limited the width of the amplifying volume to the size of the central lobe in a bare cavity. Even though this is an arbitrary option, it fairly represents several lasers. Figures 6 and 7 illustrate the new beam profiles. As expected, the secondary lobes have lower amplitude and the effective reflectivity is larger. This feature is less noticeable in the SFUR resonator for the beam is filtered twice in each round trip.

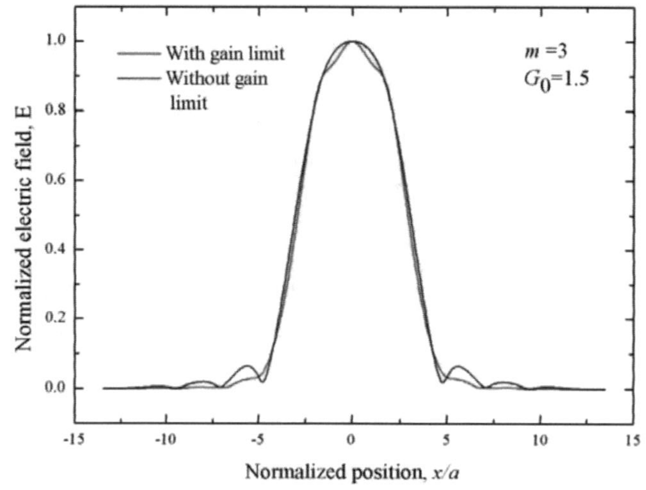

Figure 6. Field pattern with (red) and without (blue) spatially limited gain for a SFUR resonator.

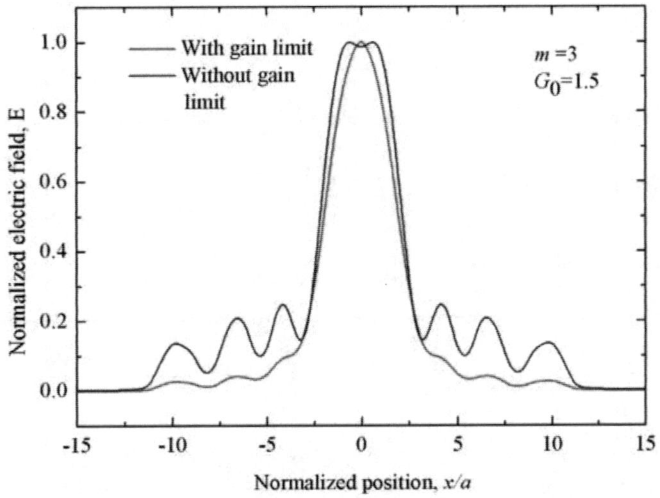

Figure 7. Field pattern with (red) and without (blue) spatially limited gain for a dot mirror resonator.

The effective reflectivity can be seen in figures 8 and 9. To ease comparison the values have been normalized to the bare cavity figure (R_{ef}/R_0). The general behavior of the curves has been commented previously. It is interesting to note that, for some values of G_0 and m, the loaded cavity reflectivity can be as low as one-half the value of the bare cavity. This reduction is smaller when the gain zone has a limited width.

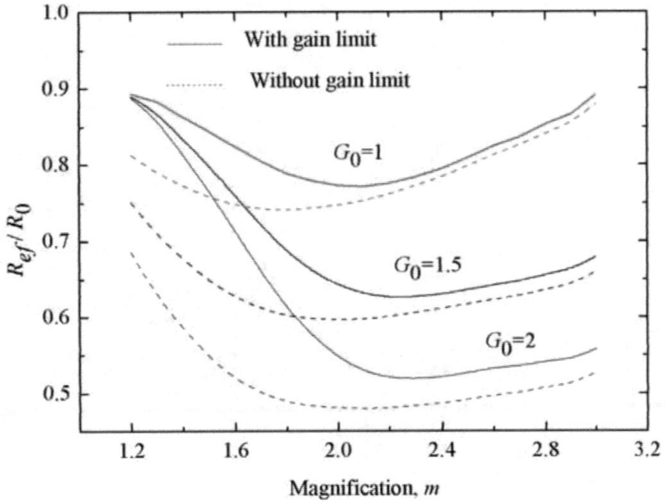

Figure 8. Normalized effective reflective versus magnification for a SFUR resonator.

These results allow us analyzing the optical power extraction process from a different perspective. In a uniform intensity laser the extraction efficiency η is computed as the ratio of the output intensity I_{out} to the maximum available intensity I_{max} from a highly saturated amplifier ($I_{max} = G_0 \cdot I_s$). In our case the extraction efficiency is better computed in terms of total optical power because the intensity pattern is not uniform. To ease comparison, we imagine a stable cavity that operates in a multimode ("flat top") regime, and whose size equals the width of the main lobe of the "unstable" resonator.

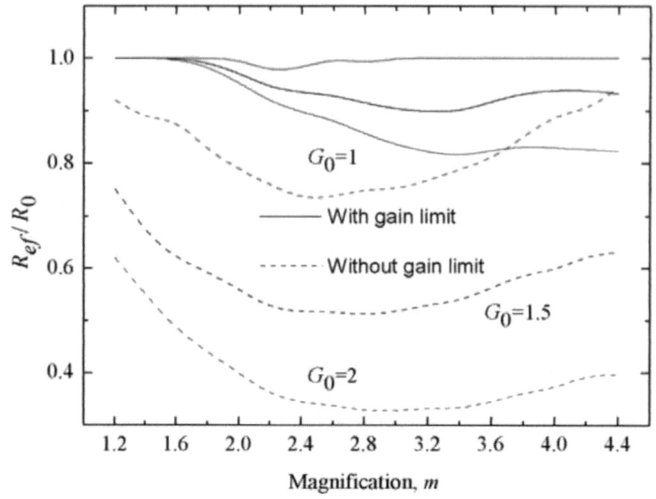

Figure 9. Normalized effective reflective versus magnification for a dot mirror resonator.

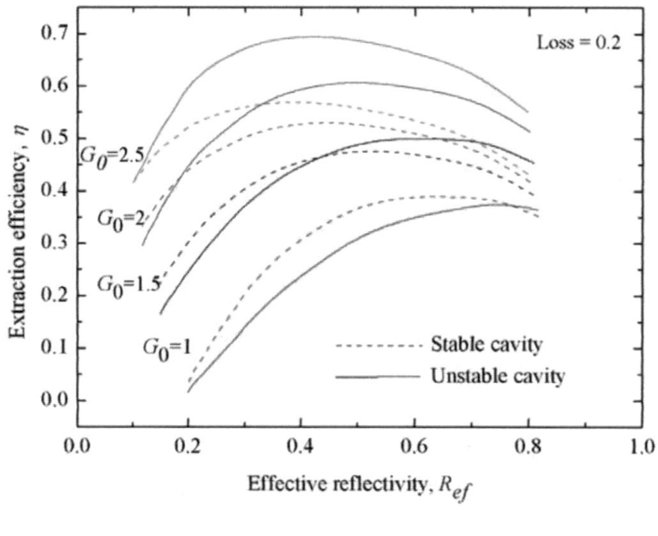

Figure 10. Extraction efficiency versus effective reflectivity for a SFUR resonator.

Figures 10 and 11 show the extraction efficiency of the SFUR (a) and the dot mirror cavity (b). To highlight the presence of a maximum, we included a non-saturable loss of 20%.

The results differ from those obtained by other authors [9] who plotted the extraction efficiency versus the bare–cavity reflectivity. "Unstable" resonators can be more efficient than the stable, multimode cavity when compared at equal effective reflectivity. Unfortunately, the latter depends, in a complex way, on geometrical magnitudes and the amplifier gain.

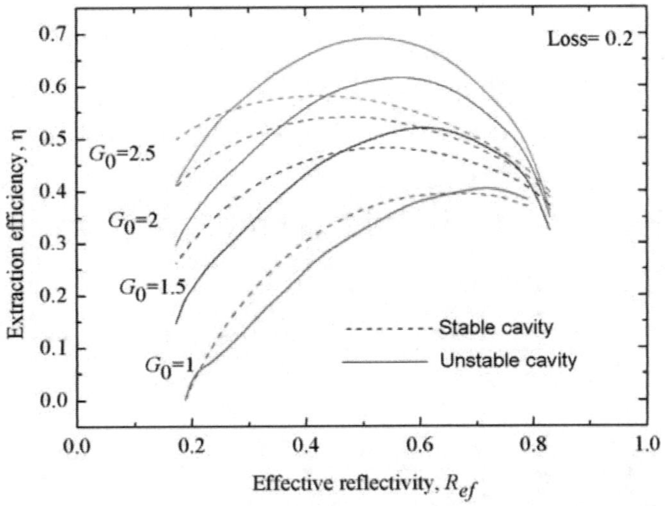

Figure 11. Extraction efficiency versus effective reflectivity for a dot mirror resonator.

3. Conclusions

Even though the effective reflectivity in a loaded unstable resonator differs from that of a stable one has been known for a long time, the analysis of the extracted energy can be carried out in different ways. The standard approach, based on the bare cavity reflectivity leads to a superiority of the stable resonator. However, when the effective reflectivity is taken into account, the unstable resonator gives better results.

Acknowledgments
This work was supported by the University of Buenos Aires, grants UBACYT-20020100100139; UBACYT-20020120100025; Consejo Nacional de Investigaciones Científicas y Técnicas (CONICET), grant PIP-112-201101-00676; and the Agencia Nacional de Promoción Científica y Tecnológica, grant PICT-2011-01211.

References
[1] Siegman A E 1964 Unstable optical resonators for laser applications *Proc. IEEE* **53** 277–267
[2] Siegman A, Sziklas E 1974 Mode calculations in unstable resonators with flowing saturable gain. 1:Hermite-Gaussian Expansion *Appl. Opt.* **13** 2775-2791
[3] Siegman A, Sziklas E 1975 Mode calculations in unstable resonators with flowing saturable gain. 2: Fast Fourier transform method *Appl. Opt.* **14** 1874-1889
[4] Rigrod W W 1965 Saturation effects in high-gain lasers *J. Appl. Phys.* **36** 2487-2490
[5] Santiago G D, Peuriot A L, Rosito C A 2001 Extension of the self-filtering unstable resonator to a ring cavity *Opt. Eng.* **40** 2215-2219
[6] Peuriot A L, Santiago G D, Rosito C A 2002 Numerical and experimental study of stable resonators with diffractive output coupling *Opt. Eng.* **41** 1903-1907
[7] Gobbi P G, Reali G C 1984 A novel unstable resonator configuration with a self filtering aperture *Opt. Commun.* **52** 195-198
[8] Pax P, Weston J 1991 Novel large mode volume resonator *IEEE J. Quantum Electron.* **27** 1242-1246
[9] Anan'ev Y A 1992 *Laser resonators and the beam divergence problem* (Bristol: Adam Hilger)

Journal of Physics: Conference Series **605** (2015) 012020 doi:10.1088/1742-6596/605/1/012020

Electromagnetic field in an active slab: conditions for spontaneous oscillation

M T Garea[1], M C Duplaá[1], F E Veiras[1,2], L I Perez[1,3], C L Matteo[1,2] and P A Sorichetti[1]

[1] *GLOmAe*, Departamento de Física, Facultad de Ingeniería, Universidad de Buenos Aires, Av. Paseo Colón 850, C1063ACV, Cdad. Autónoma de Buenos Aires, Argentina

[2] CONICET (Argentina)

[3] INTECIN-CONICET (Argentina)

E-mail: mgarea@fi.uba.ar

Abstract. In this work we analyze the conditions for the propagation of plane waves due to spontaneous oscillation within an active slab of width d and complex relative permittivity $\tilde{\varepsilon}_b$. The slab is immersed in transparent, semi-infinite media of relative permittivities ε_a and ε_c respectively, without external field sources -except, of course, the power source used for pumping the active medium in the slab. It is well known that, if there is enough gain in the active medium, it may sustain spontaneous oscillations of the electromagnetic field, which result in an impaired operation. This model gives a limiting value of the small signal gain to avoid this condition. It is applicable to disk lasers, amplifiers and other optoelectronic devices in which pumping is confined to a region with transverse dimensions much larger than its length.

1. Introduction

Spontaneous oscillation (SO) takes place in an active medium when the electromagnetic field fluctuacions are enhanced by the gain mechanism. In general, the volume occupied by the active medium is limited. When the reflections in the front, back and lateral surfaces are taken into account, it is easy to see that for some radiation trajectories the gain-length product may exceed the oscillation threshold [1], giving rise to SO. Of course, when spontaneous (alsa called "parasitic") oscillations occur the device will not work properly, if the threshold for the useful signal is higher. As the transverse dimensions of the active medium increase in comparison to its thickness, the feedback provided by the reflections from the periphery becomes increasingly relevant for the onset of SO. From the above is clear that the suppression of SO is a prime concern in the design of disk laser amplifiers and oscillators [2].Therefore, numerous authors have presented models and/or experimental results where SO control is achieved by the elimination of the reflected waves, surrounding the active medium with a cladding or an absorbing layer [3–6]. However, if the gain is high enough, SO may be originated by reflections from the front and back surfaces only. In this work we study this limiting case, where the transverse dimensions of the active medium are very large in comparison to its thickness, and there are no reflected waves from the edges.

The onset of SO is studied by treating the electromagnetic field classically. Linear propagation in active media can be assumed for small amplitudes, where the gain may be considered as independent of the electric field magnitude. Also, the relative magnetic permeability of the media is taken as unity. Under these assumptions the active media is described by a complex permittivity, with its imaginary part opposite in sign to that in usual dissipative materials [7]. The optical dispersion and gain lineshape may modeled by the frequency dependence of the real and imaginary parts of the complex permittivity. It must be remarked that, since in many cases of technological interest the gain is limited to a narrow band, in this work we will study the conditions for the onset of SO at a given frequency, considering the complex permittivity as a parameter.

2. Conditions for the existence of the spontaneous field

We will assume the existence of four plane waves (figure 1): two in the active medium and one in each of the transparent media. The wave vectors k, of the four plane waves have the same component k_z, parallel to the plane interfaces. This is due to the continuity conditions for the fields at the interfaces. Inside the active slab, the normal components of the wave vectors have the same magnitude but opposite directions. In the transparent media, the normal components of the wave vectors point away from the slab. Since the transparent media are semi-infinite and without field sources, we assume that there are no incident external waves impinging on the slab [8–11].

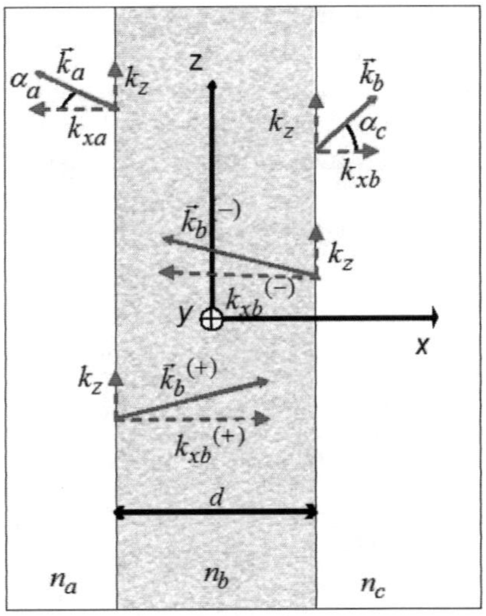

Figure 1. Coordinate system and wave vectors of the transparent – active interfaces.

In the transparent media (a and c), we define α_a and α_c as the angles between the wave vector and the normal to the interface x for each media, respectively. Also, we define η_z, a dimensionless parameter associated to the angles α_a and α_c, given by:

$$\eta_z = \sqrt{\varepsilon_l} \sin \alpha_l = n_l \sin \alpha_l \quad (l = a,\, c) \tag{1}$$

where n_a and n_c are the refractive indexes of the transparent media, respectively. It must be noted that, under the assumptions of this work, the value of the angles α_a and α_c (and therefore η_z) are limited by the ratio between the thickness and the largest tranverse dimension of the active medium. For example, for a circular slab in air of thickness d and radius R, $\eta_z \leq \sin(\arctan(R/d))$. Therefore, η_z may be large but remains bounded. If the refractive index

Journal of Physics: Conference Series **605** (2015) 012020 doi:10.1088/1742-6596/605/1/012020

of any of the transparent media were greater than η_z, this condition would not fulfilled at an interface, and the waves in the transparent medium would be evanescent (no outgoing wave propagating from the slab). This case will not be considered in this work.

The wave vector in each media will be normalized using the vacuum wavelength λ_v, as:

$$\begin{array}{ll} \vec{k}_a = \frac{2\pi}{\lambda_v}(-\eta_{ax}\hat{\mathrm{x}} + \eta_z\hat{\mathrm{z}}) & \vec{k}_b^{(+)} = \frac{2\pi}{\lambda_v}(\eta_{bx}\hat{\mathrm{x}} + \eta_z\hat{\mathrm{z}}) \\ \vec{k}_c = \frac{2\pi}{\lambda_v}(\eta_{cx}\hat{\mathrm{x}} + \eta_z\hat{\mathrm{z}}) & \vec{k}_b^{(-)} = \frac{2\pi}{\lambda_v}(-\eta_{bx}\hat{\mathrm{x}} + \eta_z\hat{\mathrm{z}}) \end{array} \tag{2}$$

where η_z and k_z are related by $\eta_z = k_z\lambda_v/2\pi$. The remaining η parameters are defined through the following relationships:

$${\eta_{lx}}^2 + {\eta_z}^2 = \varepsilon_l \;\; (l = a,\, c) \;\; , \;\; {\eta_{bx}}^2 + {\eta_z}^2 = \tilde{\varepsilon}_b \tag{3}$$

It is important to bear in mind that the relative permittivities ε_a and ε_c are real, but $\tilde{\varepsilon}_b$ is complex. In consequence, η_{bx} is the only complex η parameter (assuming outgoing waves). Thus, in the active medium, it is possible to define a complex refractive index $\tilde{n}_b = \sqrt{\tilde{\varepsilon}_b} = n_{br} - \imath\, n_{bi}$, where n_{bi} is associated to the active medium gain value. In fact, in technological applications the small signal gain coefficient G is customarily used, and it is related to the imaginary part of the refractive index by $n_{bi} = (1/2)(G\,\lambda_v/2\,\pi)$ [12].

The boundary conditions that must be satisfied by the fields at the interfaces impose constraints on the complex amplitudes of the plane waves. This leads to a homogeneous system of linear equations with non-trivial solutions only if the system determinant is zero [13]. For instance, if we consider that the electric field vector is perpendicular to the interface (perpendicular polarization) a dimenssionles magnitude ρ_s can be defined:

$$\rho_s = \frac{(\eta_{bx} + \eta_{ax})(\eta_{bx} + \eta_{cx})}{(\eta_{bx} - \eta_{ax})(\eta_{bx} - \eta_{cx})} \tag{4}$$

as a function of the independent magnitude η_z and the media permittivities. Analogously, for parallel polarization, there is an equivalent dimensionless magnitude ρ_p:

$$\rho_p = \frac{(\eta_{bx} + \frac{\tilde{n}_b^2}{{n_a}^2}\eta_{ax})(\eta_{bx} + \frac{\tilde{n}_b^2}{{n_c}^2}\eta_{cx})}{(\eta_{bx} - \frac{\tilde{n}_b^2}{{n_a}^2}\eta_{ax})(\eta_{bx} - \frac{\tilde{n}_b^2}{{n_c}^2}\eta_{cx})} \tag{5}$$

that also is a function of the same parameters. In consequence, the equations that define the existence of non-zero electromagnetic fields in the systems corresponding to perpendicular (s) and parallel (p) polarizations may be written as:

$$\rho_{s,\,p} = e^{4\pi\imath\eta_{bx}d'} \tag{6}$$

where d' is the slab thickness scaled to the vacuum wavelength: $d' = d/\lambda_v$.

Equation (6) may be recast as:

$$d' = \frac{\log|\rho|}{4\pi\,\mathrm{Im}(\eta_{bx})} \tag{7}$$

$$\arg(\rho) - \frac{\mathrm{Re}(\eta_{bx})\log|\rho|}{\mathrm{Im}(\eta_{bx})} = 2\pi\, m \qquad m \in \mathcal{Z} \tag{8}$$

where $\rho = \rho_s$ or ρ_p. Therefore, equation (8) gives the condition for non-trivial solutions that is independent of the width of the slab; moreover, for each value of η_z it depends only on the relative permittivities. For example, given the real part of the media permittivities, i. e. n_a, n_{br} and n_c, for each η_z there is a discrete set of n_{bi} values that verify equation (8), associated to different values of m. For each value of n_{bi} belonging to this discrete set, the value of d' is obtained from equation (7). It is worth mentioning that the values of m for which there are solutions of equation (8) belong to a bounded interval; the upper and lower bounds depend on the equation parameters and the interval length increases when d' decreases.

It is easy to see that equation (6) may be regarded as a generalization of the well-known threshold condition for oscillation (see for example [1] chap. 1), taking into account the polarization and the angular dependence of the phase and amplitude of the electromagnetic field. One of the most interesting consequences of equations (7) and (8) is the existence of a gain threshold for the existence of non-trivial solutions. This must be understood in the sense that the necessary condition for SO in not fullfilled for gain values corresponding to n_{bi} below those given by equations (7) and (8). It must be remarked that, for a given value of d', we can calculate for each η_z the gain value n_{bi} as a function of the real permittivities of the media, that is, n_a, n_{br}, n_c. Figures 2 and 3 show these values for several combinations of permittivities and polarizations.

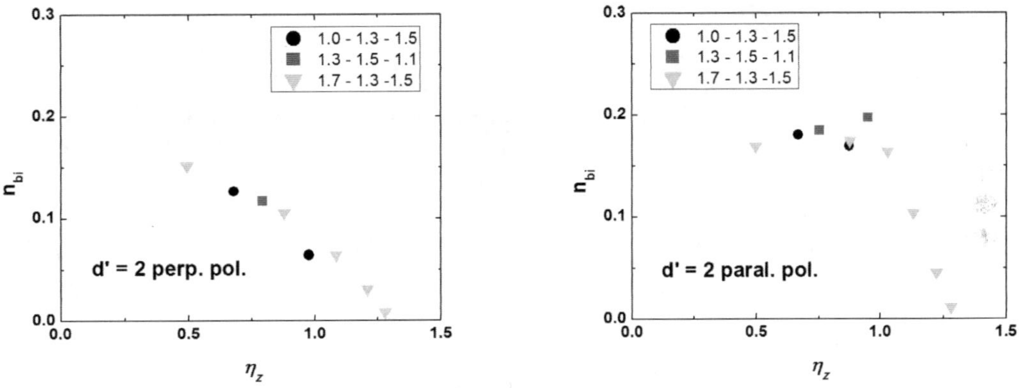

Figure 2. Values of n_{bi} from equations (7) and (8) as function of η_z for $d' = 2$ and the two polarizations. The values in the text box represent n_a, n_{br} and n_c, respectively.

It must be stressed that the non-trivial solutions of equations (6) and (7), given by condition of equation (8), are a set of discrete values; for low values of d', the solutions are few and widely separated. On the contrary, for higher values of d' there is an increasing number of closely spaced solutions, and the set of non-trivial solution resembles a continuum. From the results presented in figure 3 it is evident that, for parallel polarization, there are certain critical values of η_z where the value of n_{bi} for SO is noticeably higher. This behaviour will be studied in detail in future works.

As an application example, we will consider a Nd:Yag slab as an active medium inmersed in air, with a thickness of 2 cm and a maximum transverse dimension of 14 cm. The small signal gain coefficient at $\lambda_v = 1.06$ μm is taken as $G = 8.4\ 10^{-2}$ cm^{-1}, and the real part of the refractive index of the active medium is $n_{br} = 1.56$. These parameters correspond to the active media of the laser described by McMahon et. al. in ref. [3]. These authors analyzed in detail the detrimental effects of SO originated in each slab by the reflections on the edges, even for $G < 4\ 10^{-2}$ cm^{-1} . To supress the SO, the periphery of the active media was coated with absorbing glass, thus practically eliminating the reflections, as was confirmed by further measurements. McMahon et. al. reported that they used numerical methods based on intensity calculations only, to assure the absence of SO due to the reflections in the front and back surfaces of the slabs. Experimental data then confirmed the absence of SO even when G reached the

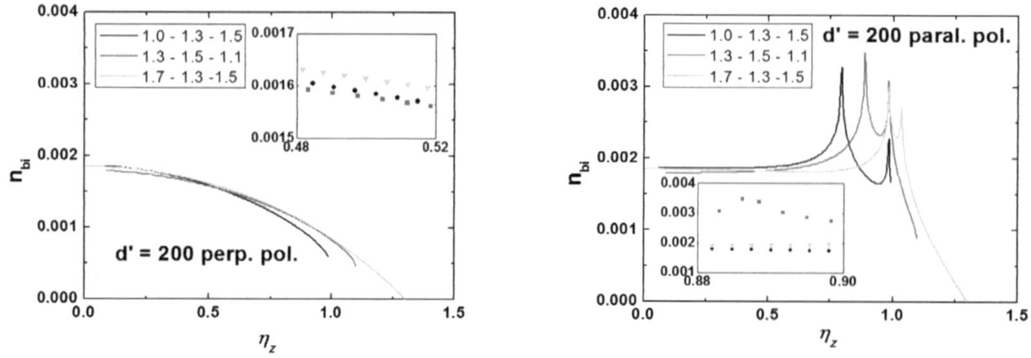

Figure 3. Values of n_{bi} from equations (7) and (8) as function of η_z for $d' = 200$ and the two polarizations. The values in the text box represent n_a, n_{br} and n_c, respectively. The insets show that the solutions are a set of discreted points.

maximum value indicated above. According to our model, the scaled slab width is $d' = 1.89\ 10^4$, the maximum value of η_z given by geometrical considerations is 0.96 and the imaginary part of the refractive index of the active medium n_{bi} has a value of $7.09\,10^{-7}$. From equations (6) and (7) the threshold values of n_{bi} for the maximum values of η_z are $3.14\,10^{-6}$ for perpendicular and $7.75\,10^{-6}$ for parallel polarization, respectively. These are well above the n_{bi} corresponding to the maximum value of G. It is clear that our analytical calculation predicts satisfactorily the absence of SO for the experimental configuration describen in ref. [3].

3. Conclusions

This work presents a study of the conditions for the onset of the spontaneous oscillations in an active slab immersed in transparent, semi-infinite media. The electromagnetic field is treated classically, and since gain saturation effects are neglected, linear propagation of electromagnetic fields is assumed. The active medium is described by a frequency-dependent complex permittivity with its the imaginary part opposite in sign to that in usual dissipative materials. Therefore, the optical dispersion and gain lineshape may modeled by the frequency dependence of the real and imaginary parts of the complex permittivity. It is assumed that the transverse dimensions of the active medium are very large and there are no reflected waves from the edges.

A system of equations describes the boundary conditions that must be satisfied by the field at the front and back interfaces. The values of thickness and gain that correspond to non-trivial solutions of this system determine the necessary conditions for the onset of SO at a given frequency. When the slab thickness is small in comparison to the wavelength, there are few solutions, widely separated; for thicker slabs the set of non-trivial solutions resembles a continuum. In practice, this implies that the gain value obtained for each thickness is in fact a threshold, since in most cases of technological interest saturation effects come into play when the gain exceeds the calculated value. The description of SO in the non-linear regime fall outside the scope of this work.

This model may be applied to estimate the maximum small signal gain without SO in a thin slab in the most favourable case, where reflections from the edges have been totally eliminated. This is a limiting condition for disk lasers, amplifiers and other optoelectronic devices in which pumping is confined to a region with transverse dimensions much larger than its length. The model predicted adequately the absence of SO in a high power device described in the literature.

Acknowledgments

This work was supported by CONICET and grants from Universidad de Buenos Aires UBACyT (2011–2014) 20020100100139, UBACyT (2011–2014) 20020100100406, UBACYT (2013–2016) 20020120100025BA and from CONICET PIP (2012–2014) 11220110100241CO. The authors also wish to thank the anonymous Reviewer for the helpful comments.

References

[1] Svelto O 2010 *Principles of Lasers, fifth Ed.* (New York: Springer) p 511
[2] Chester A N 1973 Gain thresholds for diffuse parasitic laser modes *Appl. Opt.* **12** 2139–46
[3] McMahon J M, Emmet J L, Holzrichter J F and Trenholme J B 1973 A glass-disk-laser amplifier *IEEE J. Quant. Elec.* **QE–9** 992–99
[4] Giesen A and Speiser J 2007 Fifteen years of work on thin-disk lasers: results and scaling laws *IEEE J. Selec. Topics Quant. Elec.* **13** 598–609
[5] Yagi H, Bisson J F, Ueda K and Yanagitani T 2006 $Y_3Al_5O_{12}$ ceramic absorbers for the suppression of parasitic oscillation in high-power Nd:YAG lasers *J. Lumin.* **121** 88–94
[6] Slezak O, Lucianetti A and Mocek T 2014 Efficient ASE management in disk laser amplifiers with variable absorbing clads *IEEE J. Quant. Elec.* **50** 1052–60
[7] Landau L D and Lifshitz E M 1960 *Electrodynamics of Continuous Media* (Oxford, UK: Pergamon Press)
[8] Born M and Wolf E 1999 *Principles of Optics* (Cambridge, UK: Cambridge University Press)
[9] Siegman A E 2010 Fresnel reflection, lenserf reflection and evanescent gain *Optics & Photonics News* **21(1)** 38–45
[10] Duplaá M C, Perez L I, Matteo C L and Garea M T 2011 Reflection and refraction in active dielectric materials *J. Phys.: Conf. Series* **274** 012131
[11] Callary P R and Carniglia C K 1976 Internal reflection from an amplifying layer *J. Opt. Soc. Am* **66** 775–79
[12] Siegman A E 1986 *Lasers* (Millvalley, US: University Science Books) p 269
[13] Abeles F 1967 *Optics of Thin Films en Advanced Optical Techniques* (Amsterdam: North Holland Publishing)

Journal of Physics: Conference Series **605** (2015) 012021 doi:10.1088/1742-6596/605/1/012021

Carbon/hydrogen clusters [$C_nH_x^+$] formation from laser irradiation of coronene

F Betancourt[1], J C Poveda[2], I Alvarez[1], A Guerrero[1] and C Cisneros[1*]

[1]Instituto de Ciencias Físicas Universidad Nacional Autónoma de México Av. Universidad s/n Col. Chamilpa, Cuernavaca, Mor. México
[2] Laboratorio de Espectroscopia Atómica Molecular Escuela de Química Universidad Industrial de Santander Carrera 27 Calle 9 Ciudad Universitaria, Bucaramanga, Santander, Colombia

[*]E-mail: carmen@fis.unam.mx

Abstract. This article presents the photo induced dehydrogenation of a cooled molecular jet of coronene, exposed to 266 nm laser radiation. Using unfocused laser radiation of 1064 nm, synchronously coupled with the ionization laser pulses, a system recently developed. Molecular beams were produced by laser desorption of coronene. Analysis of the photoproducts made by time-of flight mass spectrometer showed that a wide variety of ionic species were formed; more than 300 different species were observed. The results showed carbon clusters C_n^+ with n up to 24 as well as carbon/hydrogen clusters $C_nH_x^+$ with masses higher than 300 m/z. The effect on the laser irradiance on the formation of different ions, in the rage from 10^9 W/cm2 to 10^{10} W/cm2, is discussed as it is reflected on the evolution from the big ions to the smaller ones.

1.Introduction

The polycyclic aromatic hydrocarbons, PAH, constitute a family of compounds with a molecular structure base of carbon and hydrogen, characterized by sp^2 hybridization of a carbon atom, and delocalization of electrons π type. The broad spectrum of applications and their spectroscopic characteristics makes them very attractive for study. PAH are the main compounds present in crude oils and their heavy fractions. In nanotechnology, they are suggested as electric conductors by the production of molecular wires, [1] and in the case of astrophysics, they are considered to be carriers of the ubiquitous aromatic infrared bands observed at different wave lengths from interstellar molecular clouds. They may play a major role in the physics and chemistry of photo dissociation regions of the interstellar media, which has motivated a major area of studies of their properties [2-5]. In general PAH are also considered to be an important component of the extraterrestrial carbon materials. It has been proposed that many diffuse interstellar bands can be identified as result of π-π^* transitions of dehydrogenated coronene ($C_{24}H_x^+$) in the IV-VIS region or vibrational emissions in the infrared. This explains why dehydrogenated coronene is studied in order to model the evolution of PAH in photodissociation regions.

Different measurements of dehydrogenation of coronene and carbon clusters formation have been reported using, for instance Xe lamp, [6] electrons [7], or eximer laser pulses [8]. Dehydrogenated cations of coronene are supposed to be responsible of many diffuse interstellar bands [9], excellent new models of PAH chemical evolution [8] relay on the available molecular data such as the hydrogenation state of coronene ion.

In an attempt to contribute to the characterization of the response of coronene molecules to multiphoton interactions, an experimental system based on desorption preparation of a particular PAH coupled with a R-TOF (Reflectron Time of Flight) mass spectrometer has been developed and the resulting ions have been characterized. When these molecules were exposed to a laser light of 266 nm, a considerable number of ion products were observed. These ion products included type C_n^+ carbon clusters with n from 1 up to 24 and carbon/hydrogen clusters $C_nH_x^+$ with masses higher than 300 m/z .

2.Experimental

A system for desorption and ionization of solid samples using laser radiation of 1064 nm, and 266nm respectively was constructed. Desorption processes took place in a chamber at 10^{-3} torr seeded with helium gas although, other inert gases can be used if desired. Low energy laser pulses of 1024nm (Nd:YAG fundamental)were used to desorb the sample of coronene. Vapors of coronene were adiabatically expanded through a cone with pin hole of 0.2 mm into a high vacuum chamber at 10-8 torr. As a consequence of the expansion, molecules were rotationally and vibrationally cooled. The molecular jet was collimated using a skimmer of 0.2 mm previous to access the ionization region. The ionization processes were produced by multiple photon absorption using laser radiation of 266 nm, synchronously coupled with the desorption step. Delay times for desorption and ionization laser pulses were experimentally optimized to obtain the maximal total ion current, in the present case were between 20-30 microseconds. Ions resulting from ionization processes were extracted and accelerated using two continuously polarized plates at 4500 and 3500 volts respectively, with an electrolytic cooper mesh to allow the positive and negative ions to travel in opposite directions. The positive ions formed by the laser interaction, were analyzed using an R-ToF mass spectrometer analyzer; reaching resolutions up to 7000. Figure 1 shows the experimental apparatus. As shown, ions were detected using a microchannel plate detector; current signal was preamplified and digitized using a picoammeter.

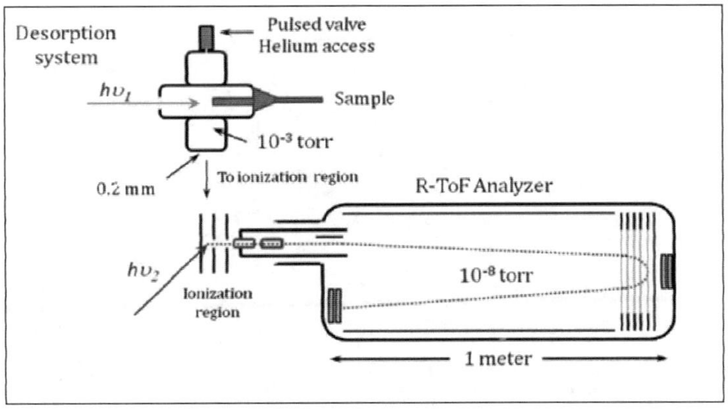

Figure 1 Experimental apparatus

3.Results and Discussion

The first step is to identify well known masses, in present experiment H^+ and C^+, and a reasonable combination of both. The correct position of each maximum on the time of flight spectrum is set by adjusting a Gaussian to each one. By using the positions of maxima obtained, a second degree curve is generated, plotting the masses (maximums) versus time of flight .Then it is obtained the equation that determines the relationship between the flight time of an ion and it's mass. Twenty mass spectra for different intensities of radiation were measured. Figure 2 shows one of the time of flight spectrum for a laser intensity of 2.70 x 10^{10} W/cm2.

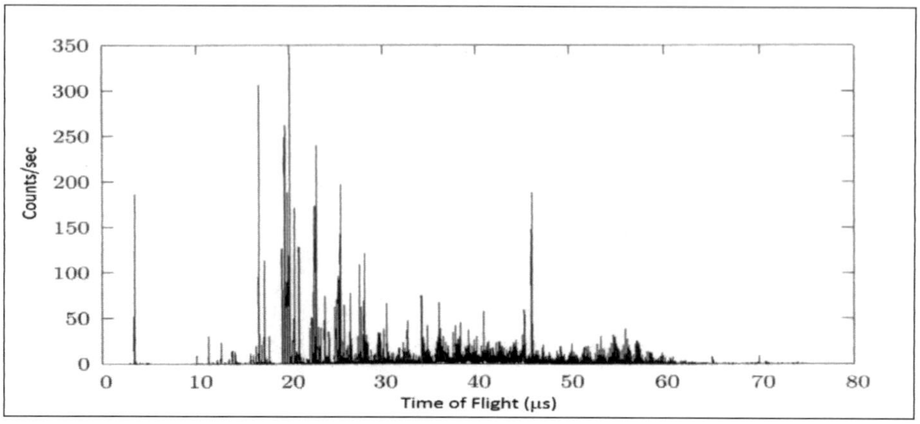

Figure 2 R-ToF mass spectra.

Figure 3a and Figure 3b focus on light masses, while Figure 3c focuses on masses around 300 amu displaying the products with masses close to that of coronene molecular ion.

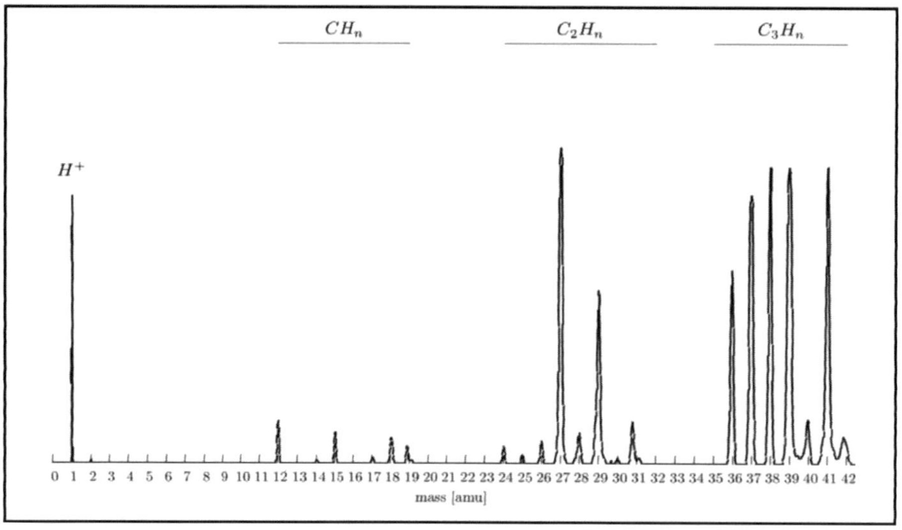

Figure 3a Ion signal versus amu for ion products from 1 to 42 amu

The loss of H_2 was observed at all the range of ions detected. Champeaux, et al. [11] suggested that, as a result on the ionization/fragmentation of isolated coronene by a 100-keV proton, loss of even numbers of hydrogen atoms are evidence of dehydrogenation processes. In our case, the loss of even numbers of hydrogen atoms is observed. The H_2 elimination path could be–a result of hydrogen transfer from one carbon to another in two possible ways to form a -CH_2 structure [12]. After this transfer a neutral H_2 is formed. Other evidences of this process [11] are the detection of CH_2+ and H_2^+, because the multiple pathways possible for the reaction. In the present experiment only the peak corresponding to the H^+ ion was detected.

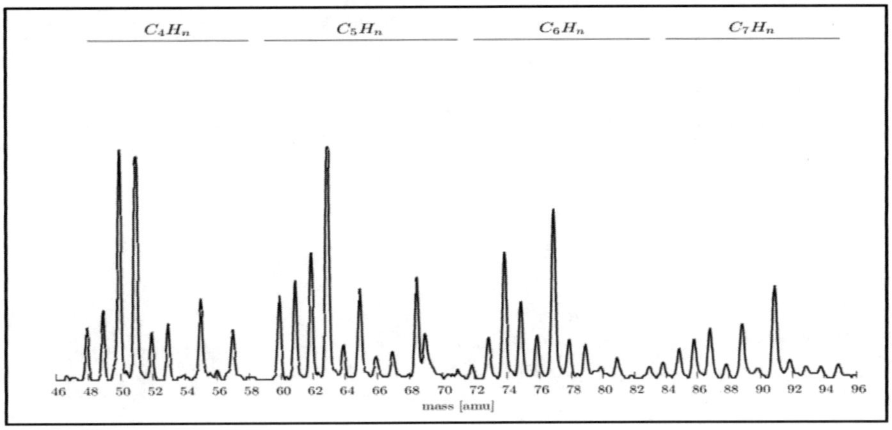

Figure 3b Ion signal versus amu for ion products from 48 to 95 amu

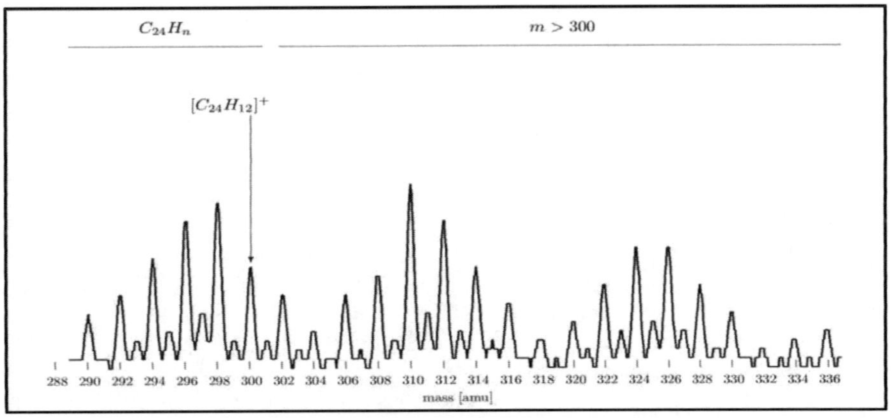

Figure 3c Ion signal versus amu for ion products from 288 to336 amu

Results of this experiment showed that at higher laser intensities light ions predominate, while at relatively lower laser intensities greater masses are more frequent. With this, it can be inferred that light or heavy ions formations depend on the laser intensity as it was expected. This is relevant for experiments dealing with multiphoton absorption since it will facilitate the possibility to concentrate the study of certain families of ions by, knowing the laser radiance at which its appearance is more favored.

From the experimental results the relative calculated currents for groups of different range of masses are presented in Figure 4. This figure shows the relationship between the ion currents and the laser radiance for different mass ranges.

From our results it is possible calculate the number of photons absorbed, leading the formation of each of the ions. The number of photons needed to ionize the coronene was calculated and subsequently the same analysis was carried out for each particular ion. For the case of coronene ion the number of absorbed photons was 1.7 consistent with ionization energy for this molecule. Knowing the number of absorbed photons needed to produce each particular ion it is possible to propose dehydrogenation paths much more complicated than one early proposed [10] for the case were electrons are used.

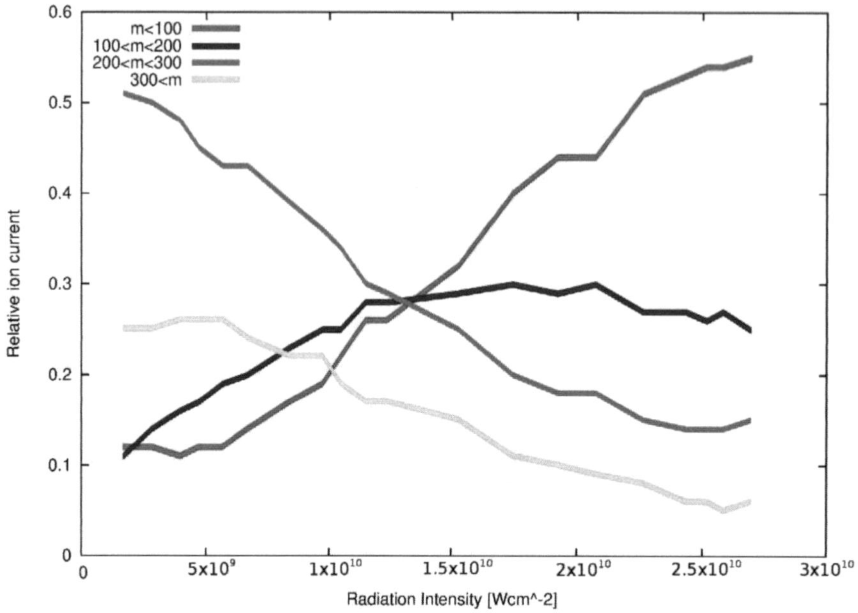

Figure 4. Relative ion currents as a function of the laser radiance

4. CONCLUSIONS

Fully dehydrogenated carbon ions, C_n^+, with n from 0 to 24 were produced. Carbon/hydrogen clusters $C_nH_x^+$ were detected, some of them for the first time. The presence of masses higher than 300 suggest that coronene clusters are form during the desorption process. All the ions observed are stable. The main coronene dissociation process is hydrogen lost. The loss of acetylene was observed however it is not predominant channel as in the case of other PHA. The dehydrogenation process, via H_2 loss, was observed. Finally the results leave several open questions on the explanation of the resulting cations formation. In particular, processes involving multiphoton ionization provide an excellent insight into ionization and dissociation dynamics. Understanding such mechanisms will be an advance in the study of PAHs important in many fields ranking from health to astrophysics.

Acknowledgments

This work was supported by CONACYT Project 165410 and PAPIIT IN107912 and IN102613

References

[1] Altgelt K H, Boduszynski M M 1994 *Composition and Analysis of Heavy Petroleum Fractions* (Marcel Dekker. New York) ps 220-231, 396
[2] Herbig G H, The diffuse interstellar bands 1995 *Annu. Rev. Astrophys.* **33** 19-73

[3] Allamandola L J, Tielins A G G M and Baker J R 1989 Interstellar polycyclic aromatic hydrocarbons the infrared emission bands, the excitation-emission mechanism and the astrophysical implication *The Astrophysical Journal Supplement* **71** 733-775

[4] Montillaud J, Joblin C and Toublanc D 2013 Evolution of polycyclic aromatic hydrocarbons in photodissociation region *Astronomy & Astrophysics 552 A* **15** 1-15

[5] Joblin C and Tielens A G G 2011 PHAS in the Universe *EAS Pub.Ser, Toulouse*.

[6] Ekern S P, Marshall AG , Szczepanski J and Vala M 1997 M Photon induced complete dehydronation of putative interstellar policycyclic aromatic hydrocarbon cations :coronene and naphto[2,3a] pyren *The Astrophysical Journal* **488**, L39-L41

[7] Kim H-S and Saykally R J 2003 An ion beam reflectron/single-photon infrared emission spectrometer for the study of gas- phase polycyclic aromatic hydrocarbons:testing proposed carriers of unidentified infrared emission bands *Rev. Sci. Instrm.***74** 2488-2494

[8] Brechignac Ph , Schmith M, Manson A, Pino T, Parmeix P and Brechignac C 2005 Photon induced products from cold coronene clusters *Astronomy & Astrophysics* **442** 239-247

[9] Duley W Dehydrogenated cations of coronene, 2006 C24Hx+: the source of many diffuse interstellar bands *The Astrophysical Journal* **643**, L21-L24

[10] Guo X, Sievers H L and Grutzmacher H F 1999 Hydrogen atom elimination from polycyclic aromatic hydrocarbons with sustained of resonance irradiation: a new approach to produce carbon/ hydrogen cluster cations *Int.Jour.Mass.Spect.***185-187** .1-10

[11] Chapeaux J, Moretto-Capelle P, Cafarelli P, Deville C. Sence M. and Casta R. 2014 Is the dissociation of coronene is stellar winds a source of molecular hydrogen? Application to the HD 44179 nebula, *Monthly Notices of the Royal Astronomical Society* **441** 1479-1487

[12] Dyakov Y, Ni C K, Lin S, Lee Y and Mebel A 2006 *Phys. Chem. Chem. Phys* **8**, 1404

Surface modification of Ti6Al4V by nanosecond laser ablation for biomedical applications

M P Fiorucci, A J López, A Ramil

Laboratorio de Aplicacións Industriais do Láser, Centro de Investigacións Tecnolóxicas, Departamento de Enxeñería Industrial II, Escola Politécnica Superior, Universidade da Coruña (UDC), Campus de Ferrol, 15403 Ferrol, Spain.

E-mail: `p.fiorucci@udc.es`

Abstract. This paper presents the surface textured process of biometal Ti6Al4V by means of 355 nm Nd:YVO_4 nanosecond laser. Our target is to create structures with sizes which favour osseointegration. In this work a pattern of parallel grooves was generated after a deep analysis of the irradiation parameters involved. Ablation modifies not only the topography but also physico-chemical properties of the metal surface. Changes in the morphology and the physico-chemical state of the laser induced groove pattern were studied by a scanning electron microscopy, X-ray diffraction and X- ray photoelectron spectroscopy, which revealed, among others, an increase of micro roughness and a oxide layer entirely formed by TiO_2, which can improve biocompatibility properties of the textured surface.

1. Introduction

Metallic materials are used in orthopedics and dentistry as biomaterials for bone-replacing applications. Titanium alloy Ti6Al4V is of great choice due to its biocompatibility, high strength and corrosion resistance. Another critical factor to be fulfilled by biometals to achieve implant success is osseointegration [1, 2]. Different studies have demonstrated that the topography of the metal surface is an important factor affecting osseointegration [3, 4, 5]. Besides mechanical considerations, the process at the metal-bone interface is fully dependent on the biological interactions as a response to surface chemistry of the implant [6].

Earlier results do not completely elucidate the influence of surface roughness in osseointegration. Both random roughness and organized patterned biometal surfaces have been tested. In any case, induced surface features must be related and comparable to the size of biological entities like proteins, osteoblast cells and/or osteons [7]. Determination of the optimal topography type and dimensions is still challenging.

Laser texturing has proved to be an adequate technique for implant modification [8, 9]. Although ultrashort pulsed lasers have been increasingly used in this field, nanosecond laser technology is widely adopted in the industry for micromachining (based on time and cost effectiveness). It has been implemented over a wide range of laser sources with different combinations of wavelength, pulse duration, energy and pulse frequency [10].

The aim of this work is the analysis of irradiation parameters involved in laser surface texturing of titanium alloy Ti6Al4V in order to obtain well organised structures with sizes in a suitable range ($\approx$ 30 μm width and $\approx$ 10 μm depth) for biomedical applications and the characterisation of the possible physico-chemical alterations which are produced at the surface due to the textured laser process.

With this objective in mind, we have generated a pattern of parallel grooves by means of different combination of laser irradiation parameters. Moreover, physico-chemical characterisation of the topographies has been performed using standard surface analysis techniques. These results will allow us to establish the most adequate laser parameters in manufacturing of the biometallic surface with a high reproducibility as well as the evaluation of the effects that the physico-chemical changes induced during the ablation process produce in the material's properties related with biocompatibility.

2. Materials and Methods

Ti6Al4V, Ti64 henceforth, in a shape of 4 mm sheet was implemented as a typical metal implant material. Samples were prepared under standard metallographic procedures. Before and after laser treatment, surfaces were ultrasonically cleaned in ethanol, acetone and distilled water.

The laser used was a Nd:YVO_4 (Coherent AVIA Ultra 355-2000) at the wavelength of 355 nm and pulse duration 25 ns. The intensity profile at the laser output was near-Gaussian ($M^2 < 1.3$) and the beam diameter at $1/e^2$ intensity level was about 2.2 mm. The laser beam presents vertical polarization ($> 100 : 1$). Pulse rate, f, can be selected from single-shot to 100 kHz, with energy per pulse around 0.1 mJ. The maximum mean power output is 2 W at $f = 20$ kHz. In order to improve the focalization, a convergent lens with focal length of 35 mm was used. The beam diameter, measured at the beam waist, was 32 ± 6 μm, which corresponds to an energy density of about 26 J/cm^2. Samples were set on XYZ translation stage Newport ILS-CC. Newport MM4006 controller was used for the movement of the stages and to *start/stop* the laser. The focused beam was impinged normal to the surface of the sample. Customised software was applied to program the controller in any arbitrary trajectory and to synchronise the laser with sample displacement. All experiments were done in air. To modify the topography of the metal surface, the sample was precisely positioned at the waist of the focused beam using the Z-direction stage.

After laser machining, samples were analysed with a scanning electron microscope (SEM) JEOL JSM 6400. Metallographic images of transverse sections were obtained by re-mounting samples at a perpendicular position to the main axis of grooves. Chemical etching with Kroll´s reagent was performed for metallographic examination. The topographical analysis was obtained by means of confocal microscopy (Sensofar PLu 2300). The crystalline structure was investigated by X-ray diffraction (XRD). Spectra were recorded with a Bruker D5000 difractometer, Cu cathode, 40 kW, 30 mA under Bragg-Brentano geometry. X-ray photoelectron spectroscopy analysis (XPS) was performed using a Thermo Scientific K-Alpha ESCA instrument equipped with Al-Ka radiation. Charge referencing was set as C1s photo peak at 285 eV.

3. Results and Discussion

3.1. Surface structuring

Polished surfaces of metal samples were textured with a pattern of parallel grooves by overlapping single laser pits to achieve the line structure. In a previous work, the systematic procedure to select the most adequate parameters to obtain topographic characteristics in the appropriate range for biomedical applications was performed [11]. The pulse rate (f) was fixed at 10 kHz because, at this value, the laser delivers its maximum output power and, moreover, this high rate allows us to optimise the process, i.e., treating large areas of the metal surface in a shorter time, which is of great interest from a practical point of view.

Regarding the scan speed (v), the selected rage was $1 - 30$ mm $\cdot$ s^{-1}. Higher values of v would result in an important loos of precision in the translation stages, and, consequently, in sample micromachining. Conversely, smaller values would needlessly increase the processing time. It was estimated that at this working v range it does not cause appreciable heat accumulation in the sample surface. The degree of overlapping depends on both the pulse repetition rate and the scan speed. Otherwise, the spacing between the grooves should be taken into account to obtain a suitable periodic pattern. This value was selected by considering the maximum width of the grooves (obtained from a Gaussian fit of the groove profile) with the criterion that the overlap is 50% of the groove depth, i.e., the spacing was approximately 1 FWHM. At the maximum value of groove width, w = 40 μm, the spacing or pitch is FWHM = 1.177.w ≈ 50 μm .

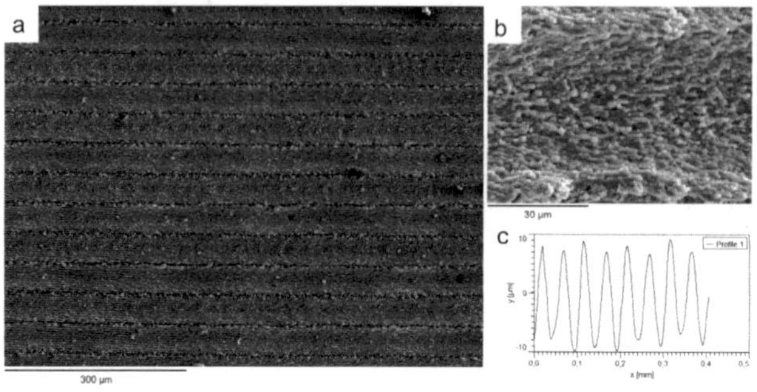

Figure 1. SEM micrograph of machined grooves in titanium alloy generated by laser irradiation in a single beam pass at $v = 25\ \mathrm{mm\cdot s^{-1}}$, $f = 10$ kHz and pitch 50 μm. (*a*) General view at 400x; (*b*) details of one groove at 1000x; (*c*) groove profiles obtained by confocal microscopy.

Figure 1 depicts SEM top view of a pattern of grooves in Ti64. In order to study the effect of scan speed on the groove depth, cross sections of samples processed at scan speed in the range $v = 1 - 30\ \mathrm{mm\cdot s^{-1}}$ were analysed by SEM. Figure 2 shows the cross sections at $v = 25$, 10 and $5\ \mathrm{mm\cdot s^{-1}}$. In the range $v = 10 - 30\ \mathrm{mm\cdot s^{-1}}$, the profile of the laser machined area is smoothly wavy. However, at $v = 5\ \mathrm{mm\cdot s^{-1}}$, the grooves become much deeper and residues debris produced during ablation remain trapped inside, which makes scan speeds below $5\ \mathrm{mm\cdot s^{-1}}$ rather ineffective for texturing the metal surfaces. The dependence of groove sizes (depth and width) with irradiation parameters (Figure 3) shows that groove depth is the most sensible characteristic; it decreases with the scan speed while the width remains almost constant. Therefore, the analysis of the different parameters involved allows us to establish the most adequate irradiation conditions to obtain grooves with features in the range of interest: at focused beam and $f = 10$ kHz the optimum value of scan speed would be in the range $25 - 30\ \mathrm{mm\cdot s^{-1}}$.

3.2. Surface characterisation

Changes of the morphology and the physico-chemical state of the laser induced groove pattern were studied. Morphological features created at the textured surface were shown in Figure 1 (*a*) and (*b*). They consisted of splattered material ejected from molten pool during ablation in the form of drops, clusters and pillar structure, which resulted in a characteristic microroughness inside the groove. Further analysis of these micro-structures developed inside the groove are being conducted in order to quantify the roughness of Ti64 bio-metal surfaces structured by laser, since it has been demonstrated that the creation of roughness and sub-micron features could be beneficial for cell proliferation and adhesion processes [12]. From transverse sections of

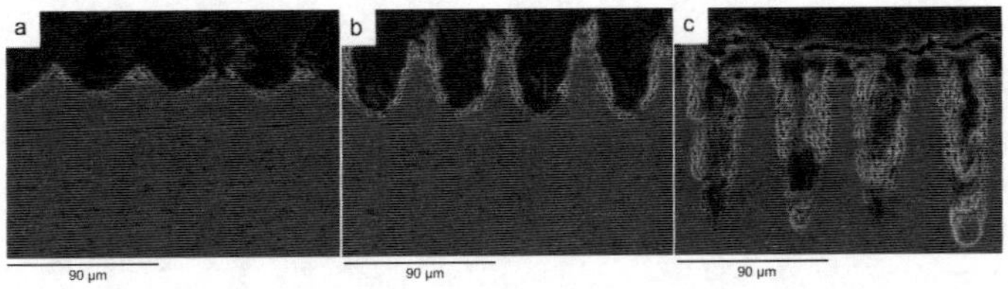

Figure 2. SEM images of cross sections of parallel grooves obtained at $v = (a)$ 25 , (b) 10 and (c) 5 $\mathrm{mm \cdot s^{-1}}$; $f = 10$ kHz and pitch 50 µm in Ti64.

the grooves (Figure 2) it can be observed that a rib of molten material was redeposited at the top of the groove walls. The rib is part of the remelted layer of bulk material which has been produced during intense laser heating and rapid cooling, conforming the heat affected zone. This layer is about 2.5 µm and exhibits a porous texture. Grain boundaries of the original crystalline phases are no longer appreciated at the remelted zone.

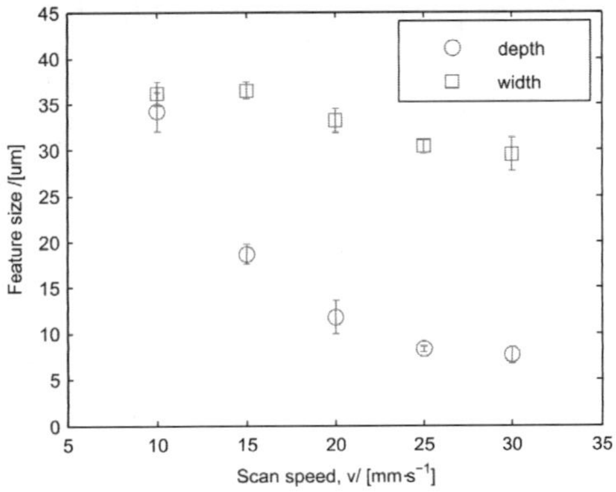

Figure 3. Sizes of the grooves (depth and width) as a function of the scan speed.

XRD analysis of treated sample (Figure 4) do not show new crystalline structures. Typical peaks corresponding to α and β tinanium phases were detected. Changes observed in the spectra of textured surfaces are regarding to the shift and broadening of the diffraction peaks which are related to a grain refinement of the metal or a strain state caused by high cooling rates [13].

Chemical characterization of the changes induced in the remelted layer by laser texturing was performed by XPS analysis. The survey spectra of the control and laser treated samples show peaks which correspond to the principal alloying constituents: titanium and aluminium. Moreover, two other dominant peaks corresponding to carbon and oxygen were detected. Figure 5 depicts high resolution spectra of samples in Ti64. In control samples, fitted peaks for high resolution O1s spectra corresponded mainly to oxide compounds and a smaller peak which corresponds to hydroxide compounds. High resolution Ti2p spectrum shows the presence of three compounds: a dominant one assign to TiO_2, and two weaker peaks for TiO and for metallic Ti. Aluminium was mainly present as oxide Al_2O_3, hydroxide and metallic state. XPS analysis of laser treated samples reflects a significant increase of the oxygen content at the surface of the samples. This increase could probably be a consequence of the surface oxidation during

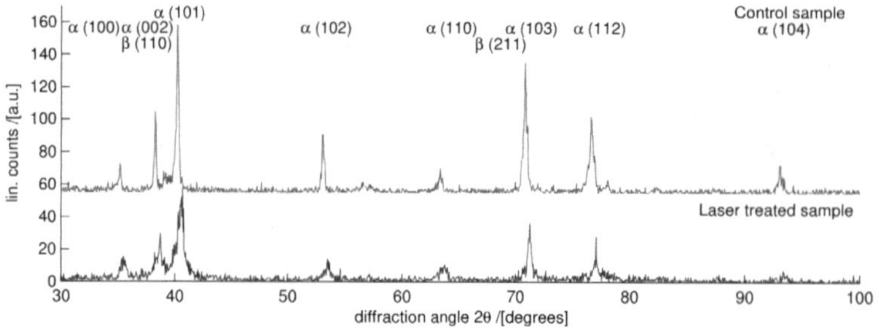

Figure 4. XRD patterns of control and laser treated samples in Ti64.

laser structuring under a non-controlled atmosphere. Laser irradiation induced an oxide layer entirely composed of TiO_2, a highly stable protective surface film, which makes titanium one of the most corrosion-resistant metals [14]. Furthermore, Ti and Al metallic peaks are not longer detected; this shows the growth in thickness of the oxide layer, as it was described in previous works [15, 16]. XPS also indicates a major rise of OH compounds after laser treatment. This surface enrichment of polar compounds could positively contribute to the surface hydrophilicity [17].

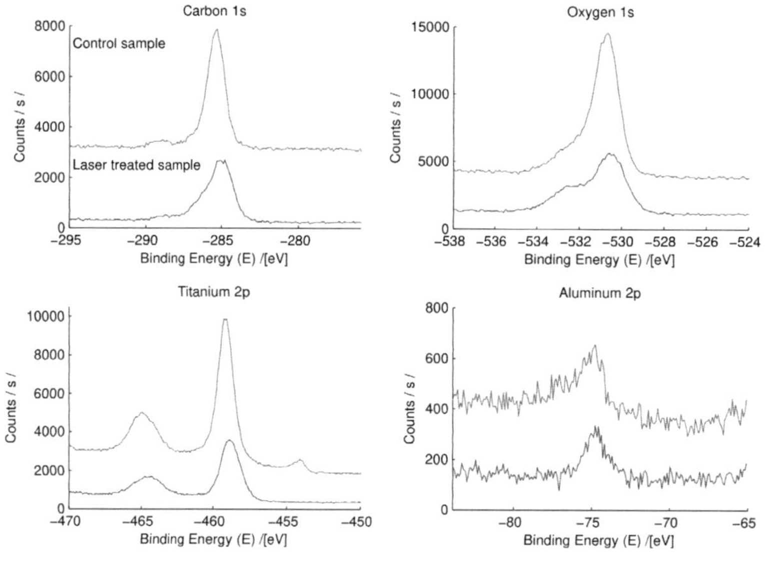

Figure 5. XPS high resolution spectra of control and laser treated surface samples with parallel grooves in Ti64.

4. Conclusion

The study of the micromachining process of Ti6Al4V by means of 355 nm Nd:YVO_4 nanosecond laser was presented. The surface was textured under a groove pattern with an appropiated set of parameter, i.e. $f = 10\,\text{kHz}$, v in the range $25 - 30\,\text{mm}\cdot\text{s}^{-1}$ and pitch is 50 μm, in order to obtain surface features in the range of ≈ 30 μm width and ≈ 10 μm depth which is reported to be suitable for biomedical applications. SEM analysis of transverse section samples revealed the presence of a thin layer of remelted material throughout the groove topography. This layer exhibited a micro-roughness and porous structure which maximise the exposure surface and, consequently, could be beneficial for cell adhesion and proliferation. A grain refinement of original crystal structure and consequently an induced residual strain in those layers could be appreciated. Regarding to the surface chemistry, XPS analysis reflected an increased surface oxidation in addition to a grow in OH which could positively contribute to the surface hydrophilicity by laser surface modification. Those findings could improve corrosion and biocompatibility properties of the metal.

Acknowledgments
M.P. Fiorucci work was supported by Contratos Predoutorais da Universidade da Coruña.

5. References

[1] Brånemark P I, Adell R, Albrektsson T, Lekholm U, Lundkvist S and Rockler B 1983 *Biomaterials* **4** 25–28
[2] Ratner B D 2004 *Biomaterials science: an introduction to materials in medicine* (Academic press)
[3] Wennerberg A and Albrektsson T 2009 *Clin. Oral Implant. Res.* **20** 172–184
[4] Frenkel S R, Simon J, Alexander H, Dennis M and Ricci J L 2002 *J Biomed. Mater. Res.* **63** 706–713
[5] Curtis A and Wilkinson C 1997 *Biomaterials* **18** 1573–1583
[6] Brunette D M 2001 *Titanium in medicine: material science, surface science, engineering, biological responses, and medical applications* (Springer)
[7] Kasemo B and Gold J 1999 *Adv. Dent. Res.* **13** 8–20
[8] Brånemark R R, Emanuelsson L, Palmquist A and Thomsen P 2011 *Nanomed.* **7** 220–227
[9] Hallgren C, Reimers H, Chakarov D, Gold J and Wennerberg A 2003 *Biomaterials* **24** 701 – 710
[10] Kurella A and Dahotre N B 2005 *J. Biomater. Appl.* **20** 5–50
[11] Fiorucci M, López A and Ramil A 2014 *The International Journal of Advanced Manufacturing Technology* 1–7 ISSN 0268-3768
[12] Ulerich J P, Ionescu L C, Chen J, Soboyejo W O and Arnold C B 2007 *Proc. SPIE - The International Society for Optical Engineering* vol 6458
[13] Waseda Y, Matsubara E and Shinoda K 2011 *X-Ray Diffraction Crystallography. Introduction, Examples and Solved Problems* (Springer)
[14] Oshida Y 2013 *Bioscience and bioengineering of titanium materials* (Elsevier)
[15] Bereznai M, Pelsöczi I, Tóth Z, Turzó K, Radnai M, Bor Z and Fazekas A 2003 *Biomaterials* **24** 4197 – 4203
[16] Reimers H, Gold J, Kasemo B and Chakarov D 2003 *Appl. Phys. A* **77** 491–498
[17] Kang B S, Sul Y T, Oh S J, Lee H J and Albrektsson T 2009 *Acta biomaterialia* **5** 2222–2229

Fabrication of a microlens array in BK7 through laser ablation and thermal treatment techniques

M Blanco, D Nieto and M T Flores-Arias

Microoptics and GRIN Optics Group, Applied Physics Department, Faculty of Physics, University of Santiago de Compostela, Santiago de Compostela, E15782 Spain

E-mail: maite.flores@usc.es

Abstract. We propose a laser-based method for fabricating microlens on borosilicate glass substrates. The technique is composed by a laser direct-write technique using a $Nd : YVO_4$ for fabricating the microlens arrays and a post thermal treatment with a CO_2 laser for improving its morphological and optical properties. The proposed technique will allow us to obtain microlenses with a broad range of diameters ($50\mu m-500\mu m$) and focal lengths ($1mm-5mm$). By combining laser direct-write and the thermal treatment assisted by a CO_2 laser, we are able to obtain good quality elements.

1. Introduction

Micro-optics plays an important role in the industrial production processes, like micro-electronics, laser micro-machining and material processing. Microlens arrays offer an unique enabling technology in critical domains such as sensors, communications, metrology and medical imaging, often providing solutions where other technologies prove unsuitable, unwieldy or cost prohibitive. It can be found in the literature the use of a big variety of methods for fabrication of microlenses: laser direct writing [1, 2, 3], molding or similar thermoforming techniques [4, 5], lithographic methods [6] and wet and dry etching [7]. All of these methods usually need to be done in very controlled enviroments and might be rather expensive. Recently, some of the authors have published a method to fabricate microlenses in commercial substrates using a combination of a $Nd : YVO_4$ laser with a post-thermal oven treatment [8]. This work proposes an improvement of the post-thermal treatment by using a CO_2 laser.

In this paper we report a laser based technique for fabricating a microlens arrays in BK7 glass which combines a Q-switched $Nd : YVO_4$ laser and a thermal treatment CO_2 laser roller furnace [9] that reduces the thermal shock and avoid the complications related to the conventional thermal treatments to be able to melt only the top surface of the glass, on which the array is located.

2. Experimental procedure

The microlenses are fabricated via laser ablation using a Q-switched $Nd : YVO_4$ laser in combination with a galvanometer system for adressing the output laser beam. A flat-field lens, with an effective focal lenght of $100mm$, provides an uniform irradiance distribution over an area of $80 \times 80mm^2$. The most suitable laser parameters for this material and purpose have been found to be a repetition rate of $10kHz$, a wavelength of $1064nm$, a pulse width of 20 ns and a scan speed of $50mm/s$. The structure designed for fabricating the microlenses is composed by a set of concentric rings. Each ring is obtained by a circular trench separated $11\mu m$ from the following one. The laser beam is moved, using a galvanometer system, relatively to the glass sample, being maintained in the same position. The laser fluence used for manufacturing each ring varies from $F = 72(14)J/cm^2$ for the last stage to $F = 189(38)J/cm^2$ for the first stage (see figure 1). After the promoting glass structures are generated in the sample we use a roller furnace combined with a CO_2 laser (see figure 2), for melting the glass and reshape its surface, reducing the damage generated before and softening the morphology of the microlenses.

The scheme used for fabricating the microlens (distance between consecutive rings and laser fluence) are selected individually to avoid extremal thermal effects that lead to breakage of the material.

A visual explanation of the processes involved in the fabrication is shown in figures 1 and 2. As it was said before, figure 1 depicts the experimental setup together with the scheme of fluences applied to the sample, figure 2 shows how the thermal process is done.

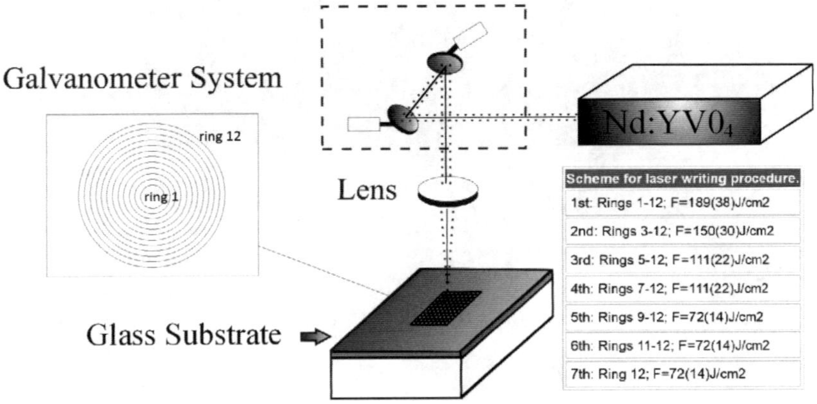

Figure 1. Setup of the ablative process. The applied structure is depicted by the rings on the left side. The scheme of fluences is shown in a table on the right, in which we explain for every step of the process what is the fluence and over which rings it acts.

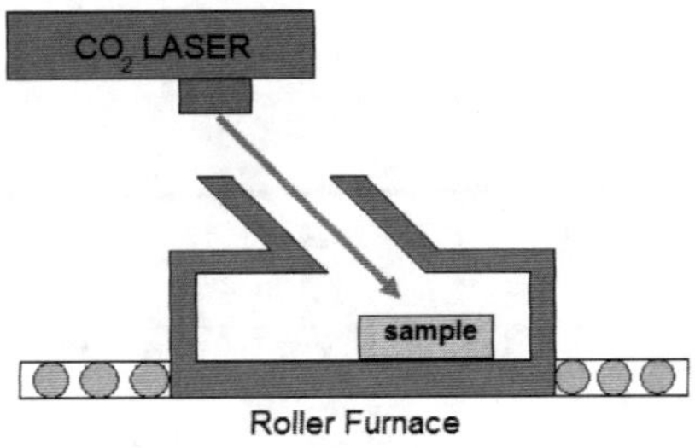

Figure 2. Representation of the thermal treatment composed by a dinamic furnace and a CO_2 laser.

3. Results and characterization

In this section we analize the fabricated microlens array. Measurements of diameter and height have been obtained using an optical microscope Nikon MM400. Images of a top and lateral view of the carved structure can be seen in 3 and 4 respectively. In figure 3 it can be seen the annular structure defined in the previous section and how the separation between rings is optimal to avoid walls of material between two consecutive lines, figure 4 supports this conclusions and also gives a measurement of the height of the profile. Microlenses with diameter $290\mu m$ and height of $\sim 7\mu m$ are presented.

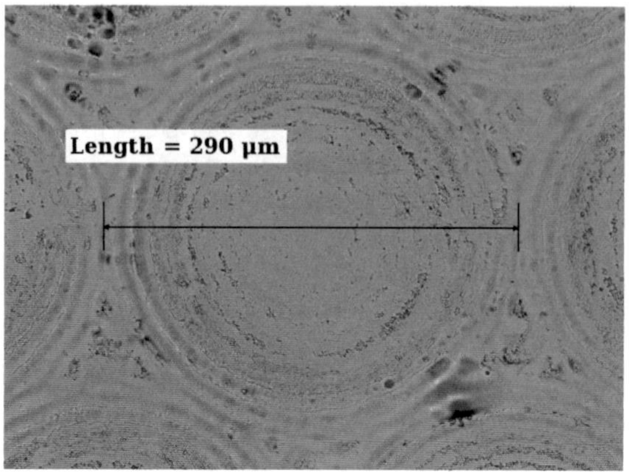

Figure 3. Microscope image of the top view of the direct carving process. The ring structure can be noticed perfectly in the outer regions. The diameter of the microlens was measured using the microscope commercial software. The measurement precision is $\sim 10\mu m$.

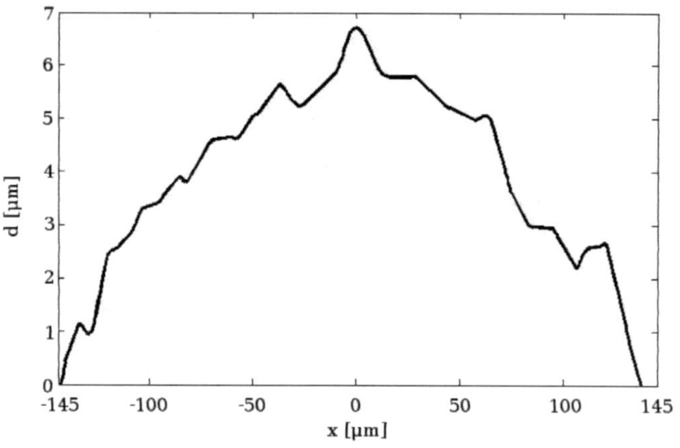

Figure 4. Lateral view of the carved structure. It can be seen that the shape is very irregular but it can be improved by a melting procedure.

It can be seen in the images 1 and 2 the obtained annular structure, as well as the total abscense of crackings or breakage of any kind on the material.

The obtained array has an efficiency (ratio between the power in the focuses and the total incident power) of $\varepsilon = 69.2(6.3)\%$ at $\lambda = 633nm$. The focal length obtained is $5.2(1.2)mm$ via direct measurement, which is in accordance with the curvature ($R = 2066.1(0.1)\mu m$) obtained via measurement with a Zeta-20 True Color 3D Optical Profiler, shown in figure 5 according to:

$$f = \frac{R}{n-1} \tag{1}$$

In this equation f is the focal length, R is the curvature radius and n the refractive index.

In figure 5 we can observe how a central region ($\sim 200\mu m$) of the lens is effectively spherical with a small deviation in the edges.

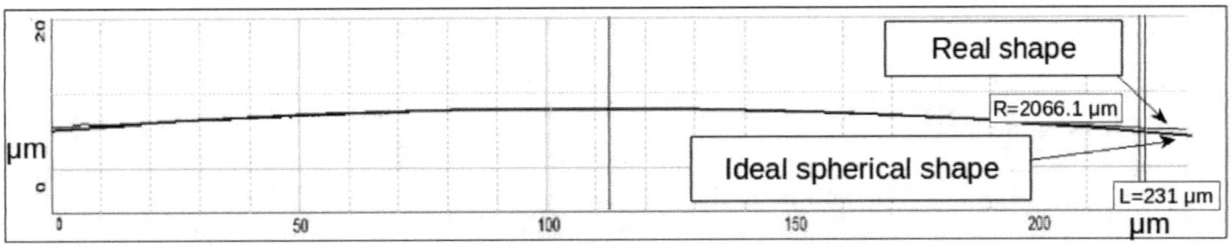

Figure 5. Lateral profile of a microlens obtained with a Zeta-20 True Color 3D Optical Profiler. It can be seen how a central region of $\sim 220\mu m$ clearly fits to a theoretical parabolical/spherical shape with a small deviation on the edges.

Analizing this profile we can obtain a diameter value of $\sim 220\mu m$ and a curvature radius of ($R = 2066.1(0.1)\mu m$), which in combination with the equation 1 and assuming a refractive index of $n = 1.5$ returns a focal length of $4.1322(0.0001)mm$.

The dimensions of the spot of light are wider than the real ones due to magnification because of the use of a microscope objective in our measurements. Taking this into account, the spot can be fitted to the theoretical Airy disk formula (equation 2) obtained for a circular lens. The irradiance distribution of a spot produced by a microlens in its focal length has been measured by a Thorlabs BP100VIS beam profiler (see figure 6). The fit presents a regression coefficient of $r^2 = 0.92$.

$$I = I_0 \left(\pi r^2\right)^2 \left[\frac{2J_1\left(\frac{2\pi x r}{\lambda f}\right)}{\frac{2\pi x r}{\lambda f}}\right]^2 \tag{2}$$

In this equation, I_0 is the peak irradiance, r the microlens radius, x is the position coordinate, f the focal length of the microlens, λ the wavelength of the light source and $J_1(x)$ is the Bessel spherical function of first order.

Figure 6 shows a fit of the theoretical formula in equation 2 to the measured data, it can be seen the very good agreement between them, reflected on the previously mentioned regression coefficient of $r^2 = 0.92$.

Journal of Physics: Conference Series **605** (2015) 012023 doi:10.1088/1742-6596/605/1/012023

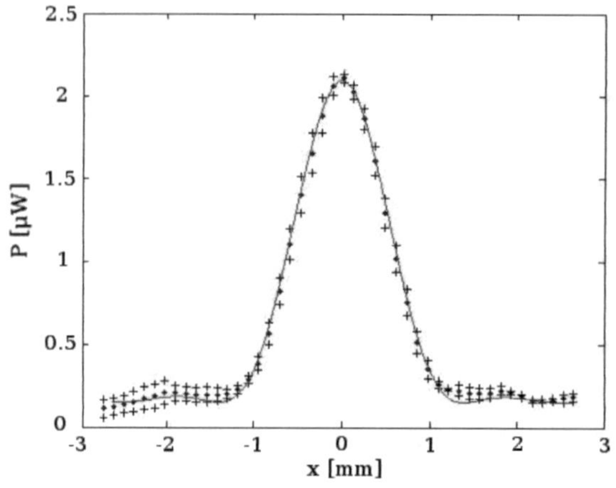

Figure 6. Fit of one of the spots to the Airy disk formula. The fit is made numerically in Matlab with the *cftool* package, leaving as parameters the diameter and the focal length in the Airy equation in 2. The quality of the fit is ensured by the squared regression factor $r^2 = 0.92$.

The fit of this equation is made numerically in Matlab, and it returns a value for the diameter of $264(12)\mu m$ and a value of focal length of $f = 4.48(0.95)mm$.

One last method to measure the diameter and the focal length of the fabricated microlenses is based on the use of a He-Ne laser ($\lambda = 632.8nm$), and a microscope objective, as depicted in figure 7. In order to determine the focal distance of the tested microlenses, we focused the microscope objective on the surface of the microlenses. Then, this was moved until observing the focal plane of the microlens array, the difference between the two positions gives us the focal distance of the fabricated microlenses. Figure 7 shows a close look of the experimental setup.

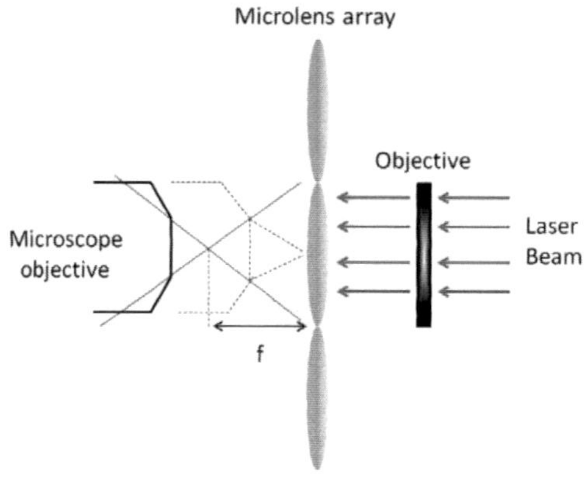

Figure 7. Setup for measuring the microlens focal length.

The value measured for the focal length was $5.2(1.2)mm$, that seems to be in good agreement with the previous ones.

4. Conclusions

The proposed technique will allow us to obtain microlenses with a broad range of diameters ($50\mu m \rightarrow 500\mu m$) and focal lengths ($1mm \rightarrow 5mm$) and opens a new window for generating different kind of surfaces in optical materials through a cheap laser process. By combining laser direct- write and the thermal treatment, we are able to obtain good quality elements using a common laser in industrial processes, which makes this technique competitive with other methods.

Acknowledgments

This work was supported by Xunta de Galicia, Spain, under project EM2012/019.

References

[1] Nieto D, Flores-Arias M T, O'Connor G M and Gomez-Reino C 2010 *Appl. Op.* **49** 4979
[2] Nieto D, Arines J, Gomez-Reino C, O'Connor G M and Flores-Arias M T 2011 *Proc. SPIE.* **8011** 80110T-6
[3] Smuka A Y and Lawandy N M 2000 *J. Appl. Ph* **87** 4026
[4] Huang C Y, Hsiao W T, Huang K C, Chang K S, Chou H Y and Chou C P 2011 *J. Micromech. Microeng.* **21** 085020
[5] Wu M H and Whitesides G M 2002 *J. Micromech. Microeng.* **12** 747
[6] Kim J, Kim K M, Jeon S K, Moonjung K, Park E H and Kwon Y S 2003 *Int. Conf. on Indium Phosphide and Related Materials.* 175
[7] Kunnavakkam M V, Houlihan F M, Schlax M, Liddle J A, Kolodner P, Nalamasu O, and Rogers J A 2003 *Appl. Phys. Lett.* **82** 1152
[8] Nieto D, Arines J, Gomez-Reino C, O'Connor G M and Flores-Arias M T 2010 *J. Appl. Phys.* **110** 023108
[9] Estepa L C and De la Fuente G F 2006, "Continuous Furnace with Coupled Laser for the Surface Treatment of Materials", Patent No. 200600560

Photobleaching effect in azo-dye containing epoxy resin films: the potentiality of carbon nanotubes as azo-dye dispensers

Guadalupe Díaz Costanzo[1,2], Silvia Goyanes[2], Silvia Ledesma[1]
[1]Image Processing Laboratory, Physics Department, School of Sciences, University of Buenos Aires, Pabellón I, Ciudad Universitaria, C1428EGA, Buenos Aires, Argentina.
[2]Nanomaterials Group, LP&MC, Physics Department- IFIBA, School of Sciences, University of Buenos Aires, Pabellón I, Ciudad Universitaria, C1428EGA, Buenos Aires, Argentina.
E-mail: ledesma@df.uba.ar

Abstract. Azo-dye molecules may suffer from bleaching under certain illumination conditions. When this photoinduced process occurs, it generates an irreversible effect that is characterized by the loss of absorption of the dye molecule. Moreover, the well-known isomerization of azo-dye molecules does not occur anymore. In this work it is shown how the addition of a small amount of multi-walled carbon nanotubes (MWCNTs) helps to decrease the bleaching effect in a photosensitive guest-host azo-polymer film. Two different systems were fabricated using an epoxy resin as polymer matrix. An azo-dye, Disperse Orange 3, was used as photosensitive material in both systems and MWCNTs were added into one of them. The optical response of the polymeric systems was studied considering the degree of photoinduced birefringence. Photobleaching of the azo-dye was observed in all cases however, the effect is lower for the composite material containing 0.2 wt % MWCNTs. The weak interaction between MWCNTs and dye molecules is less favorable when the material is heated. The optical behavior of the heated composite material suggests that carbon nanotubes can be potentially used as azo dye dispensers. The results are interpreted in terms of the non-covalent interaction between azo-dye molecules and MWCNTs.

1. Introduction

The combination of carbon nanotubes and azo-dye chromophores has recently arisen as a new nanoarchitecture for optoelectronics applications. Particularly, the electronic structure of carbon nanotubes and azo-dye chromophores allows them to weakly interact via a non-covalent binding [1, 2].

Among the many possible ways of working with carbon nanotubes and azo-dye chromophores, one alternative turns out from the addition of both carbon nanotubes and azo-dye molecules into a polymeric matrix [3, 4]. As a result, this nanocomposite material can exhibit new properties given by the presence of the dye and the nanotubes. Particularly, azo-dye molecules are well-known for their light-triggered behavior, the photoisomerization: azo-dye molecules have two isomers, trans and cis, being the trans state the energetically most stable one. Exposure to light of the proper wavelength photoinduces cis isomers and that then relax to trans isomers. Azo-dye molecules photoisomerize if their dipole moment has a non-vanishing component in the direction of the light polarization. Therefore, azo-dye chromophores accumulate in the direction perpendicular to the polarization axis of light giving rise to some degree of birefringence in the host material. Since the photoisomerization process is reversible and sensitive to the light polarization, azo-dyes have been used for the fabrication of a broad class of photosensitive materials with varied applications including waveguiding [5], optical switching [6] and optical information storage and processing [7, 8]. However, irreversible photoinduced bleaching (photobleaching) can be suffered by the azo-dye molecules [9, 10]. If this occurs, it means that the molecule loses absorption intensity and thus the activity of the chromophore is lost. Even though it has been demonstrated that this effect can be useful for the development of waveguide devices [10], for most practical purposes photobleaching is usually an undesired effect.

The aim of this work is to demonstrate that the addition of a small quantity of multi-walled carbon nanotubes (MWCNTs) into a photosensitive guest-host azo-polymer system helps to decrease the photobleaching effect. In order to show this, the optical behavior of a nanocomposite material using an epoxy resin as matrix, Disperse Orange 3 as azo-dye and MWCNTs as filler was compared to the behavior of a material with only azo-dye. In both materials the photoinduced birefringence was explored. Successive photoinduced processes led to decreasing birefringence that can be related to photobleaching in the both photosensitive materials. The systematic study of the photoinduced birefringence reveals that the photobleaching process is diminished in the material with MWCNTs and that temperature could be used as an external trigger to modify its optical response. Results suggest that MWCNTs may work as azo-dye dispensers that can be activated by heating the material.

2. Materials and methods

2.1. Materials

One of the most common epoxy-based materials is epoxy resin and at the same time, the most important class of epoxy resins is diglycidyl ether of bisphenol A (DGEBA). DGEBA resins are transparent to visible light. The epoxy pre-polymer and curing agent used in this work were DGEBA LY 556 and IPDA HY 2962, both from Ciba Geigy. The employed azo-dye was Disperse Orange 3 (DO3) from Sigma-Aldrich. Its chemical structure is shown in figure 1. Commercially available MWCNTs were purchased from Nanocyl (NC3100). Their length is about 1.5 μm and their diameter is around 20-40 nm [1]. MWCNTs were dried in a vacuum oven at 120° C during 3 h (in order to remove adsorbed water) and they were storage in vacuum at room temperature until used. Reagent grade tetrahydrofuran (THF) from Biopack (Argentina) was used as solvent.

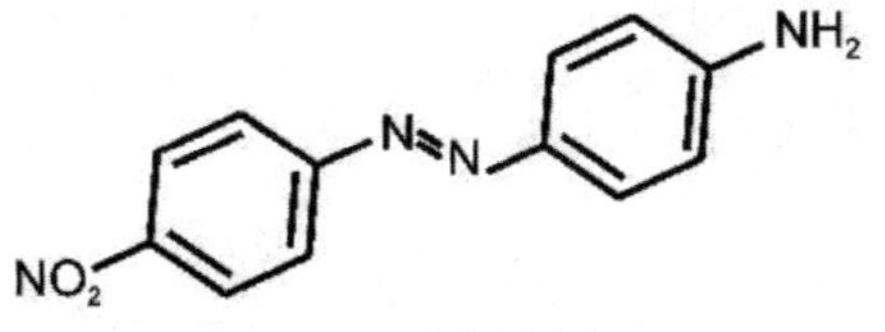

DO3

Figure 1. Chemical structure of the azo-dye chromophore Disperse Orange 3 (DO3).

2.2. Preparations of films

Two different kinds of films were prepared: films of epoxy resin and DO3 (DO3 epoxy film) and films of epoxy resin, DO3 and MWCNTs (DO3-MWCNTs epoxy film). For both kinds of films, a stoichiometric ratio (aminohydrogen/epoxy) of 1 was used. The followed protocol was used in order to prepare the photosensitive films: 0.05 g of DO3 were dissolved in 1 mL of THF and incorporated into 0.8 g of the pre-polymer DGEBA. The mixture was stirred at room temperature during 5 min. Then, 0.2 g of hardener were added and mixed again during 5 min. Immediately after that, the DO3 films were prepared by spin-coating technique. DO3 films were first storage at room temperature for 20 h and then were dried at 120 °C for 3 h in an air flow conventional oven.

DO3-MWCNTs films were done following the same protocol as for the DO3 films. In this case, 0.002 g of MWCNTs were bath sonicated together with the DO3 and THF during 30 min in order to have a stable dispersion of MWCNTs. Then, this dispersion was mixed with the pre-polymer DGEBA at room temperature. The following steps were the same as for DO3 films. After these processes, DO3 films with 5 wt % dye concentration and DO3-MWCNTs films with 5 wt % dye and 0.2 wt % carbon nanotubes concentration were obtained.

2.3. Characterization

The photosensitive materials were characterized by measuring the time dependence of the transmittance. The experimental setup for this kind of measurements is shown in figure 2. Taking into account the absorption wavelength of the dye, birefringence was induced with a writing laser beam at 473 nm. The incidence was almost normal to the sample (about 5°) in order to avoid asymmetries in the spot. The response was monitored using a probe laser beam at 633 nm in a configuration of polarizer/sample/analyzer (P1, S and A in figure 1). The writing beam was also linearly polarized (s polarization) after passing through a polarizer (P2) which axis was set to 45° with respect to the polarizer/analyzer orientation in order to measure maximum response. Even though there are some phenomena, for instance mass transport, that are dependent on the state of polarization of the writing beam, enough general results are expected for the bleaching phenomenon from a linear polarization state. The power of the writing beam was set to 75 mW/cm^2 in all experiments and it was controlled using a neutral filter (F). A red filter (RF) was placed right before the photodiode (PD) to avoid detecting light from the writing beam. Misalignment of dye molecules can be photoinduced by irradiating the sample with circularly polarized light. To do this, a quarter wave plate was placed after P2 when necessary. For both kinds of samples, DO3 epoxy films and DO3-MWCNTs epoxy films, seven cycles to induce and erase birefringence were done successively on the same area of each film. After 60 min of photoinducing birefringence, the erasing process was done for five minutes, in order to ensure that the birefringence turns to its minimum value and then the re-writing process began again. For experiments done at a temperature other than room temperature, a hollow cylindrical-shaped oven was used as it is drawn with a dotted line in figure 2. The main axis of the oven was placed in the direction of the optical axis. The temperature was set to 85 °C for photoresponsive nano-composite. Films were placed inside the oven with a film holder specially designed for this purpose. The oven plateau was 5 cm guaranteeing thermal stability in that region where the film was placed. The temperature was determined with variations less than 1 °C using a temperature controller (TC). The experiment was done as follows. First, birefringence, relaxation and erasure of birefringence were induced for a total of 70 min at room temperature (25 °C). Immediately after that, temperature was raised up to 85 °C until thermal stabilization occurred. Once the stabilization was reached, the same process for inducing, relaxing and erasing birefringence was repeated on the same area of the sample.

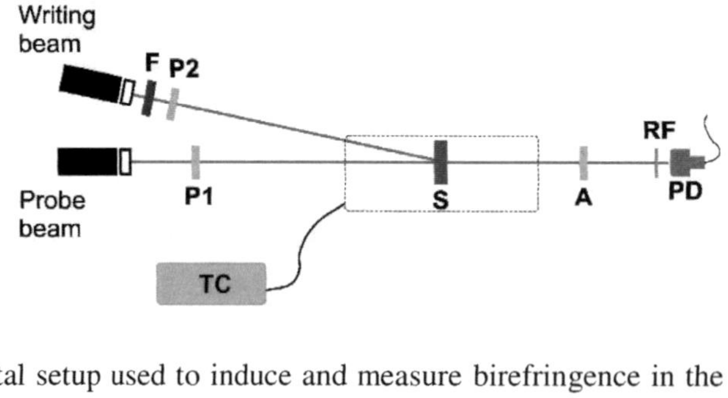

Figure 2. Experimental setup used to induce and measure birefringence in the materials. The writing beam, first attenuated by filter F and polarized by P_2, impinges on the sample S; the probe beam, polarized by P_1, impinges on the sample S, is then polarized by A and finally collected by the photodetector PD. RF is a red filter and then dotted line indicates the cylindrical oven.

Journal of Physics: Conference Series **605** (2015) 012024 doi:10.1088/1742-6596/605/1/012024

3. Results and discussion

Figure 3 shows the birefringence curves for DO3 epoxy films (Figure 3a) and for DO3-MWCNTs epoxy films (Figure 3b). For both kinds of materials, inducing and erasing cycles of birefringence were done successively on the same area of the film. In order to easily appreciate the obtained effect, only three curves of inducing birefringence corresponding to cycles 1, 3 and 7, were plotted in each figures 3a and 3b, respectively. For both materials the final value of birefringence decreases with the successive cycles indicating that during each cycle there are some dye molecules that lose their optical response. This behavior can be explained in terms of photobleaching effects. Moreover, it is clear that the growth of the signal for the first cycle for DO3 films and for DO3-MWCNTs films is quite different: the detected signal for DO3-MWCNTs film is slower. This suggests that the process to photoinduce birefringence is affected by the addition of MWCNTs. In particular, the change in the rate at which birefringence is photoinduced could be due to an increment in the steric hindrance introduced by the addition of MWCNTs [3].

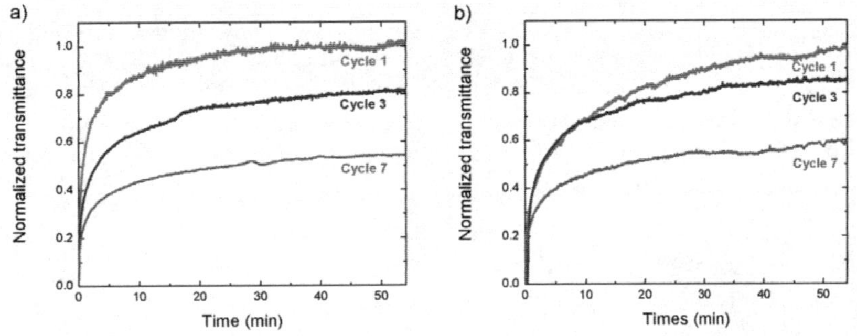

Figure 3. Photoinduced birefringence for (a) DO3 film and (b) DO3-MWCNTs film.

The maximum levels of birefringence for all the measured cycles are plotted in Figure 3 for both kinds of films: DO3 films and DO3-MWCNTs films. In figure 4a the asymptotic values of the signal were plotted as a function of the number of cycles. These values of birefringence seem to follow a linear trend for both kinds of samples. Of course, negative slopes are expected to be found in both cases as it reflects the losses in the values of birefringence. The asymptotic values belonging to the DO3 films and to the DO3-MWCNTs films were fitted with linear functions in order to quantify the losses rate for successive cycles. First, it should be noted that the maximum transmittance values for DO3-MWCNTs films is always lower than those for DO3 films. Second, the slope obtained is 10 times smaller for DO3-MWCNTs films than for DO3 films. In figure 4b the same values of maximum reached birefringence were normalized to the highest level of birefringence reached in the first cycle. This could be interpreted as a normalization to the dye molecules initially available to undergo photoisomerization. When the results are plotted in this way, the normalized maximum signal for both kinds of samples follows a similar trend. This reflects that if there were initially the same amount of available DO3 molecules, then the response for both materials would be similar.

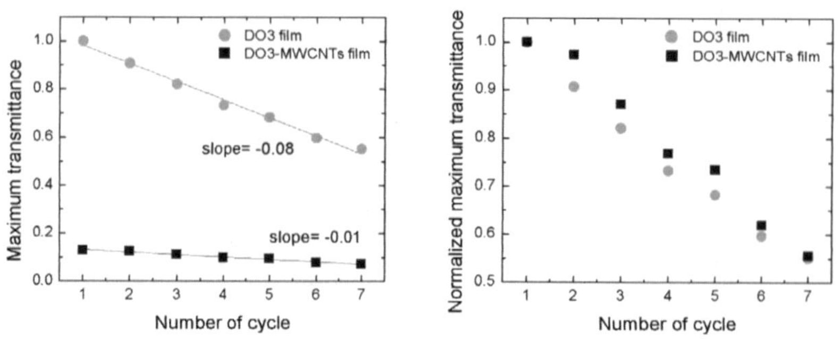

Figure 4. (4a) Maximum level of birefringence reached for the successive cycles on both films. Solid lines are fitted linear functions. (4b) The same values normalized to the maximum level of birefringence of the first cycle.

The progressive losses in the optical response of the materials can be related to the morphological changes observed with naked eye. In fact, DO3 epoxy films and DO3-MWCNTs epoxy films showed photobleaching effects after exposure to light at 473 nm. Picture in figure 5 is showed in order to illustrate the bleaching process in a DO3 film. This image was obtained after irradiation of the sample at 473 nm (75 mW/cm^2) during 6 h. The circular shape of the bleached spot (~ 2 mm) follows the beam shape.

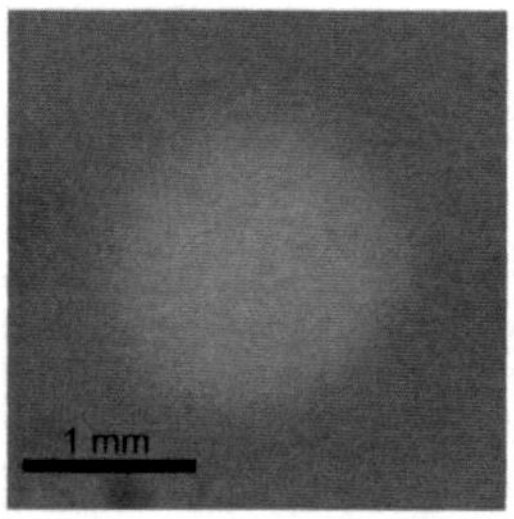

Figure 5. Illustrative picture of a photobleached effect in a DO3 epoxy film.

According to the previous results, the material with MWCNTs presents a different behavior with respect to the DO3 film reflecting that the presence of MWCNTs plays an important role in the optical response. In fact, the two observations previously mentioned in the description of Figure 3a could be explained in terms of the π-π interaction between DO3 molecules and MWCNTs [1, 4]. This interaction generates that some fraction of DO3 molecules are preferentially in the vicinities of MWCNTs and less molecules are available for photoisomerization and thus for photobleaching.

A birefringence experiment at higher temperature was performed in order to create an environment where the weak, non-covalent π-π interaction between MWCNTs and dye molecules is less favorable. Figure 6 shows two curves of induction, relaxation and erasure of birefringence for DO3-MWCNTs epoxy film at two different temperatures: the first one at 25 °C and the second one at 85 °C. The sample was kept fixed between consecutive measurements in order to irradiate the same area of the film. The time interval between each measurement corresponds to the required time to increase and stabilize the temperature. As it can be seen from figure 6 the measurement at 85 °C presents a similar temporal behaviour to the previous measurement at room temperature. Moreover, biexponential functions were used to fit each measurement (dotted line in Fig 6) and the asymptotic values for both curves are similar. Contrary to what happens when two successive cycles are done at room

Journal of Physics: Conference Series **605** (2015) 012024 doi:10.1088/1742-6596/605/1/012024

temperature, in this case there is no decrease in the transmittance for the second cycle. This result indicates that some dye molecules can be activated by heating the material. A plausible explanation for this effect to occur is that thermal agitation could increase the number of available azo-dye molecules as it overcomes the non-covalent interaction between the azo-dye molecules and MWCNTs. Presumably, these dye molecules initially do not participate in the photoisomerization and photobleaching process for being in the vicinities of the carbon nanotubes. This characteristic of the DO3-MWCNTs epoxy films suggest that carbon nanotubes could be used as DO3 dispensers that can be activated by increasing temperature.

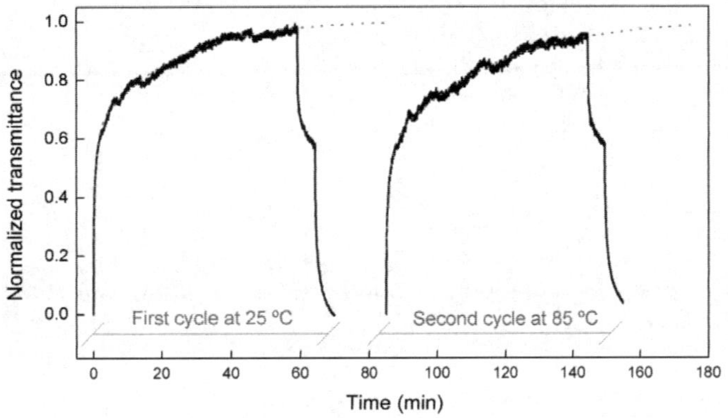

Figure 6. Curves of the induction, relaxation and erasure of the birefringence for DO3-MWCNTs films performed for two successive cycles: first at 25 °C and second 85 °C.

4. Conclusions

The study of the induced birefringence at room temperature in guest-host azo-epoxy resin materials reveals that the addition of a low concentration of MWCNTs decreases the photobleaching effect suffered by the azo molecules. The bleaching effect, which can be seen with the naked eye, can be detected by successively measure the photoinduced birefringence. Successive measurements for DO3 epoxy films and DO3-MWCNTs epoxy films revealed that some azo-dye molecules lost their absorption. However, MWCNTs epoxy films showed that the rate at which the material photobleaches is 10 times slower than for the materials without MWCNTs. Measurements of photoinduced birefringence at higher temperature were done in order to evaluate if the rate at which the DO3-MWCNTs epoxy film photobleaches was the same. As revealed by the experiments, not only the asymptotic value did not decrease but it was the same after successive cycles. A possible explanation for this behavior could be that the non-covalent interaction between MWCNTs and azo-dye molecules keeps azo molecules in the vicinities of the nanotubes preventing the azo molecules from bleaching. The fact that a higher level of birefringence was obtained by heating the sample indicates that there exist more azo-dye molecules contributing to the final birefringence. Thermal agitation could be responsible for a higher availability of azo-dye molecules overcoming the non-covalent interaction between azo-dye molecules and MWCNTs. These results suggest that MWCNTs may work as azo-dye dispensers that can be activated by heating the material.

Acknowledgments
This work was supported by UBACyT 20020100100689, CONICET PIP 112-200801-03047, and ANPCYT PICT 2010-02179.

References

[1] Díaz Costanzo G, Ledesma S, Mondragon I, Goyanes S 2010 *J. Phys. Chem. C* **114** 14347–14352.
[2] Huang C, Wang R K, Wong B M, McGee D J, Léonard F, Kim Y J, Johnson K F, Arnold M S, Eriksson M A, Gopalan P 2011 *ACS nano* **5** 7767-74
[3] Vijayakumar C, Balan B, Kim M-j, Takeuchi M, 2011 *J. Phys. Chem. C* **115** 4533–4539
[4] Díaz Costanzo G, Ribba L, Goyanes S, Ledesma S 2014 *J. Phys. D: Appl. Phys.* **47** 135103
[5] Stockermans R J, Rochon P L 1999 *Appl. Opt.* **38** 3714–3719
[6] Luo Y, She W, Wu S, Zeng F, Yao S 2005 *Appl. Phys. B: Lasers Opt.* **80** 77-80
[7] Hvilsted S., Sanchez C, Alcala R J 2009 *Mater. Chem.* **19** 6641–6648
[8] Huang W H, Hu Y L, Zhang Z S, Chen Y H, Zhang Q J 2010 *Opt. Lett.* **35** 46–48
[9] Galvan-Gonzalez A, Canva M, Stegeman G I, Sukhomlinova L, Twieg R J, Chan K, Kwok P, Kowalczyk T C, Lackritz H S 2000 *J. Opt. Soc. Am. B* **17** 1992-2000
[10] Hattori T, Shibata T, Onodera S, Kaino T 2000 *J. Appl. Phys.* **87** 3240-3244

Effect of surface plasmon resonance in TiO_2/Au thin films on the fluorescence of self-assembled CdTe QDs structure

I Moura[1]*, M F Cerqueira[1], D Melnikau[2], D Savateeva[2], Y Rakovich[2,3], J Borges[1], F Vaz[1] and M Vasilevskiy[1]

[1]Centro de Física, Universidade do Minho, Campus de Gualtar, 4710-057 Braga, Portugal

[2]Centro de Física de Materiales (MPC, CSIC-UPV/EHU) and Donostia International Physics Center (DIPC), E-20018 Donostia-San Sebastian, Spain

[3]IKERBASQUE Basque Foundation for Science, 48013, Bilbao, Spain

* E-mail: mouraisabel1@gmail.com

Abstract. The exceptional properties of localised surface plasmons (LSPs), such as local field enhancement and confinement effects, resonant behavior, make them ideal candidates to control the emission of luminescent nanoparticles. In the present work, we investigated the LSP effect on the steady-state and time-resolved emission properties of quantum dots (QDs) by organizing the dots into self-assembled dendrite structures deposited on plasmonic nanostructures. Self-assembled structures consisting of water-soluble CdTe mono-size QDs, were developed on the surface of co-sputtered TiO_2 thin films doped with Au nanoparticles (NPs) annealed at different temperatures. Their steady-state fluorescence properties were probed by scanning the spatially resolved emission spectra and the energy transfer processes were investigated by the fluorescence lifetime imaging (FLIM) microscopy. Our results indicate that a resonant coupling between excitons confined in QDs and LSPs in Au NPs located beneath the self-assembled structure indeed takes place and results in (i) a shift of the ground state luminescence towards higher energies and onset of emission from excited states in QDs, and (ii) a decrease of the ground state exciton lifetime (fluorescence quenching).

1. Introduction

Surface plasmon resonance (SPR) typically arises in nanostructures of noble metals. Their unique properties such as strong field confinement, resonant behavior and local field enhancement effects make them ideal candidates to control the emission of luminescent nanoparticles (NPs) [1, 2]. Among these NPs, quantum dots (QDs) are highly versatile and suitable for a wide range of applications, due to their single optical properties and tunable emission, as well as high quantum yield (QY) and photostability, making them good energy donors and acceptors [3]. Förster resonant energy transfer (FRET) is an important microscopic mechanism of excitation transport between QDs, which is based on dipole-dipole interactions between energy donors and acceptors [4]. This energy transfer process can be implemented in bio-sensing and light-harvesting applications [5, 6]. It has been demonstrated that the QDs luminescence is either enhanced or quenched when they interact with localized SPR [7-9]. At the same time, the energy transfer through a SPR-mediated coupling between metal nanostructures and light-emitting nanomaterials can change the luminescence of light-emitting acceptors [9, 10]. Notwithstanding, until now there are only a few experimental reports on SPR effects on FRET, especially for QDs systems [11].

The ability to self-assemble nanostructures into functional devices, where these assembled systems take advantage of the unique properties of the colloidal QDs becomes a new challenge in the field of nanotechnology. However, QD's environment has to be taken into account during their organization into self-assembled structures and under pulsed optical excitation, optimising them for applications. Therefore, the development of a novel material combining QDs and dielectric/plasmonic thin films is of considerable interest.

In this work we investigate the effect of localized SPR, supported by nanocomposites consisting of a dielectric matrix (TiO_2) with embedded Au nanoparticles (NPs), on the steady-state luminescence, its time-resolved kinetics and resonant energy transfer in self-assembled dendrite structures consisting of CdTe mono-size QDs. The results are discussed in view of their relation to surface plasmons localised in gold NPs.

2. Experimental methods

Negatively charged water-soluble CdTe QDs, capped with thioglycolic acid (TGA) were synthetized according a procedure previously reported [12, 13]. CdTe QDs have a maximum photoluminescence emission at 597 nm, Stokes-shifted with respect to the absorption peak associated with the exciton ground state in the dots (see figure 1). From the absorption peak we evaluated an average size of ≈ 4 nm, according to the calculated data presented in Ref. [14]. The quantum yield (QY) of colloidal CdTe-TGA was determined using Rhodamine B in ethanol (QY=70%) as a reference, resulting in a value of 36% [15].

Self-assembling technique [16] was used to prepare QDs dendritic structures. First, a diluted solution of QDs was prepared by the addition of sodium hydroxide pH =11, to an aqueous solution of CdTe QDs with an initial concentration of 10^{-5} mol L^{-1}. Subsequently, aliquots of prepared solution were deposited on the surface of a conventional substrate (glass) and on TiO_2 films doped with Au NPs produced by co-sputtering and subsequently annealed at different temperatures (see [17] for details of the co-sputtering and annealing regimes) and allowed to dry at room temperature.

A Shimadzu UV-3101PC UV-vis-NIR spectrophotometer was used to record the absorption spectra of TiO_2/Au films and the absorption spectrum of colloidal CdTe QDs in solution. The photoluminescence (PL) spectrum of the QDs solution was recorded using a Fluorolog 3 fluorescence spectrometer with an excitation wavelength of 360 nm provided by a Xenon lamp. These and all other optical measurements were performed under ambient conditions.

Steady state emission of the dendrite samples was investigated by using the Nd:YAG frequency-doubled 532 nm laser line (P_{Laser}=1.5mW at the sample) for excitation and equipped with a motorized sample positioning XY-stage (positioning accuracy better than 1 μm). In particular, the 800^{0}C annealed sample (the map is shown below) was studied within a square (120 × 120 μm^2) area representing a typical dendrite. The measurements were performed along 120 parallel lines, each 120 μm long, measuring 120 spectra at each line. Using Witec software tools, the 120 × 120 spectra were then converted into a "Raman map" of the studied area, which consisted of a single spectrum acquired at each pixel of the area. An appropriate averaging procedure was applied afterwards.

Fluorescence lifetime mapping (FLIM) measurements were recorded using a PicoQuant Micro-time200 time resolved confocal microscope system, equipped with an Olympus IX71 inverted microscope. The samples were excited by a 485 nm picosecond laser pulse (PicoQuant DH485 laser head controlled by Sepia II driver). The system has an overall resolution of 100 ps. Fluorescence lifetime maps were elaborated on a per pixel basis with respect to the logarithm of the emission intensity.

3. Results and discussion

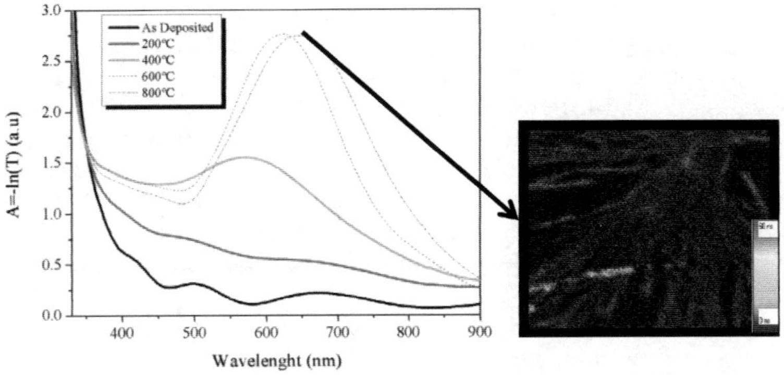

(a) (b)

Figure 1. (a) Absorption spectra of the TiO_2/Au films on SiO_2 substrates annealed at different temperatures and (b) QDs structure/TiO_2/Au thin film substrate annealed at 800°C.

Prior to the QDs deposition, absorption spectroscopy measurements were carried out in order to evaluate the SPR effect in TiO_2/Au films annealed at different temperatures. Figure 1(a), shows the onset of the SPR-related absorption band with the increase of the annealing temperature, caused by the formation of Au NPs. No sign of SPR is seen in the optical spectra of the as-deposited and 200 °C annealed samples, with only a weak and broad absorption band in the visible range for samples annealed at 400 °C. In a previous study it was reported that TiO_2/Au films produced by co-sputtering show SPR-related optical response only after annealing at sufficiently high temperatures (above 500°C), as also can be observed in our figure 1. Torrell et al. [17] demonstrated that high annealing temperatures are required to produce Au NPs inside the dielectric matrix and consequently, the SPR effect on the optical spectra. Such an annealing affects significantly both Au NP's shape and size and also the phase composition of the TiO_2 matrix modifying its dielectric constant.

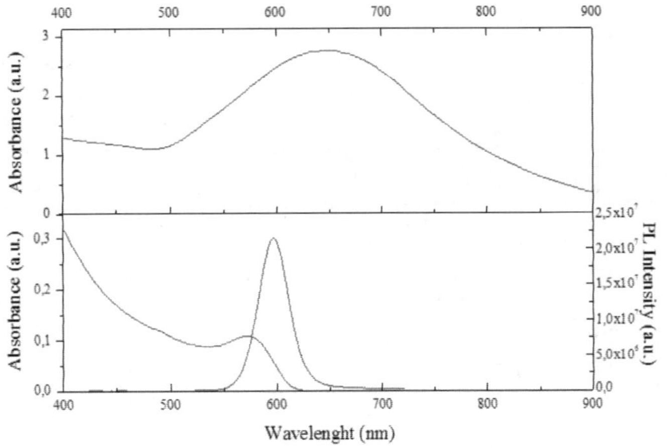

Figure 2. Absorption spectrum of a co-sputtered TiO_2 film doped with Au NPs (top) and absorption and photoluminescence spectra of CdTe QDs in solution (bottom).

The absorption spectrum of a TiO_2/Au thin film annealed at 800ºC is also shown in figure 2, together with the absorption and photoluminescence spectra of colloidal CdTe QDs in solution. As it can be seen from figure 2, the spectra of TiO_2/Au film show a pronounced SPR-related absorption band with a maximum at ≈ 650 nm, which encompasses the absorption and emission peaks of the CdTe QDs. Therefore, one can expect a resonant coupling between excitons confined in QDs and surface plasmons in Au NPs. This could open an additional channel of energy transfer between the dots (QDs1→NP→QDs2) [9, 18], which would enhance the FRET probability. On the other hand, the QDs luminescence can be quenched because surface plasmons are damped and the energy transfer QDs→NP may be irreversible, since the energy is dissipated within the metallic particle [7, 18]. Beyond these dynamical effects, the exciton-plasmon coupling should modify the energy spectrum of the hybrid system and, consequently, its steady-state optical properties, as it has been shown both experimentally [19] and theoretically [20] for structures composed of a QDs monolayer and a silver film.

To assess the LSP effect on the steady-state photoluminescence spectra of the QDs structures, we performed spatially resolved scanning spectroscopy studies using the Witec Raman alpha300M+ apparatus. By using the Witec tools Cluster and Basis analysis, the emission maps were decomposed into three typical bands (see figure 3) and their relative intensities were compared for different pixels of the map. For QDs samples deposited on "non-plasmonic" substrates (TiO_2/Au film annealed at 200ºC) the spectra (not shown) were composed of almost solely green band corresponding to the QDs ground state emission, similar to the same dots in solution.

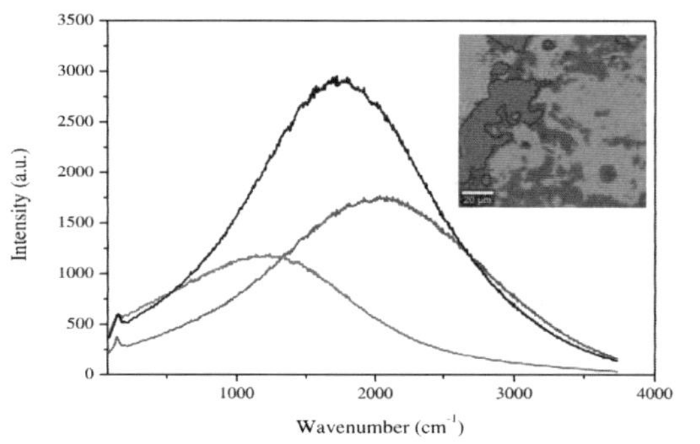

Figure 3. Emission bands for CdTe QDs dendritic structure fabricated on a TiO_2/Au thin film substrate annealed at 800ºC. The wavelengths of the maxima of the red, blue and green bands are 566, 587 and 596 nm, respectively. The small peak at very low wavenumbers is the phonon-related Raman scattering signal from QDs. The inset shows a digital map where the color corresponds to the dominating emission band.

In contrast, the spectra of the dendritic structure developed on the surface of TiO_2/Au thin film substrate annealed at 800ºC are inhomogeneous and contain the three emission bands (figure 3). The green band corresponds to the highest wavelength of ≈ 596 nm, characteristic of the QDs emission in solution. The blue band is only slightly shifted towards higher photon energies with respect to the former (the shift is ≈ 30 meV). The origin of this shift could be the electromagnetic coupling between Au NPs and CdTe QDs since it results in a mode anticrossing and an anti-Stokes shift of the upper coupled mode [19, 20]. Another possible explanation is the formation of QDs molecules with inter-dot-coupled exciton states [21] and/or phonon-assisted (up-converted) emission [22]. Indeed, our analysis indicates that this emission comes mostly from those zones of the self-assembled structure where the QDs density is higher. Then the question about the role of gold NPs arises, namely, whether they can stimulate the aggregation of QDs at the self-assembly stage.

Finally, the "red" band is shifted towards higher energies by more than 100 meV with respect to the "fundamental" emission mode (596 nm). Its origin is not completely clear to us at the moment but it might be an emission owing to the recombination of excited state (possibly $1S_e$–$1S_{3/2}$) excitons. Such a "hot" luminescence has sometimes been observed for resonantly excited QDs [23]. The role of Au NPs here could consist in quenching of the ground state luminescence (see below), which helps revealing the weaker emission owing to the direct recombination of the excited state excitons. We also investigated fluorescence decay kinetics using pulsed laser excitation (λ_{exc} =485 nm) and FLIM

Journal of Physics: Conference Series **605** (2015) 012025 doi:10.1088/1742-6596/605/1/012025

mapping of the emission from self-assembled dendritic structures on plasmonic and non-plasmonic platforms. FLIM images of the QDs dendritic structures deposited on a conventional substrate and on the surface of a TiO_2/Au thin film annealed at 200ºC and 800ºC are presented in figure 4.

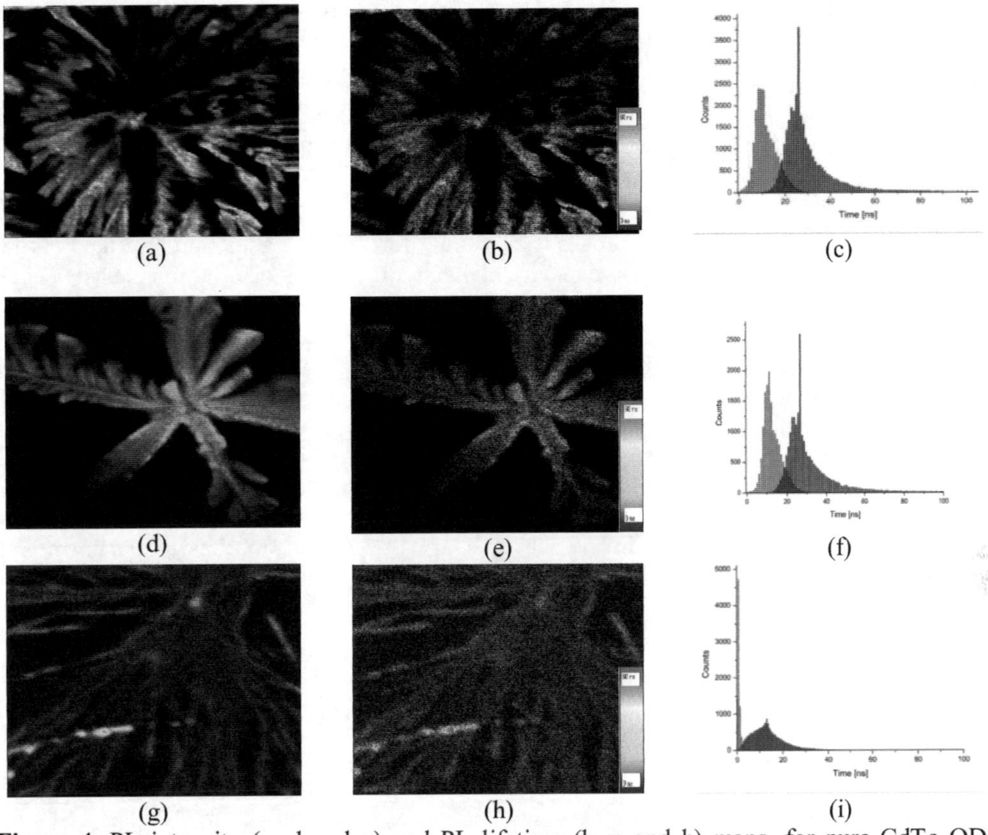

Figure 4. PL intensity (a, d and g) and PL lifetime (b, e and h) maps, for pure CdTe QDs dendrite structures fabricated on glass (a and b) and on the surface of TiO_2/Au thin films annealed at 200 ºC (d and e) and 800 ºC (g and h). The corresponding PL lifetime histograms (c, f and i) were obtained by means of two exponential fitting of the PL decay kinetics at each image spot while scanning over an 82 x 82 μm, 78 x 78 μm and 67 x 67 μm, respectively. The bar heights represent the relative statistical weights of the two components (τ_1, and τ_2, red). The laser excitation wavelength is 485 nm.

According to the results of the analysis performed on the PL decay kinetics, it is possible to distinguish two different time scales with characteristic times τ_1 and τ_2 ($\tau_1 > \tau_2$). These characteristic times are represented by blue and red colors, respectively, in the histograms of figure 4. Presumably τ_1 represents the radiative recombination of a QDs exciton affected by non-dissipative processes like FRET (it would be just a radiative lifetime for an isolated QDs), while τ_2 describes non-radiative dissipation processes. Figure 4 shows that τ_1 and τ_2 remain unchanged for the QDs dendritic structure on a TiO_2/Au substrate annealed at 200 ºC compared to those fabricated on glass surface. However, for the structure fabricated on the surface of a TiO_2/Au film annealed at 800 ºC τ_2 shortens dramatically, meaning that the plasmonic substrate leads to the fast fluorescence quenching. The longer characteristic time (τ_1) is also decreased, which means that it is not just the intrinsic radiative lifetime of a single quantum dot and a more complex analysis involving a continuous distribution of lifetimes would be appropriate [24].

Table 1. Characteristic decay times (τ_1 and τ_2) for QDs structures fabricated on TiO_2/Au substrates annealed at different temperatures.

Sample	Reference	TiO_2/Au 200°C	TiO_2/Au 400°C	TiO_2/Au 600°C	TiO_2/Au 800°C
τ_1	29.3	27.9	21.0	21.0	13.0
τ_2	12.2	12.5	5.3	7.7	0.8

We notice that the decrease of τ_1 and τ_2 is observed only for TiO_2/Au thin film annealed at 800°C. According to the FLIM maps, a good homogeneity in the lifetime spatial distribution is observed, even though a slightly higher luminescent core, with a longer τ_1 is seen for the QDs dendrite on the TiO_2/Au thin film surface annealed at 800 °C. This observation can be explained by the higher QDs density in the core region. Nevertheless, there is no any pronounced inhomogeneity along the structure in the lifetime map, unlike it was found in our recent study of fractal aggregates of CdTe dots [24]. This indicates that the lifetime is almost independent of the local environment in the self-assembled structure and therefore, energy transfer process does not play an important role within the QDs dendritic structure but rather it takes place in the direction from the dots to the plasmonic substrate. From table 1 we can see that the characteristic time of the slower non-dissipative kinetic processes decreases with the onset of the SPR in the substrate. It can be understood as (plasmon-mediated) exciton energy transfer to larger dots from which it is later emitted at a somewhat different wavelength.

4. Conclusions

In summary, we investigated the effect of SPR supported by nanocomposites consisting of a dielectric matrix (TiO_2) with embedded Au NPs annealed at different temperatures, on the steady-state luminescence and resonant energy transfer in self-assembled structures consisting of CdTe mono-size QDs.Our results indicate that a resonant coupling between excitons confined in QDs and LSPs in Au NPs located beneath the self-assembled structure indeed takes place and results in a shift of the ground state luminescence peak towards higher energies, as well as an onset of a "hot" emission from excited states in QDs. Depending on the morphology, spatial distribution, shape and size of the Au NPs, determined by the annealing temperature, the characteristic fluorescence decays show some SPR-related changes in lifetimes. Indeed, we can conclude that the changes are correlated with the onset of the SPR in the NPs. They possibly include both dissipative quenching of the emission and non-dissipative FRET-type processes. We continue our studies of the microscopic mechanisms operating in these hybrid systems but our results already show the potential of TiO_2/Au platforms for environmental sensing applications based on the coupling of localised surface plasmons to the elementary excitations responsible for the light emission, such as QDs excitons.

Acknowledgments
Financial support from the European Science Foundation (ESF) through PLASMON-BIONANOSENSE programme and Portuguese Foundation for Science and Technology (FCT) and FEDER through Projects PTDC-FIS-113199-2009 and PEst-C/FIS/UI0607/2013 are gratefully acknowledged. DM, DS and YR acknowledge support from Project FIS2013-41184-P of the Spanish Ministry of Economy and Competitiveness MINECO.

References

[1] Raether H, Heitmann D and Pockrand I 1998 *Surface plasmons on smooth and rough surfaces and on gratings* (Verlag: Springer)
[2] Barnes W, Dereux A and Ebbesen T 2003 *Nature* **424** 824
[3] Clapp A, Medintz L and Mattoussi H 2006 *ChemPhysChem* **7** 47
[4] Lunz M and Bradley A 2014 *In: Cadmium Telluride Quantum Dots: Advances and Applications* ed J. Donegan and Y. Rakovich (USA: Pan Stanford Publishing) p 87
[5] Medintz I, Konnert J, Clapp A, Stanish I, Twigg M, Mattoussi H, Mauro J and Deschamps J 2003 *Nat. Mater.* **2** 630
[6] Franzl T, Klar T, Schietinger S, Rogach A and Feldmann J 2004 *Nano Lett.* **4** 1599
[7] Matsumoto Y, Kanemoto R, Itoh T, Nakanishi S, Ishikawa M and Biju V 2007 *J. Phys. Chem. C* **112** 1345
[8] Pompa P, Martiradonna L, Torre A, Sala F, Manna L, Vittorio M, Calabi F, Cingolani R and Rinaldi R 2006, *Nanotechnology* **1** 126
[9] Lunz M, Zhang X, Gerard A, Gunko Y, Lesnyak V, Gaponik N, Susha A, Rogach A and Bradley A 2012 *J. Phys. Chem.* C **116** 26529
[10] Moerland R, Rekola H, Sharma G, Eskelinen A.-P, Väkeväinen A and Törmä P 2012 *Appl. Phys. Lett.* **100** 221111
[11] Gryczynski I, Malicka J, Jiang W, Fischer H, Chan W, Gryczynsk Z, Grudzinski W and Lakowicz J 2005 *J. Phys. Chem. B* **109** 1088
[12] Byrne S, Corr S, Rakovich T, Gun'ko Y, Rakovich Y, Donegan J, Mitchell S and Volkov Y 2006 *J. Mater. Chem.* **16** 2896
[13] Rogach A, Franzl T, Klar T, Feldmann J, Gaponik N, Lesnyak V, Shavel A, Eychmüller A, Rakovich Y and Donegan J 2007 *J. Phys. Chem. C* **111** 14628
[14] Fernández Y and Vasilevskiy M 2014 *In: Cadmium Telluride Quantum Dots: Advances and Applications* ed J. Donegan and Y. Rakovich (USA: Pan Stanford Publishing) p 1
[15] López F, Ojeda P and Arbeloa I 1989 *J. Lumin.* **44** 105
[16] Volkov Y, Mitchel S, Gaponik N, Rakovich Y, Donegan J, Kelleher D and Rogach A 2004 *ChemPhysChem* **5** 1600
[17] Torrell M, Kabir R, Cunha L, Vasilevskiy M, Vaz F, Cavaleiro A, Alves E and Barradas N 2011 *J. Appl Phys.* **109** 074310
[18] Novotny L and Hehct B 2006 *Principles of Nano-Optics* (Cambridge: Cambridge University Press)
[19] Gomez D, Vernon R, Mulvaney P and Davis T 2010 *Nano Lett.* **10** 274
[20] Bludov Y and Vasilevskiy M 2012 *J. Phys. Chem. C* **116** 13738
[21] Kiravittaya S, Rastelli A and Schmidt O *Rep. Prog. Phys.* 2009 **72** 046502
[22] Santos J, Vasilevskiy M and Filonovich S 2008 *Phys. Rev. B* **78** 245422
[23] Hessman D, Castrillo P, Pistol M, Pryor C and Samuelson L 1996 *Appl. Phys. Lett.* **69** 749
[24] Bernardo C *et al.* 2014 *J. Phys. Chem. C* **118** 4982

Time-domain terahertz spectrometer for angular reflection measurements

José Antonio Nóvoa[1], Ismael Ordoñez[1], Carmen Rial[2], Jorge Martín[1], Álvaro Gil[2] and José Ramón Salgueiro[1]

[1]Departamento de Física Aplicada, Universidade de Vigo, As Lagoas, s/n, 32004 Ourense, Spain

[2]Instituto de Cerámica de Galicia, Universidade de Santiago de Compostela, Campus Sur, 15782 Santiago de Compostela, Spain

E-mail: joseantonionolo@gmail.com

Abstract. We describe a terahertz time-domain spectrometer developed to perform measurements by reflection under a wide range of angles. The different parts of the device are described as well as the process of frequency-scale calibration and signal linearity evaluation. Also the main parameters that characterize the system and account their performance are measured. Finally an example of application to the measurement of the bandgap of a photonic crystal designed for the terahertz band is presented.

1. Introduction

The terahertz (THz) electromagnetic band, in the range between 0.1 and 10 THz which corresponds to wavelengths between 3 mm and 30 μm respectively, has recently been the object of an intense research activity because of the finding of novel and interesting applications [1–5]. It had been to great extent neglected up to know because of the strong absorption of the atmosphere at these frequencies what made them useless for telecommunications and also because the available sources were scarce and inefficient. The traditional interest on this radiation came on one hand from radioastronomers who wanted to work with increasingly shorter wavelengths, and from spectroscopists who wanted to work at increasingly longer wavelengths in the infrared.

The recent development of efficient sources of such radiation has made possible to apply THz technology to a number of fields, remarkably tomography and spectroscopy. In fact, THz techniques are useful for non-destructive diagnostic as for instance early detection of internal caries or tumoral cells [6–8]. This is due to the fact that water strongly absorbs at this band and so it results easy to distinguish between tissues with slightly different water content. On the other hand this radiation is very useful for the structural analysis of materials and complex molecules [9]. Many molecules that are key components of drugs or explosives have a characteristic spectrum at these frequencies and so the detection of such substances becomes possible, leading to the use of this technology for safety applications [10–12]. Besides, it allows to distinguish among different types of materials, measure density of dopants and detect and measure pollution gases [13, 14].

One of the cheapest ways to generate THz radiation is by means of the photoconductive effect[5]. It requires from an ultrashort pulsed laser source that is used to irradiate an antenna made of two close conductors (forming a gap) attached to a semiconductor material. A bias voltage is applied to the conductors in order to generate an electric field. The laser pulse promotes charge carriers from the valence band to the conduction band which are then accelerated by the field producing the THz radiation. Using a pulsed laser of a particular repetition rate results in a train of THz pulses known as T-rays. The detection of the THz radiation can be carried out using the same effect. In this case the incoming THz pulse, impinging on the antenna synchronously with the laser pulse, moves the promoted carriers generating a small current that can be detected, amplified and measured. A controlled time delay in this laser pulse allows to scan the THz pulse and directly measure the electric field as a function of time.

The production and detection mechanisms described above are the basis of the successful time domain spectroscopy technique (THz-TDS) [3, 5], which is now one of the most versatile and cheapest for many applications, particularly for material analysis. A number of spectrometers were already described in the literature and the THz-TDS is at the moment a recent but quite well established technique [5, 15–17]. Most of the described spectrometers, however, are based on measuring radiation transmitted by the sample under analysis. THz-TDS technique based on measurements by reflection is quite scarcely used [18–20] because it results technically more complicated since the emitter and/or receiver must be mobile devices and that makes alignment more difficult. Reflection measurements under different angles are, however, interesting for many applications as they provide information that transmission not always can provide, for instance when the material under study is deposited on a substrate surface or for high absorbent bulk materials. This makes a reflection spectrometer a very useful tool.

In this work we describe our developed spectrometer for making reflection measurements for a broad range of angles. In the following section, a description of the set up is carried out. In section 3 we present the results of the calibration process as well as the parameter evaluation that accounts for the device performance. Finally, section 4 is devoted to show an example of application to the measurement of the bandgap of a photonic crystal.

2. Description of the spectrometer

A scheme of the developed set up is shown in figure 1. In order to make possible the recording of measurements under different angles, the receptor is mounted on a bench attached to a rotating stage. On the other hand the sample holder is also mounted on another rotating stage to properly orientate the sample respect to the impinging THz beam. A beam from a Ti-Za pulsed laser (wavelength $\lambda = 810\,\mathrm{nm}$, pulse duration 75 fs, and repetition rate 80 MHz) is divided in two beams to feed both emitter and receiver photoconductive antennas. In order to assure the mobility of the receiver, the corresponding beam is coupled to a 1 m long optical fiber. The output of the fiber is coupled to a beam collimator installed on the receiver head, properly aligned to feed the lens that focuses it on the antenna gap. On the other hand, the other beam is also directed to the emitter using an identical piece of fiber in order to compensate for optical path and pulse widening due to dispersion. In such a way two similar synchronized pulses are used for both emitter and receiver. In order to scan the THz pulse a delay line, consisting in a motorized stage that displaces a mirror, is installed in the path of the beam directed to the receiver. In such a way, it is possible to change the time delay between both pulses directed to emitter and receiver. Programming the scanning stage to run at a fixed velocity a set of measurements of the THz-electric field are taken for different delay times. A polarizing beam splitter and a combination of a half and a quarter wave plates are used to properly separate the two fractions of the laser beam used for the emitter and receiver.

The output of the receiving antenna is a current in the range of the tenths of nA which is transformed into a voltage using a low-noise transimpedance preamplifier with a gain about $10^5\,\mathrm{V/A}$ and then further enhanced using a lock-in amplifier. The output of the lock-in amplifier

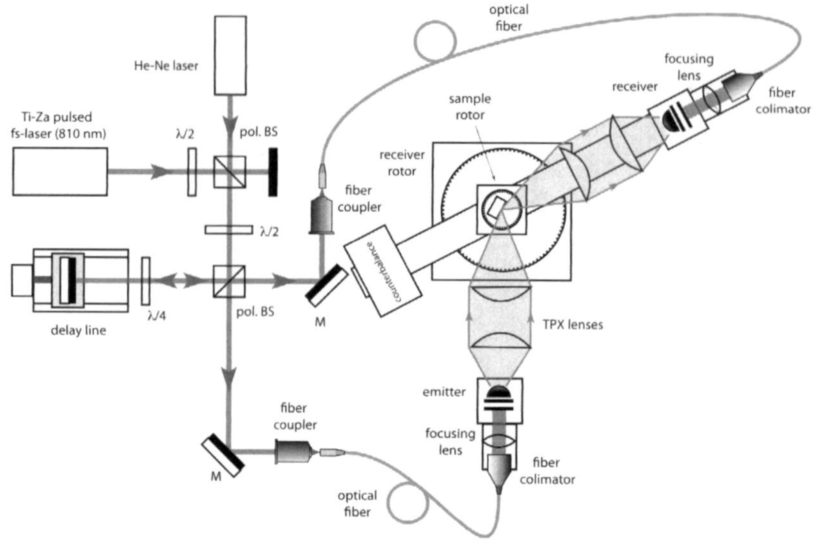

Figure 1: Sketch of the spectrometer set up.

is read by a data acquisition device on the same computer controlling the delay line. In this way, measurements are automatically taken. In combination with the lock-in amplifier, an electronic chopper is used to drive the emitter bias voltage as an square signal of frequency about 15 kHz and amplitude of 30 V (the required by the antennas used). To maximize the efficiency, instead of periodically interrupting the bias voltage the polarity is inverted at every half period, so that emission of THz radiation is never interrupted and a periodic phase change is introduced in the signal, enough to make the lock-in amplifier to work. This solution also introduces periodic phase changes in the THz radiation but they have no effect as those are produced at a rate between four and five magnitude orders slower than the pulse repetition rate.

The set up also includes an He-Ne laser whose beam is superposed to the Ti-Za laser in order to perform the system alignment more comfortably and safely. Besides, a half wave plate and a polarizing beam splitter are used at the beam input to adjust power which should not rise over 10 mW to avoid damage of the antennas.

3. Spectrometer evaluation

In order to tune the spectrometer and assess its performance a calibration process as well as an evaluation of the main characteristic parameters is necessary. To be precise, the frequency scale has to be calibrated to assure accurate measurements and also the linearity of the output checked. Considering that our assembly is a reflection spectrometer, both frequency and linearity calibrations will be defined for zero angle. The process is completed with the measurement of the signal to noise ratio, dynamic range, spectral resolution and THz electric field of the generated THz radiation.

3.1. Frequency calibration

As we have described before, a THz-TDS spectrometer works in the time domain and the use of an FFT algorithm is needed to get the frequency spectrum. Therefore, the frequency scale must be ensured externally. Many methods and techniques have been developed but most of them are based in a contrast pattern, consisting in the use of a material with a known THz spectrum which is compared to the spectrum obtained from the set up. Two common possibilities are the use of the absorption spectrum from a gas or the frequency pattern of an etalon [21]. The first one involves the use of a external database and is more expensive than the etalon. Thus, we chose an etalon-based calibration. Our etalon is a plane parallel plate of pure silicon with both faces optically polished that forms a Fabry-Perot interferometer and generates a peak pattern when it is introduced in the path of the THz beam.

If n is the refractive index of the etalon material, l its thickness and N an integer describing the order of the peak, corresponding to the number of wavelengths in the etalon, the frequencies for the peaks and valleys are given by the expressions

$$f_N = \frac{c}{2nl}N; \qquad \text{peaks} \tag{1}$$

$$f_{N+\frac{1}{2}} = \frac{c}{2nl}\left(N+\frac{1}{2}\right); \qquad \text{valleys.} \tag{2}$$

According to the Fabry-Perot interferometer theory, the model for reflectance R and transmittance T to compare with the measured spectrum is mathematically described by

$$R = \left(\frac{n-1}{n+1}\right)^2; \qquad \delta = 4\pi nl\frac{f}{c}, \tag{3}$$

$$T = \left(1 + F\sin^2\frac{\delta}{2}\right)^{-1}; \quad F = \frac{4R}{(1-R)^2}, \tag{4}$$

$$\tag{5}$$

where δ is the phase difference induced by the etalon and F the finesse coefficient. Using the characteristics of the etalon, we will get a theoretical model for the frequency transmitted through the etalon. With this model we can estimate the correlation between the theory and the spectrum obtained from THz-TDS. We used an etalon with a thickness of $l = 0.5\,\text{mm}$ and refractive index $n = 3.42$. In figure 2 it is shown the transmittance (angle zero) obtained from this theoretical model (dashed line) superposed to the measured signal (continuous line). There is a pretty close match between model and experimental results, and the measured peaks can be identified with those of the model to calibrate the frequency scale.

3.2. Linearity calibration

Ensuring a linear response of the spectrometer is very important because the frequency spectrum of a substance should be independent of the sample thickness. So, if the spectrometer has a lineal response, samples of a same substance with different thickness will produce the same frequency spectrum. The linearity of the spectrometer can be affected, among other factors, by the response of the antennas and the signal amplification system.

To check the linearity we have chosen a method based in the measurements using a different number of silicon plates placed into the THz-beam path [22]. Even though silicon is quite transparent in the THz range, there is some radiation reflected at the plate polished faces which can be evaluated using the Fresnel coefficients. Every silicon plate will introduce the same losses, so measuring the peak maximum for a different number of interposed plates should produce a linear decrease of its amplitude.

We used 3 mm thick silicon plates separated the same distance of 3 mm. This plates are thicker than the etalon used for frequency calibration to avoid the formation of stationary waves as happens for the etalon measurements. In figure 3-a we show temporal pulses measured for an increasing number of plates interposed in the THz beam. There is an increasing delay of the maximum amplitude peak due to the optical path change introduced by the silicon plates

Journal of Physics: Conference Series **605** (2015) 012026 doi:10.1088/1742-6596/605/1/012026

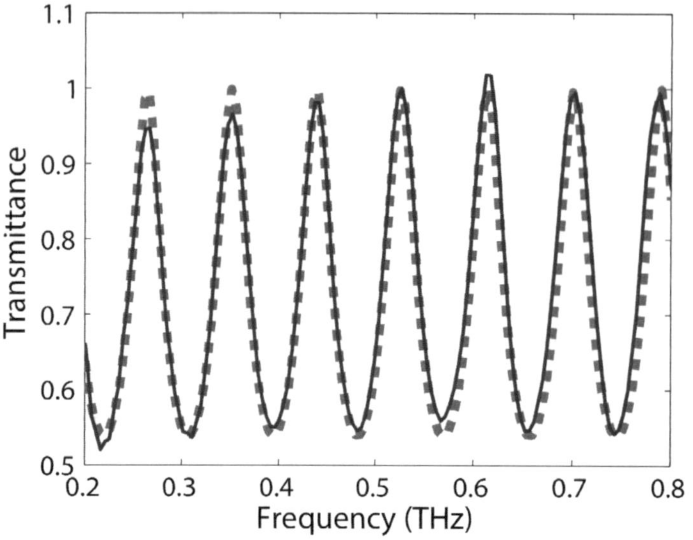

Figure 2: Transmittance measured by the spectrometer (continuous blue) and transmittance obtained with the Fabry-Perot model (dashed red).

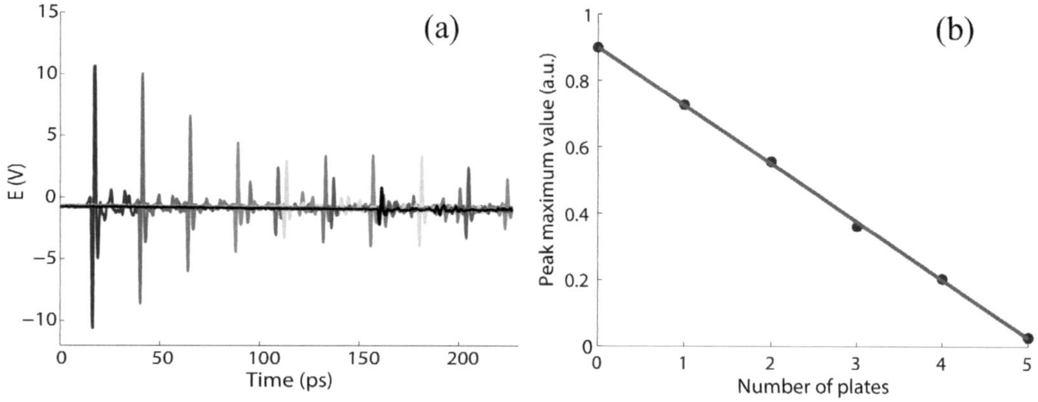

Figure 3: (a) Time domain measurements for different levels of attenuation of the THz radiation, obtained with the interposition of an increasing number of silicon plates. (b) Linear fit of the peak maximum value obtained for different number of silicon plates.

and also a progressive decrease of the peak amplitude due to the increasing reflection losses. In figure 3-b we plot the maximum amplitude value versus the number of plates. The points fit very well to a straight line revealing that the response is definitively linear.

3.3. Signal-to-noise ratio and dynamic range

Signal-to-noise ratio (SNR) and dynamic range (DR) are two parameters that determine the performance of the spectrometer. The SNR is defined as the ratio between the signal power and the noise power and measures the minimum signal change that it is possible to detect. On the other hand the DR is the ratio from the maximum peak of the signal to the background noise and is a count of the maximum detectable signal change. Both parameters should be evaluated in both the spectral and time domains. These parameters are calculated, respectively, by using the expressions[23]:

$$SNR = \frac{\text{average value of the amplitude}}{\text{standard deviation of the amplitude}}, \quad (6)$$

$$DR = \frac{\text{maximum of the amplitude}}{\text{root mean square of noise}}. \quad (7)$$

To calculate the SNR and DR in both time and spectral domains, ten measurements in vacuum (no sample) are recorded and the spectral signal obtained with a fast Fourier transform (FFT). For each point of the time and spectral signal the average and standard deviation over the ten measurements are obtained. Also, a set of measurements were also done without THz signal in order to evaluate the root mean square of the background noise. This allows to calculate, using equations (6) and (7), both SNR and DR. For the SNR it was taken the value correspondent to the point of maximum signal amplitude. For the DR, it was taken the maximum value obtained all over the signal domain. Those values are given in table 1.

	SNR	DR (maximum)
Time domain	55	1100
Spectral domain	66	9066

Table 1: *Signal to noise ratio and dynamic range obtained for the spectrometer.*

3.4. Bandwidth and spectral resolution

The bandwidth and the spectral resolution are two parameters that determine the operating range of any spectrometer. Bandwidth is the frequency range in which most of the signal power is concentrated and the spectral resolution determines the minimum frequency shift that can be detected by the spectrometer. This parameters, respectively, are calculated by using the equations [5]:

$$\Delta\Omega = \frac{2\pi}{\delta t}, \quad (8)$$

$$\delta\omega = \frac{2\pi}{t_{sg}}, \quad (9)$$

where $\Delta\Omega$ is the bandwidth, δt is the sampling interval of the measured signal, $\delta\omega$ is the spectral resolution and t_{sg} is the duration of the measured temporal signal. For a typical measurement we acquired 1700 samples equally distributed in a scanning length of 3.4 cm (the maximum possible for motorized stage we employed). In this way, we have $\delta t = 1.3 \times 10^{-13}$ s and $t_{sg} = 2.3 \times 10^{-10}$ s and we consequently obtain $\Delta\Omega = 47$ THz and $\delta\omega = 0.027$ THz. These values are ideal but useful for a coarse estimation of the system performance. Other (non considered) factors have also influence on this parameters, limiting the performance of the system. Some of these factors are the stability and straightness of the stage, the laser beam quality, the efficiency of emitter and receiver antennas or the laser pulse duration.

3.5. Electric field

The electric field generated by the terahertz radiation can be calculated from the equation [5]

$$\overline{J} = \overline{N} e \mu E_{THz} \tag{10}$$

where $\overline{J}$ is the density of current, e is the electron charge, μ is the mobility of the charge carriers, $\overline{N}$ is the average number of charge carriers and E_{THz} is the THz electric field. For the calculation we take $\overline{N} = 2.5 \times 10^{22}\,\mathrm{m}^{-3}$, $\mu = 3000\,\mathrm{cm}^{-2}\,\mathrm{Vs}$ and $e = 1.602 \times 10^{-19}\,\mathrm{C}$ [24]. The amplification factor for the transimpedance amplifier is $10^5\,\mathrm{V/A}$ and the gain factor used in the lock-in amplifier is 3×10^3. The value of the photocurrent induced by the THz beam in the receiving antenna after passing through the plano-convex lens system is $I_{ph} = 33.3\,\mathrm{nA}$ for a focused laser beam of radius $r = 38\,\mu\mathrm{m}$. So, we get a value $E_{THz} = 0.6\,\mathrm{V/cm}$ for the electric field.

4. Application example

As an example of application of the reflection spectrometer, we present measurements intended to detect and evaluate the bandgap of a photonic crystal of the woodpile type. This type of 3D periodic structure consists of a number of layers each one formed by a set of long elements (cylinders) located parallel to each other in the plane. Consecutive layers are disposed with the elements relatively perpendicular, i. e. both planes are rotated 90 degrees, and cylinders belonging to every second plane present a relative shift of half a period [25]. This structure presents a tetragonal symmetry and if properly designed can be made to present a diamond-like face centered cubic (FCC) structure which corresponds to the case when the bandgap is the widest. For the existence of the bandgap, however, it is necessary a minimum value for the refractive index of the material.

Woodpile samples were fabricated of alumina (Al_2O_3), a highly transparent material at THz frequencies and presenting a dielectric constant of $\epsilon = 9.8$, using the direct ink writing (DIW) technique [26]. They present network constants of $d = 368\,\mu\mathrm{m}$ (separation between elements in plane) and $w = 153\,\mu\mathrm{m}$ (separation between planes of same-oriented cylinders). They were placed on the spectrometer and the THz beam focused on the face parallel to the piled planes. Spectrum measurements for incidence angles of zero (transmission), 45, 60 and 75 degrees (reflection) were recorded. Also, the same measurements were made on a homogeneous block sample for the sake of comparison. Results are plotted in figure 4.

The bandgap width depends on the tridimensional wavevector modulus and so the incidence angle of the radiation influences the values obtained. In the figure it is shown the presence of the bandgap centered on 0.3 THz approximately, where a deep valley in transmission is obtained respect to the block transmission. The bandgap is also detected by reflection as a hump in the reflectance curves. Maximum values of reflectance reach only values around 0.1 due to the high power loss as the samples measured where quite small and offered a small reflecting section to the THz beam.

5. Conclusion

In this work we have described a set up for TDS-THz spectroscopy by reflection. Details about the system and characteristics derived from its evaluation are presented. Finally, as an application example it was demonstrated its usefulness for the detection and measurement of the bandgap of a woodpile photonic crystal.

Acknowledgments

IFO acknowledges financial support from Xunta de Galicia, Spain (project EM2013/002).

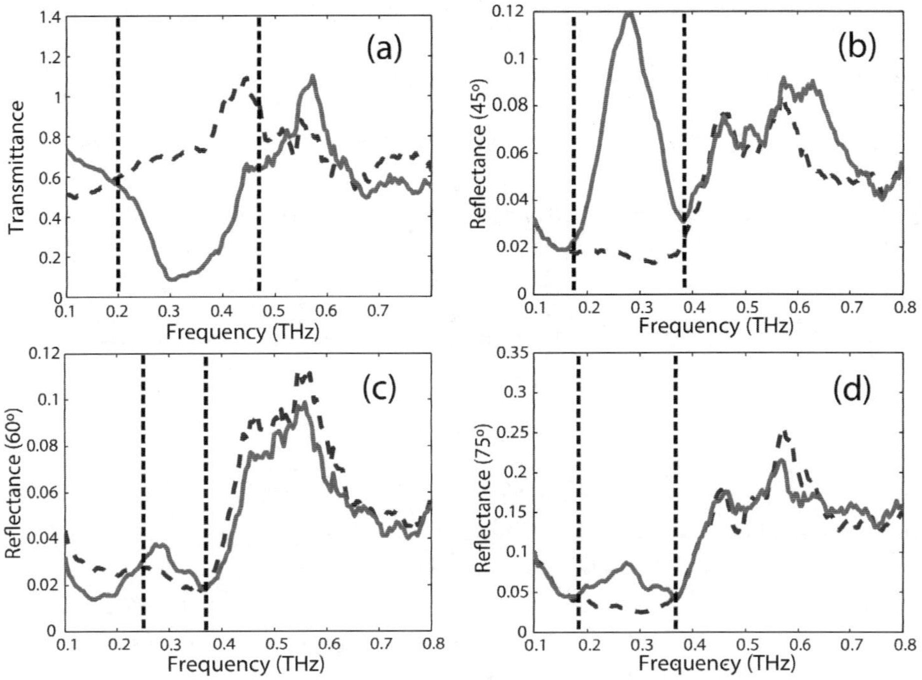

Figure 4: Measurements for a periodic structure sample of the woodpile type (continuous red line) as well as for an homogeneous block (dashed blue), both made of Al_2O_3. (a) Transmission (0 deg), (b) reflection at 45 deg, (c) reflection at 60 deg, (d) reflection at 75 deg. Vertical lines mark the bandgap region.

References

[1] Siegel P H 2002 *IEEE Trans. Microwave Theory Tech.* **50** 910–928

[2] Ferguson B and Zhang X C 2002 *Nature Mater.* **1** 26–33

[3] Tonouchi M 2007 *Nature Photon.* **1** 97–105

[4] Mueller E 2003 *The industrial physicist* 27–29

[5] Zhang X C and Xu J 2010 *Introduction to THz Wave Photonics* (New York: Springer)

[6] Pickwell E and Wallace V P 2006 *J. Phys. D: Appl. Phys.* **39**(17) R301–R310

[7] Oh S J, Kang J, Maeng I, Suh J S, Huh Y M, Haam S and Son J H 2009 *Opt. Express* **17** 3469–3475

[8] He M, Azad A K, Ye S and Zhang W 2006 *Opt. Commun.* **259** 389–392

[9] Strachan C J, Rades T, Newnham D A, Gordon K C, Pepper M and Taday P F 2004 *Chem. Phys. Lett.* **390**(1) 20–24

[10] Kurmbholtz N and et al 2009 *Milimetre Wave and Terahertz Sensors and Technology II* (*Proc. SPIE* vol 7485) p 748504

[11] Dickinson J C, Goyette T M, Gatesman A J, Joseph C S, Root Z G, Giles R H, Waldman J and Nixon W E 2006 **6212**

[12] Yamamoto K, Yamaguchi M, Miyamaru F, Tani M, Hangyo M, Ikeda T, Matsushita A, Koide K, Tatsuno M and Minami Y 2004 *Jpn. J. Appl. Phys.* **43**(No. 3B) L414 – L414–L417

[13] Redo-Sánchez A, Salvatella G, Galceran R, Roldós E, García-Reguero J A, Castellari M and Tejada J 2011 *Analyst* **136** 1733–1738

[14] Mittleman D, Jacobsen R, Neelamani R, Baraniuk R and Nuss M 1998 *Appl. Phys. B* **67** 379–390

[15] Dexheimer S L (ed) 2008 *Terahertz Spectroscopy: Principles and Applications* (Boca Raton: CRC Press)

[16] K E Peiponen Axel J and M K (eds) 2013 *Terahertz spectroscopy and imaging* first edition ed (New York: Springer)

[17] Lewis R A 2012 *Terahertz Physics* 1st ed (New York: Cambridge University Press)

[18] Matsuoka T, Fujimoto T, Tanaka K, Miyasaka S, Tajima S, Fujii K, Suzuki M and Tonouchi M 2009 *Physica C: Superconductivity* **469** 982–984

[19] Palka N 2011 *Acta Phys. Pol. A* **120** 713–715

[20] Pashkin A, Kempa M, Nemec H, Kadlec F and Kuzel P 2003 *Rev. Sci. Instrum.* **74** 4711–4717

[21] Naftaly M, Dudley R and Fletcher J 2010 *Opt. Commun.* **283** 1849 – 1853

[22] Naftaly M and Dudley R 2009 *Opt. Lett.* **34** 674–676

[23] Naftaly M and Dudley R 2009 *Opt. Lett.* **34**(8) 1213–1215

[24] Beard M C, Turner G M and Schmuttenmaer C A 2001 *J. Appl. Phys.* **90** 5915–5923

[25] Joannopoulos J D, Johnson S G, Winn J N and Meade R D 2008 *Photonic Crystals: Molding the Flow of Light* 2nd ed (Princeton, New Jersey: Princeton University Press)

[26] Smay J, Gratson G, Shepherd R, Cesarano J and Lewis J 2002 *Adv. Mater.* **14** 1279–1283

Journal of Physics: Conference Series **605** (2015) 012027 doi:10.1088/1742-6596/605/1/012027

Interferometric angular characterization of ion-exchanged glass binary phase plates

Vicente Moreno[1], Xesús Prieto-Blanco[1], Carlos Montero[1], José Crespo[1], David Barral[1], Dolores Mouriz[1], María C Nistal[1], Eduardo F Mateo[2] and Jesús Liñares[1]

[1]Área de Óptica, Departamento de Física Aplicada, Facultade de Física e Facultade de Óptica e Optometría, Universidade de Santiago de Compostela, Campus Vida s/n, E-15782 Santiago de Compostela, Galicia, Spain

[2]NEC Labs America, 4 Independence Way, Princeton, NJ 08540, USA

E-mail: suso.linares.beiras@usc.es

Abstract. An interferometric characterization of the phase angular shifting produced by rotation of an ion-exchanged glass binary phase plates is presented. The inverse WKB method is used as a starting point to fabricate the phase plates, because such a method can only characterize the phase shift for normal incidence of the light on the plate. A complete phase angular shifting characterization is made by a Mach-Zehnder interferometer where a four-step phase shifting method is used for acquisition of data and a modified Carré algorithm is applied to data processing. The theoretical phase shifting is calculated by using the phase accumulated by a plane wavefront propagating in an axial graded-index media modelling an ion-exchanged glass phase plate. Experimental results present a good agreement with theoretical predictions.

1. Introduction

Binary phase plates are key optical elements in many branchs of optics and photonics such as optical imaging, optical metrology [1], optical communications [2], and so on. In order to check the importance of such plates, we briefly centre our attention, for instance, on optical communications where significant efforts are being made to increase optical network capacity as data traffic continues to grow in a amazing way. One promising candidate to further increase fiber capacity is mode-division multiplexing in few-mode fibers, which uses transverse space modes as a new information-bearing dimension [2, 3]. A fundamental optical element for spatial multiplexing is a modal transformer, which converts the fundamental mode LP01 of a single mode fiber into the excited modes (LP11, LP21) of a few modes fiber. Modal transformation can be implemented in free-space by using a multi-region phase plate which is one of the main building blocks of these modal transformers. Such plates present two values (binary) of the phase in two or more regions. To prevent modal cross-talk at the modal multiplexing stage, high quality modal transformation is desired. In consequence, fabrication of the mentioned phase plates must provide an accurate, stable and steep phase-shift between regions. Other examples in optical imaging, such as diffractive elements (Fresnel lenses and so on), optical metrology and so on can be found, and again phase plates must also provide such an accurate, stable and steep phase-shift between regions.

Journal of Physics: Conference Series **605** (2015) 012027 doi:10.1088/1742-6596/605/1/012027

In a recent work we presented the fabrication and characterization of phase plates fabricated by highly-uniform Ag^+/Na^+ ion-exchange (IE) glass but without paying attention to their angular properties, that is, the phase shift undergone by an ion-exchanged glass binary phase plate under rotation. These angular properties can be interesting to get a more accurate phase or even to change the phase with a specific purpose. In this work we present a preliminary study of these angular properties of IE plates. On the other hand, ion-exchange in glass is a well-established technique for materials processing [4], integrated optics [5], and so on, and it presents a high versatility and precision. The first advantage of this technique is that IE phase plates are monolithic. This ensures that the change of phase is achieved inside the material, which provides an excellent uniformity and robustness. Likewise ion-exchange technique allows to easily control the fabrication parameters such as time (t), temperature (T) and ionic concentration (c), which results in a very accurate phase step. In fact, a precision in the phase step as high as $\lambda/60$ can be achieved [8]. Moreover, under different models for the (gradual) refractive index change, compatible with the ion-exchange physical processes, we can determine and optimize the fabrication parameters (t,T,c) to obtain IE binary phase plates for several wavelengths, in particular, in soda-lime glasses by Ag^+/Na^+ ion-exchange. Although phase plates can be characterized by the Inverse WKB (IWKB) method [5], the angular properties only can be obtained by using an interferometric technique to which a phase recovery algorithm is applied, in particular a modified Carrè algoritm. We have designed the fabrication of ion-exchanged glass phase plates by using the IWKB method [8] in order to obtain, for instance, a modal transformer to LP11 mode from a LP01 mode. The results show that a rotation can compensate small errors of the phase step due to different reasons, therefore to characterize and analyze the angular behaviour of a ion-exchanged glass binary phase plate is an important goal. The plan of the paper is as follows: in section 2 the theoretical angular phase shift of a ion-exchanged (graded-index) glass plate is obtained. In section 3 the fabrication of ion-exchanged glass plates is described and fabrication parameters are obtained in order to produce particular phase plates, that is, binary plates with two regions with phases shifted a certain amount. In section 4 the interferometric device, the acquisition data and the results and their analysis are presented. In section 5 conclusions are presented.

2. Phase Angular Shift of an Ion-Exchanged Glass Plate

The well-developed thermal ion-exchange technique in a glass substrate of refractive index n_s and thickness h produces an axial graded-index profile $n(z)$, with z the normal direction to the substrate. It is common to choose a normalized ionic concentration $c(z,t)$, that is, $c(0,t) = 1$ and the obvious boundary condition $c(\infty,t) = 0$. The ionic concentration $c(z,t)$ is obtained by solving the unidimensional nonlinear diffusion equation [5], and solutions can be written in the form $c(z/d)$, where $d = 2\sqrt{Dt}$ is the socalled effective depth and D the diffusion coefficient, with $h \gg d$, in general. Accordingly, the refractive index profiles achieved by a thermal ion-exchanged process, after a particular time t, can be written, in a good approximation, as

$$n(z) = \begin{cases} n_s + \Delta n\, c(z/d) & z \in [0,h] \\ 1 & z \notin [0,h], \end{cases} \tag{1}$$

where Δn is the change of index at surface and the following condition $\Delta n \ll n_s$ is assumed. Eq.(1) represents a z-axial index profile of a graded-index medium, in particular, of an IE plate.

We are interested in obtaining the phase $\Delta\Phi$ introduced to a plane wavefront under propagation in such a graded-index IE plate. For that purpose, let us consider, a light beam with a plane wavefront incident with an angle ϕ_0 (with wavevector included in the XZ plane) on a plane-parallel IE plate of axial refractive index profile $n(z)$ and thickness h as shown in Fig.1. The plane wavefront at the exit of the plate is defined by the pointed line joining points C and D, then if we consider light of wavelength $\lambda_0 = 2\pi/k_o$ the phase accumulated for such a

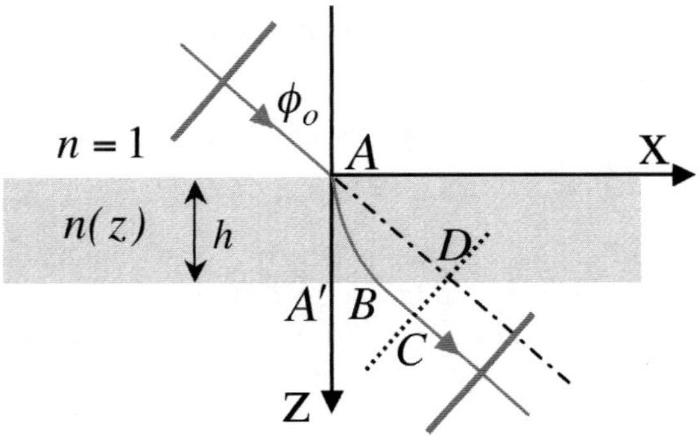

Figure 1. Geometrical and optical characteristics of a dielectric graded-index plate.

plane wavefront from the point A up to the point C is given, according to the Fermat principle, by the expression

$$\Phi_{AC} = \Phi_{AB} + \Phi_{BC} = k_o \int_0^h n(z)\sqrt{1+\dot{x}^2}\, dz + k_o BC, \tag{2}$$

where $\dot{x} \equiv dx/dz$. From the first term of the right hand member of Eq.(2) the differential equation for the ray trajectory from A to B can be obtained (Euler-Lagrange equation), that is,

$$\dot{x}(z) = \frac{\sin\phi_o}{(n^2 - \sin^2\phi_o)^{1/2}}. \tag{3}$$

An approximation to order $O(\Delta n^2)$ can be made in the ray trajectory by taking into account the assumption $\Delta n \ll n_s$, that is, by integrating Eq.(3) we obtain

$$x(z) \approx \frac{\sin\phi_o}{(n_s^2 - \sin^2\phi_o)^{1/2}} z - \frac{\sin\phi_o\, n_s \Delta n}{(n_s^2 - \sin^2\phi_o)^{3/2}} \int_0^z c(z'/d)dz' + O(\Delta n^2). \tag{4}$$

Note that $A'B \equiv x(h)$, where $A' = (0, 0, h)$. Next, by inserting Eq. (3) into Φ_{AB}, that is, in the first term of the right hand member of Eq.(2), and after a long but straightforward calculation, the following result is obtained:

$$\Phi_{AB} = \frac{k_o n_s}{(n_s^2 - \sin^2\phi_o)}\{n_s h + \Delta n \int_0^h c(z/d)dz\} - \Phi_o, \tag{5}$$

where $\Phi_o = k_o n_s \Delta n \sin^2\phi_o (n_s^2 - \sin^2\phi_o)^{-3/2} \int_0^h c(z/d)dz$. Likewise, the phase between point B and C is given by $\Phi_{BC} = k_o(A'D - A'B)\sin\phi_o = k_o(h\tan\phi_o - x(h))\sin\phi_o$, therefore by using Eq.(4) at $z = h$, the following result is obtained:

$$\Phi_{BC} = k_o h \tan\phi_o \sin\phi_o - \frac{k_o \sin^2\phi_o h}{(n_s^2 - \sin^2\phi_o)^{1/2}} + \Phi_o. \tag{6}$$

Finally, by adding Eqs.(5) and (6) we obtain the phase Φ_{AC} accumulated by the incident plane wave after propagation through the graded index plate. In the same way, the phase accumulated for $\Delta n = 0$ can be obtained. Therefore the phase angular shift $\Delta\Phi$ produced by the graded-index plate is given by the following expression

$$\Delta\Phi(\phi_o) = \Phi_{AC} - \Phi_{AC}(\Delta n = 0) = \frac{k_o n_s \Delta n}{(n_s^2 - \sin^2\phi_o)^{1/2}} \int_0^h c(z/d)dz. \tag{7}$$

Journal of Physics: Conference Series **605** (2015) 012027 doi:10.1088/1742-6596/605/1/012027

By taking into account the phase shift for normal incidence, that is, $\Delta\Phi(\phi_o = 0)$, we obtain the following phase shifting angular law

$$\Delta\Phi(\phi_o) = \frac{n_s\,\Delta\Phi(0)}{(n_s^2 - \sin^2\phi_o)^{1/2}} = \frac{\Delta\Phi(0)}{\cos\phi_s}. \tag{8}$$

where ϕ_s is the angle formed by the wavevector (or ray) in the substrate, that is, far from the surface $z = 0$ where $n(z) \approx n_s$. Expression (8) shows that the phase step $\Delta\Phi(\phi_o)$ can be changed by a simple rotation of the plate. Note that for normal incidence, that is, $\phi_0 = 0$, a minimum value of the optical phase is obtained. Therefore, it can not be reduced, which indicates that the fabrication process must be designed to make sure that a phase value equal or less to the needed one is achieved. Indeed, by increasing the incidence angle ϕ_0 the phase step can be enlarged to the required value. In particular for a glass with $n_s \approx 1.5$ the maximum theoretical increasing ($\phi_0 = \pi/2$) is about 34%, and, for instance, if $\phi_0 = \pi/3$ the increasing is about 22%. Obviously if the phase take values $\Delta\Phi = \delta + 2m\pi \equiv \delta\,(\mathrm{mod}\,2\pi)$, with m a positive integer and $\delta \in [0, 2\pi]$, then the increased phase step due to rotation will be much larger according to the value of m.

3. Fabrication of Ion-Exchanged Glass Plates

The ion-exchange process in the most common glasses is nonlinear, that is, the diffusion coefficient of the ions depends on the concentration. In these cases solutions of the nonlinear diffusion equation are needed. We have proven [8] that the following function provides an approximate solution fitting much better to the experimental results, that is,

$$c(z,t) = \frac{1}{\alpha}\log[1 + (\mathrm{e}^{\alpha} - 1)\mathrm{erfc}(z/d)],\quad d = 2\sqrt{Dt} \tag{9}$$

where α accounts for the degree of nonlinearity. Let us return to the problem of obtaining the phase introduced by a graded-index plate. From equation (7) the general expression of the phase shift is given by

$$\Delta\Phi(\phi_o, t) = \frac{2\pi}{\lambda_0}\frac{n_s \Delta n\, d}{(n_s^2 - \sin^2\phi_0)^{1/2}} I(\alpha, h), \tag{10}$$

where

$$I(\alpha, h) = \int_0^{h/d} \frac{\log(1 + (\mathrm{e}^{\alpha} - 1)\mathrm{erfc}(z')}{\alpha}\mathrm{d}z' \tag{11}$$

with $z' = z/d$. Note that the phase grows linearly with the effective depth d, that is, with the ion-exchange time. Moreover, in most practical cases, we can consider $h \gg d = 2\sqrt{Dt}$, that is, we can formally take $h/d \approx \infty$, but still this integral $I(\alpha)$ does not have a closed-form solution. However, it can be seen by numerical integration that $I(\alpha, \infty) \approx a\,\alpha + b$. For typical glasses $3 < \alpha < 7$ and $I(\alpha, \infty) \approx 0.1306\,\alpha + 0.5725$ with an accuracy better than 0.6%. Taking into account this result equation (10) can be rewritten as follows

$$\Delta\Phi(\phi_o, t) \approx \frac{2\pi}{\lambda_0}\frac{n_s \Delta n\, d}{(n_s^2 - \sin^2\phi_0)^{1/2}}(0.1306\,\alpha + 0.5725). \tag{12}$$

This equation gives a good approximation (better than 1%) of the phase step for typical nonlinear ion-exchange processes. Therefore we will design the fabrication of the guide by using the refractive index profile given by equation (1) when the concentration function given by equation (9) is taken into account. For that, we have to use the so called IWKB method [5, 6], that is, the phase plate can be also considered as an integrated multimode waveguide, that is, whith several guided modes. Thus, by a standard technique based on coupling guide-prism we measure the effective indices and hence, by using the IWKB method, the index profile is obtained [5, 7]. We must stress that the phase plates can only be characterized by the IWKB method for normal incidence. Moreover, it is also important to check the IWKB method by an alternative method as the interferometric one, which will be described in the next section. In short, the IWKB method allows us to have a starting point for the angular characterization of the plates.

The plate that we have characterized angularly was fabricated by following the procedure described in a previous work [8]. The substrates were plane-round soda-lime glasses of 25.3 mm of diameter and 1 mm thickness (refractive index n_s=1.5106 for λ=632.8 nm). After lithography they were immersed in a 5% $AgNO_3/NaNO_3$ salt melt at 340 °C for ion-exchange took place during t=1624.5'. By one hand, we chose this time because it leads to a large phase shift for λ=632.8 nm and therefore a more convenient angular phase shift characterization can be made. In effect, by using the IWKB method we obtain an index profile which has Δn=0.087303, α=3.527 and d=40.925 μm. Inserting this values in Eq. (12) it gives $\Delta\Phi=11.67\pi\simeq 11\pi+(2/3)\pi$, for normal incidence. Such a large phase shift will reduce the signal-to-noise ratio in the subsequent angular interferometric measures. On the other hand, we have also chosen the mentioned time of fabrication because the phase shift reaches a value close to $\Delta\Phi\sim 5\pi$ for λ=1550 nm, which can be used, for instance, to implement a mode converter to LP11 from a LP01 mode for optical communications.

4. Angular characterization by optical interferometry

In the above section, by using the IWKB method [8] we have designed the fabrication of an ion-exchanged glass phase plates in order to obtain a binary phase plate with a large phase shift, in particular, $\Delta\Phi = 11.67\pi \simeq 12\pi - (1/3)\pi$ for λ=632.8 nm. As commented, the phase shift of the phase plates can be characterized by the inverse IWKB method only for normal incidence, therefore their angular phase shifting properties will be obtained by using an interferometric method to which a phase recovery algorithm is applied. To that end, we have used a Mach-Zehnder interferometer to analyze such angular phase shifting properties of IE plates by measuring fringe shifting.

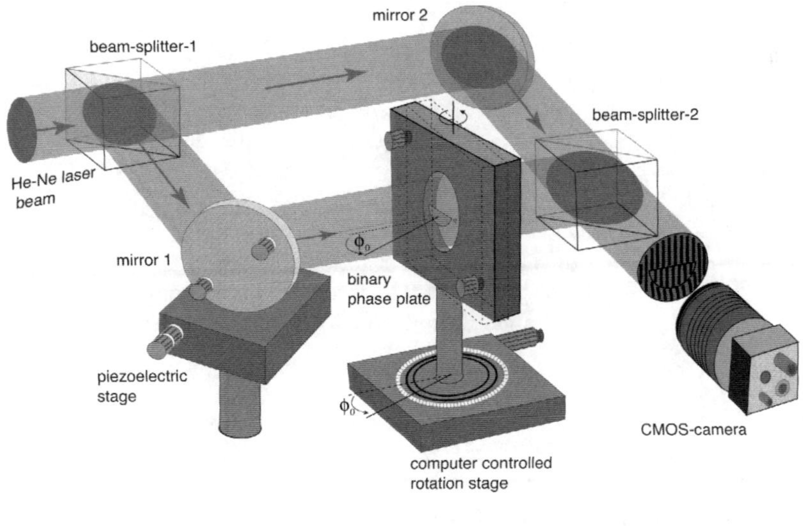

Figure 2. Mach-Zehnder interferometer for angular characterization of a IE phase plate

Journal of Physics: Conference Series **605** (2015) 012027 doi:10.1088/1742-6596/605/1/012027

4.1. Interferometric device

The experimental set-up is a Mach-Zehnder interferometer in a rectangular configuration (32 cm long arm, 23 cm short arm) as shown in Fig. 2. The illumination with gaussian beam from a He-Ne laser (λ=632.8 nm) collimated with transversal section of 2". The alignment of this interferometer was performed using an Equilateral Hyperbolic Zone Plate designed specifically for this task [8]. In the first step two focal crosses should overlap and next, in a second step, fine alignment is achieved by observing in a plane out of focus and minimizing the number of the fringe in the lobes of the interference pattern, provided the rest of the field is free of fringes. Next, when the IE plate is inserted linear fringes are obtained. In Figure 3 we show a particular interferogram where we can see the phase step produced by the IE plate.

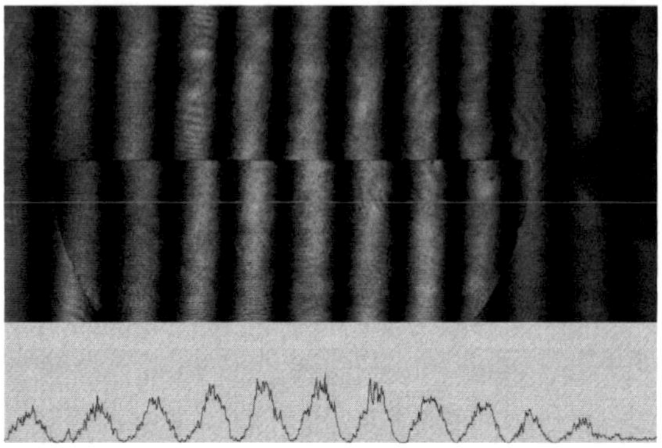

Figure 3. Example of an interferogram produced by a IE phase plate.

The determination of angular properties of the phase plates, requires the proper alignment of it with respect to the incident beam. In order to achieve this, the back reflected beam, produced for a small surface of the plate coated by aluminium (and with triangular shape for indicating the border of the ion-exchanged zone) is used. This reflected beam is carefully aligned back through all the previous optical elements till the laser window. In this point the use of a piezo electric stage (Thorlabs NF5DP20S), in closed-loop with piezo controller and strain gauge reader, situated in a corner of the interferometer, is necessary to introduce the proper nano displacement to produce a controlled phase change in the optical path between the two arms of the interferometer. After three steps of this stage, and four interferograms, captured by a CMOS camera (Thorlabs DCC1545M $1280x1024 pixels$) and a macro lens (Canon EF 100mm 1:28 USM), it is possible recover the phase map of the image using a proper algorithm.

The angular displacement was carried out by means a rotational stage (Newport 495 Series with the PMC200 digital controller) in steps of fraction of degree till 15 degrees. After every rotation the process to recover the phase map is repeated again, by using the piezo electric stage, between every interferogram of the four ones which are necessary for every angular position.

4.2. Data adquisition

As commented, the phase step $\Delta\Phi$ was measured for each orientation of the plate by an interferometric analysis based on the Carré algorithm [8]. The measurements have been made directly on the interferferograms. The algorithm was applied to four interferograms obtained after respective phase steps Φ_R of the reference beam. Although Φ_R was adjusted with the piezo-electric stage, it was also recovered in a more accurate way from the interferograms themselves [9]. Then Φ_R was used to calculate the phase (mod 2π) between the object and reference beams at each point of the plate, which is shown as a false color image in Fig. 4. In order to

Figure 4. Phase (false color) difference (mod 2π) for an incidence angle of $4°$.

make easier both the unwrapping of this phase and the interpretation of the interferograms, a slight horizontal tilt was introduced between the interfering beams, which results in the vertical fringes of the interferograms. That is, inside the exchanged region, the unwrapped phase is a ramp across each row of the image and a constant along each column. The same happens outside this region, but a discontinuity occurs at the transition. By a suitable average of the phase in two rectangular regions at both sides of the straight frontier, the height of the discontinuity $\Delta\Phi(\text{mod}2\pi)$ was obtained. The above procedure was repeated for a range of incidence angles from zero to 15 degrees. In order to obtain the real value of the phase step at normal incidence $\Delta\Phi(0)$, a multiple of 2π was added to the interferometric result to make it coincide with the WKB estimate. The same value was added to the phase steps for the remaining incidence angles, since $\Delta\Phi$ does not cross any multiple of 2π.

4.3. Results and Analysis

As, for sake of simplicity, we have made the angular characterization for 633.8 nm (He-Ne Laser), then a IE plate with phase $\Delta\Phi(\text{mod}\,2\pi)$ has to be considered. In particular, for normal incidence $\Delta\Phi(0) = (1.67 \pm 0.02)\pi\,(\text{mod}\,2\pi) \simeq -(1/3)\pi\,(\text{mod}\,2\pi)$ is the measured interferometric phase. Likewise, by using the IWKB method we had estimated, for normal incidence, a total phase equal to $\Delta\Phi(0) = 12\pi - (1/3)\pi$.

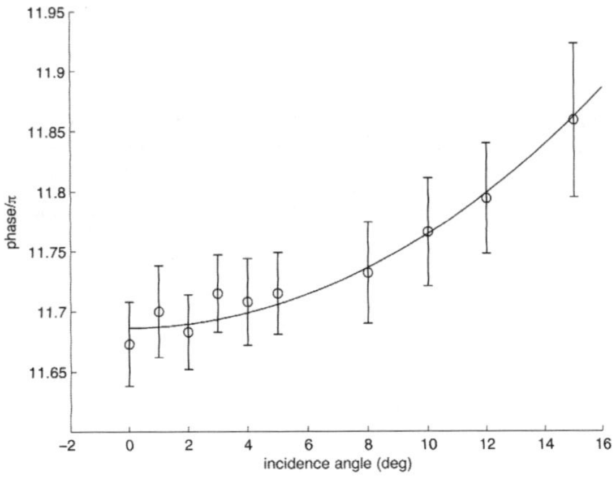

Figure 5. Phase *versus* incidence angle for a IE phase plate.

In Figure 5 we present the experimental results: phase shift *versus* incidence angle ($\Delta\Phi - \phi_o$), and the theoretical curve corresponding to equation (7). The agreement between theoretical and experimental values is quite satisfactory. Note, for instance, that a rotation up to, approximately, $\phi_o \simeq 14^o$ provides an additional phase shift $\pi/6$, therefore the new phase shift has been increased up to $\Delta\Phi(14^o) = 12\pi - (1/6)\pi$. Likewise, note that very precise phase change can be made by rotating, and such precision is increased with the wavelength because the phase under normal incidence is reduced.

5. Conclusions

As a conclusion, we have shown by a interferometric technique that the phase shift of a IE phase plate is increased under rotation. Moreover, it is a simple way to introduce a precise phase change as the requires angle can be adjusted with a great accyracy. Accordingly, when small errors of the phase step has to be corrected then a rotation of the plate can be used whenever the experimental phase is lower than the required one. On the other hand, if the phase takes values $\Delta\Phi(\text{mod}\, 2\pi)$ then the increased phase step due to rotation is much larger which can be of interest in optical metrology applications. It has been proven that the agreement between the theoretical results provided by the phase shifting angular law and the experimental ones show a good agreement, therefore the IWKB method can be used to determine the normal incidence phase $\Delta\Phi(0)$ and next the phase shift angular law can be applied to predict the phase shift.

Acknowledgments

Authors from University of Santiago de Compostela wish to acknowledge the financial support from Ministerio de Economía y Competitividad, Central Government of Spain, Project Ref. FIS2013-46584-C2-1-R and FEDER 2007-2013.

References

[1] Shamir J 2012 *Opt. Engineering* **51** 073605-1
[2] Li A, Amin A A, Chen X and Shieh W 2011 *Opt. Express* **19** 8808
[3] Bai N, Ip E,Huang Y, Mateo E, Yaman F, Li M, Bickham S, Ten S, Liñares J, Montero C, Moreno V, Prieto X, Tse V, Chung K.M., Lau APT, Tam H, Lu C, Luo Y, Peng G, Li G, and Wang T 2012 em Opt. Express **20**, 2668
[4] Liñares J, Montero C and Prieto X 1995 *Pure Appl. Opt.* **4**, 695
[5] Najafi S I, ed., 1992 *Introduction to glass integrated optics* (Artech House, Boston-London)
[6] D. L. Lee D L 1986 *Electromagnetic Principles of Integrated Optics* (John Wiley & Sons, New York)
[7] Liñares J, Montero C, Moreno V, Nistal M C, Prieto X, Salgueiro J R and Sotelo D 2000 *Proc. SPIE* **3936** 227
[8] Montero-Orille C, Moreno V, Prieto-Blanco X, Mateo E F , Ip E, Crespo J and Liñares J 2013 *Appl. Opt.* **52**, 2332
[9] Kreiss T 2005 *Handbook of Holographic Interferometry* (Wiley-VCH, Weinheim)

Transparent Boundary Condition for Oseen-Frank Model. Application for NLC Cells With Patterned Electrodes

I Orquín-Serrano[1], J Vijande[2], F R Villatoro[3], A Ferrando[4], P Fernández de Córdoba[1], and H Michinel[5]

[1]Instituto Universitario de Matemática Pura y Aplicada (IUMPA), Universidad Politécnica de Valencia (UPV), Camino de Vera s/n 46022 Valencia, Spain

[2]Departamento de Física Atómica, Molecular y Nuclear, Universidad de Valencia (UV) and IFIC (UV-CSIC), Valencia, Spain.

[3]Departamento de Lenguajes y Ciencias de la Computación, Universidad de Málaga (UMA), Málaga, Spain.

[4]Departament d'Óptica, Universitat de València (UV), 46100 Burjassot, Valencia, Spain.

[5]Optics Lab. University of Vigo, Facultade de Ciencias de Ourense, As Lagoas s/n. Ourense, 32004. Spain.

E-mail: isorser@doctor.upv.es

Abstract. In the present work a novel application of Transparent Boundary Conditions (TBC) to nematic liquid crystal cells (NLCC) with planar alignment and a patterned electrode is studied. This device is attracting great interest since it allows soliton steering by optically and externally induced waveguides. We employ the continuum Oseen-Frank theory to find the tilt and twist angle distributions in the cell under the one-constant approximation. The electric field distribution takes into account the whole 2D permittivity tensor for the transverse coordinates. Standard finite difference time domain methods together with an iterative method is applied to find an approximate solution to our coupled problem. A novel class of TBC is used to correctly define the boundary for both the distortion angle problem and the electric field distribution when using patterned electrodes. Thus, we achieve an important decrease of computational needs when solving this kind of problems and we are also capable of exploring weak anchoring conditions for NLCC.

1. Introduction

Nematic Liquid Crystal (NLC) Cells for lateral propagation are attracting much attention due, among other reasons, to their capability to form tunable reconfigurable waveguides. They can be formed by an external electric field [1] or by creating a self-consistent waveguide in bulk NLC [2]. These soliton induced waveguides can be used to implement all-optical switching and logic gating [3]. Tunable waveguide arrays can also be obtained in these kind of devices, in which all-optical switching phenomena [4] and optical multiband vector breathers [5] appear.

With respect to electrically induced waveguides, several NLC devices are being investigated. Many of them are based on using a patterned electrode to affect the molecules of the liquid crystal in such a way that the effective refractive index profile obtained yields a graded index

Journal of Physics: Conference Series **605** (2015) 012028 doi:10.1088/1742-6596/605/1/012028

waveguide inside the nematics [6]. Other dispositions, where two electrodes at different voltages are employed, are used for soliton routing [7, 8].

The problem of simulating the optical evolution in a stripped electrode device is a difficult one and involves necessarily a $2D + 1$ anisotropic beam propagation method that at least takes into account the transverse anisotropy induced by the external electric field. Moreover, vectorial effects appear when the y component (see Fig. 1) of the external electric field is not negligible and pushes the molecules of the liquid crystal out of the propagation plane, making them twist. This implies changes in the polarization of the optical beam and also losses of the extraordinary solitonic beam. Some approximations are usually employed to avoid this complexity, obtaining good agreement with the experiments. One of these approximations is not considering the twist movement of the molecules. This approximation holds for low voltage values and for those cases where, although voltages are high, light moves far away from the vicinity of the patterned electrode where this twist is noticeable. Hard anchoring conditions for the orientational problem are usually employed. This boundary condition is realistic since the glass plates confining the NLC can be treated in such a way that tilt and twist angle are fixed at the boundary. However we can also work with a non-fixed value for the twist angle next to the glass plates allowing the tilt angle to be fixed. This is also realistic since a suitable treatment of the glass plates that allows this twist movement of the director at the boundary can be found.

In a weak anchoring situation the elastic constant depends on the position, in the proximity of the boundaries [9] and to determine the value of this surface (or anchoring) energy, small deviations close to the Friedericksz transition are usually considered [10]. The anchoring energy for deformations will be neglected as it is much lower than the corresponding energy for tilt deformations (tipically, one or even two orders of magnitude)[11]. We will show the numerical boundary treatment in this case, what could be used for further numerical research looking for better characterizing the boundary influence on the elastic constants under weak anchoring conditions as it was done in some studies in the past[12].

This is the regime we want to show in this paper, where transparent boundary conditions for the twist angle will be necessary in the boundary next to the glass plates. These simulations widens our knowledge of what is actually happening in the area closest to the electrode.

This paper is organized as follows: section 2 presents the physical model, section 3 shows the numerical treatment that allows us to approximate the solution of the problem using the boundary condition explained in section 4. Some results of our numerical model are presented in section 5 for different electrode configurations and finally our conclusions are presented in section 6.

2. Physical Model

We want to simulate a NLC device made of two glass plates containing commercial NLC E7 in between indium tin oxide transparent electrodes connected to the glass plates to apply an external voltage to the liquid crystal. It can be seen in Fig. 1 the geometrical properties of such a device; z coordinate stands for the propagation axis while x and y represent transverse spatial dimensions. Tilt angle (θ) appears with any movement of the director along a plane perpendicular to the yz plane. Twist angle (φ) is defined in zy plane and holds for the twist deformations of the director. Planar alignment imposes director orientation nearly parallel to the glass plates, i.e., small tilt angle ($\theta = 2°$). $V = 0$ is applied to the lower electrode affecting the whole glass plate. Upper one is thinner and doesn't affect the whole upper glass plate, so we will refer to it as the stripped electrode, where $V \neq 0$ is applied. This physical configuration makes the numerical boundary difficult to work with because of its undefinition where the upper electrode doesn't influence the liquid crystal orientation. This discontinuity on the upper boundary, forced by the discontinuous electrode makes appear an y component for the electric field, such a component being responsible for the twist angle to appear in the proximities of

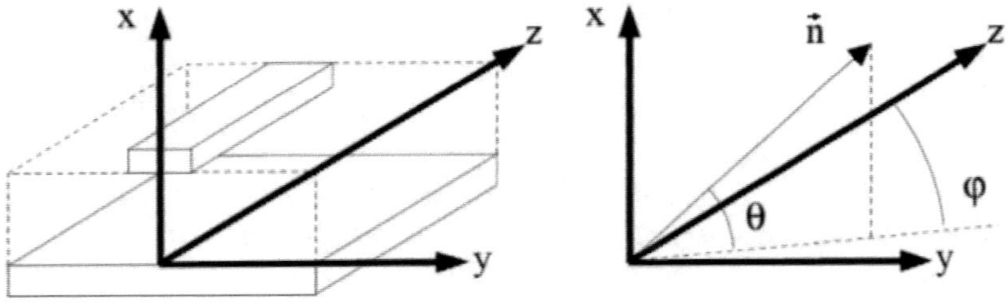

Figure 1. Sketch of the nematic liquid crystal device with patterned electrode (left) and notation for the planar alignment in the cell (right).

the stripped electrode. So a complete molecular reorientation is needed to explain both angle distributions. The whole orientational problem has already been solved in an iterative way for different NLC configurations where no discontinuous electrode is taken into account [13]. We extend this iterative method to the two dimensional case and we also present the two dimensional electric field distribution in those cases where tilt and twist angle distributions affect the electric field imposed by the electrodes, i.e., taking into account the whole 2D permittivity tensor.

2.1. *Electric field distribution*

We start from the Maxwell equations

$$\left.\begin{array}{c}\nabla D = 0\\ D = \epsilon E\\ E = -\nabla V\end{array}\right) \rightarrow \nabla\left(\bar{\epsilon}\nabla V\right) = 0, \tag{1}$$

where ϵ_{ij} with $i = x, y, z$ and $j = x, y, z$ are permittivity tensor terms, given by

$$\epsilon = \begin{pmatrix} \epsilon_\perp + \Delta\epsilon\sin^2\theta & \Delta\epsilon\sin\theta\cos\theta\sin\varphi \\ \Delta\epsilon\sin\theta\cos\theta\sin\varphi & \epsilon_\perp + \Delta\epsilon\cos^2\theta\sin^2\theta \end{pmatrix}, \tag{2}$$

where $\epsilon_\perp$ is the perpendicular (to the glass plates) component of the permittivity and $\Delta\epsilon = \epsilon_\parallel - \epsilon_\perp$ is the birefringence of the medium. Tilt ($\theta(x, y)$) and twist ($\varphi(x, y)$) angles depend only on these transverse coordinates. By doing so we are uncoupling the z coordinate therefore obtaining a static transverse distribution. We also consider negligible any dependence on z of the functions, i.e. $V(x, y, z) = V(x, y)$, $\frac{\partial V}{\partial z} = 0$, $\frac{\partial^2 V}{\partial z^2} = 0$, $\frac{\partial^2 V}{\partial y \partial z} = 0$, $\frac{\partial^2 V}{\partial x \partial z} = 0$, $\frac{\partial \theta}{\partial z} = 0$, $\frac{\partial \varphi}{\partial z} = 0$.

Substituting the dielectric tensor expressions in Eq. (1) and not considering any dependence on the z coordinate one obtains for the static two dimensional problem in the transverse plane,

$$a_1\frac{\partial^2 V}{\partial x^2} + a_2\frac{\partial^2 V}{\partial y^2} + a_3\frac{\partial^2 V}{\partial x \partial y} + a_4\frac{\partial V}{\partial x} + a_5\frac{\partial V}{\partial y} = 0, \tag{3}$$

with

$$a_1 = \frac{1}{\Delta x^2}\left(\epsilon_\perp + \Delta\epsilon\sin^2\theta\right), \tag{4}$$

$$a_2 = \frac{1}{\Delta y^2}\left(\epsilon_\perp + \Delta\epsilon\cos^2\theta\sin^2\varphi\right), \tag{5}$$

$$a_3 = \frac{1}{4\Delta x\Delta y}\Delta\epsilon\sin(2\theta)\sin\varphi, \tag{6}$$

$$a_4 = \frac{1}{2\Delta x}\left(\Delta\epsilon\left(\cos(2\theta)\sin\varphi\frac{\partial\theta}{\partial y} + \cos\theta\sin\theta\left(\cos\varphi\frac{\partial\varphi}{\partial y} + 2\frac{\partial\theta}{\partial x}\right)\right)\right), \tag{7}$$

$$a_5 = \frac{1}{2\Delta y}\left(\Delta\epsilon\left(\left(\cos^2\theta\sin\varphi - \sin^2\theta\sin\varphi\right)\frac{\partial\theta}{\partial x}\right.\right. \tag{8}$$

$$\left.\left. - \sin(2\theta)\sin^2\varphi\frac{\partial\theta}{\partial y} + \cos\theta\cos(\varphi)\sin\theta\frac{\partial\varphi}{\partial x} + \cos^2\theta\sin(2\varphi)\frac{\partial\varphi}{\partial y}\right)\right). \tag{9}$$

This is the equation we solve to obtain the electric field distribution. Notice that the dielectric tensor is responsible of the graded index profile which appears in the Beam Propagation Method, this is the way this whole static problem is coupled with the optical problem, which we are not studying here.

Finite Difference method is used to approximate this equation. Centered second order differences are employed for approximating the differential operators. Newton-Raphson method is used to solve the nonlinear equations of the finite difference scheme.

2.2. Elastic distortion angles distribution

As it was mentioned before, Oseen-Frank theory is employed to describe the molecular orientation in the NLC cell [14]. The total free energy of the liquid crystal is calculated to obtain its minimum via the Euler Lagrange theory. All three energy terms are considered in this model: the electric energy due to the external voltage, the electric energy due to the optical beam propagating inside the cell, and the elastic distortion energy.

The elongated molecules from the NLC feel any electric field due to its electric polarity, so they react changing its orientation so that it is aligned with the total electric field. Thus, molecules orientation is defined by the director vector $\overline{n} = (\sin\theta, \cos\theta\sin\varphi, \cos\theta\cos\varphi)$.

The distortion energy per volume unit considers three different types of distortion, each one with different elastic constants K. Under *one constant approximation*, all three elastic constants, related to different molecular distortions, are considered to be equal resulting in a more simplified equation. Minimizing the energy density can be done by transforming this problem to the solution of a partial differential equation via Euler-Lagrange theory, yielding:

$$\begin{aligned} -K\left(\frac{\partial^2\theta}{\partial x^2} + \frac{\partial^2\theta}{\partial y^2}\right) + \sin(2\theta)\left[-\frac{K}{2}\left(\frac{\partial\varphi}{\partial x}^2 + \frac{\partial\varphi}{\partial y}^2\right)\right. \\ +\epsilon_0\left(-\frac{1}{2}\left(\Delta\epsilon^s|E_x^s|^2 + \Delta\epsilon^o|E_x^o|^2\right) + \frac{\Delta\epsilon^s}{4}|E_y^s|^2(1-\cos(2\varphi))\right. \\ \left.\left.+\frac{1}{2}\Delta\epsilon^o\sin^2\varphi|E_y^o|^2\right)\right] - \frac{1}{2}\epsilon_0\cos(2\theta)\sin\varphi\left(\Delta\epsilon^o\left(E_y^{o*}E_x^o + E_x^{o*}E_y^o\right)\right. \\ \left.+2\Delta\epsilon^s E_x^s E_y^s\right) = 0, \end{aligned} \tag{10}$$

and

$$\begin{aligned} K\cos\theta\left[\cos\theta\left(\frac{\partial^2\varphi}{\partial x^2} + \frac{\partial^2\varphi}{\partial y^2}\right) - 2\sin\theta\left(\frac{\partial\theta}{\partial x}\frac{\partial\varphi}{\partial x} + \frac{\partial\theta}{\partial y}\frac{\partial\varphi}{\partial y}\right)\right] \\ +\frac{1}{2}\epsilon_0\cos^2\theta\sin(2\varphi)\left(\Delta\epsilon^o|E_y^o|^2 + \Delta\epsilon^s|E_y^s|^2\right) \\ +\frac{1}{4}\epsilon_0\cos\varphi\sin(2\theta)\left(\Delta\epsilon^o\left(E_y^{o*}E_x^o + E_x^{o*}E_y^o\right)\right) \\ +2\Delta\epsilon^s E_x^s E_y^s = 0, \end{aligned} \tag{11}$$

that will allow us to solve $\theta(x, y)$ and $\varphi(x, y)$ respectively. Typical values we use for the magnitudes involved are $K = 12 \times 10^{-12}$ N, $\Delta\epsilon^s = \epsilon_\parallel^s - \epsilon_\perp^s = 19.6 - 5.1$, $\Delta\epsilon^o = \epsilon_\parallel^o - \epsilon_\perp^o = n_\parallel^2 - n_\perp^2 = 1.69^2 - 1.50^2$.

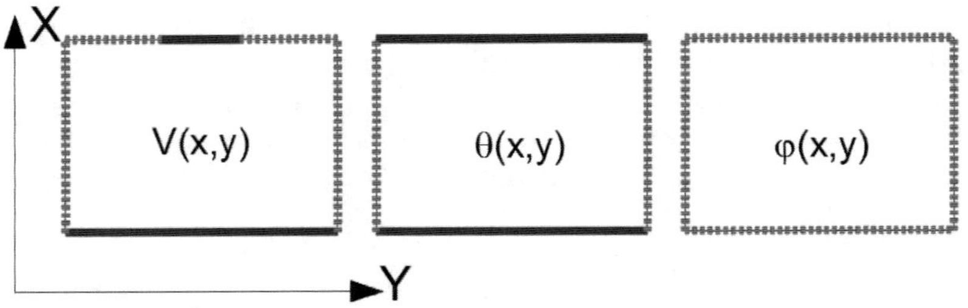

Figure 2. Numerical domain for all three differential equations. Blue solid lines represent hard anchoring condition while red dotted lines hold for transparent boundary condition.

3. Numerical Model

All three equations (3, 10 and 11) are approximated by a finite difference method consisting of difference operators which approximates the differential operators to second order, both for first and second derivatives. The corresponding nonlinear boundary value problems require the use of an iterative method such as Newton-Raphson ([15]) in order to find an approximate solution. Following the notation of [16] we write the difference equation which approximates eq. (3)

$$\begin{aligned} a_{1_{j,k}}\delta_x^2 V_{j,k} + & \; a_{2_{j,k}}\delta_y^2 V_{j,k} + a_{3_{j,k}}(\delta_{0y}(\delta_{0x}V_{j,k})) \\ + & \; a_{4_{j,k}}\delta_{0x}V_{j,k} + a_{5_{j,k}}\delta_{0y}V_{j,k} = 0, \end{aligned} \tag{12}$$

being $\delta_x^2 V_{j,k} = V_{j-1,k} - 2V_{j,k} + V_{j+1,k}$ and $\delta_y^2 V_{j,k} = V_{j,k-1} - 2V_{j,k} + V_{j,k+1}$ the second order approximations to the second spatial derivatives and $\delta_{0x}V_{j,k} = V_{j+1,k} - V_{j-1,k}$ and $\delta_{0y}V_{j,k} = V_{j,k+1} - V_{j,k-1}$ the second order approximations to the first spatial derivatives. Notice that difference operators δ_{0x} and δ_{0y} commute. Coefficients $a_{i_{j,k}}$ represent the discretized version of the respective coefficients of eq. (3). It can be seen that coefficients $a_{1_{j,k}}, a_{2_{j,k}}$ and $a_{3_{j,k}}$ are point approximations while $a_{4_{j,k}}$ and $a_{5_{j,k}}$ require the application of difference operators to the angle derivatives appearing in those coefficients. Thus the boundary condition will appear also in these coefficients.

We can rewrite eq. (12) in a more clear way

$$\begin{aligned} & A_{NW_{j,k}}V_{j-1,k-1} + A_{N_{j,k}}V_{j,k-1} + A_{NE_{j,k}}V_{j+1,k-1} \\ + & \; A_{W_{j,k}}V_{j-1,k} + A_{P_{j,k}}V_{j,k} + A_{E_{j,k}}V_{j+1,k} \\ + & \; A_{SW_{j,k}}V_{j-1,k+1} + A_{S_{j,k}}V_{j,k+1} + A_{SE_{j,k}}V_{j+1,k+1} = 0 \end{aligned} \tag{13}$$

where the coefficients stand for the location of the nodes in the discrete molecule of the difference method. This is a nine point molecule whose coefficients are $A_{NW_{j,k}} = a_{3_{j,k}}$, $A_{N_{j,k}} = a_{2_{j,k}} - a_{5_{j,k}}$, $A_{NE_{j,k}} = -a_{3_{j,k}}$, $A_{W_{j,k}} = a_{1_{j,k}} - a_{4_{j,k}}$, $A_{P_{j,k}} = -2a_{1_{j,k}} - 2a_{2_{j,k}}$, $A_{E_{j,k}} = a_{1_{j,k}} + a_{4_{j,k}}$, $A_{SW_{j,k}} = -a_{3_{j,k}}$, $A_{S_{j,k}} = a_{2_{j,k}} + a_{5_{j,k}}$, $A_{SE_{j,k}} = a_{3_{j,k}}$ accounting for the north-west, north, north-east, west, central, east, south-west, south and south-east nodes respectively. This is a very convenient notation since the matrix of coefficients of eq. (13) and the matrix representing the jacobian of that difference functional coincides.

Equations (10) and (11) are solved analogously using second order difference operators and Newton-Raphson method. Notice that the numerical scheme of these equations yields a five-point molecule instead of the nine-point molecule appearing in eq. (13) due to the crossed second derivative approximation.

The main drawback of this formalism is that an iterative scheme is needed to reach an autoconsistent solution since equations (3), (10) and (11) are coupled. An iterative scheme can

Journal of Physics: Conference Series **605** (2015) 012028 doi:10.1088/1742-6596/605/1/012028

be used to obtain the solution to this coupled problem. First an iterative loop will find the auto consistent solution of the approximated equations of (10) and (11) and then use this solution to find the electric field distribution using (3). Then other iterative loop is needed for finding an autoconsistent solution of both problems the electric and the angular one. Thus, two nested iterative loops scheme have been used to find the whole autoconsistent solution.

4. Transparent Boundary Condition

All three differential equations we solve have different boundary conditions as it can be seen in Fig. 2. First, the calculation for the external electric potential applied over the nematics requires hard anchoring condition wherever we have an electrode fixing the voltage. Where no electrode is located we use transparent boundary condition since it allows no restrictions over the voltage value at the boundary, which is a realistic boundary. Second, with respect to the tilt angle distribution, $\theta(x, y)$ we use hard anchoring condition for the x boundary, since that is the area where the nematics is limited by the glass plate, which usually has a rubbing treatment to fix the tilt angle to a low value, typically $\theta = 2°$, with no possibility of movement in this boundary. Finally, twist angle $\varphi(x, y)$ is not fixed in the x boundary, since the glass plates allows twist movement of the molecules, parallel to the glass plates, after appropriate treatment of the surface. It can be seen in Fig. 2 that TBC are used for all three problems in the y boundary allowing a finite extent in this transverse coordinate of such device. Gratings can be simulated by employing periodic boundary conditions in y boundary.

TBC allow us to predict which has to be the value of the function at the boundary using the information we have in the interior domain. From the middle 70's on much work has been done to develop a total non reflecting boundary condition for the Schrödinger equation ([17, 18, 19, 20, 21, 22, 23, 24, 25]). Absorbing boundary conditions are still being studied and its application to dispersive waves phenomena [26] or its behavior with semilinear parabolic equations [27] have been published recently. Different approaches have been proposed, both analytical and numerical, but the work of G. R. Hadley has become one of the most popular results published in this topic because of its easy implementation at the time it offers very good results. Hadley's work presents the derivation of a TBC for the Schrödinger equation, based on physical considerations about the energy flow through the numerical boundaries [28, 29]. We use similar mathematical considerations which allow us to predict the value of the function at the boundary. In this work we extend the use of this TBC to other differential equations. Let us show the calculation of this transparent boundary condition with an example for the external electric potential $V(x, y)$. Let $V_{j,k}$ be the discretized values $V(j\Delta x, k\Delta y)$ where $2 \leq j \leq N_X - 1$ and $2 \leq k \leq N_Y - 1$, N_X and N_Y being the number of unknowns in the x and y direction respectively.

Consider a fixed value for j in the external electric potential case, say $j = N_X/2$. Calculate the set of values $c_{y_{j,k}} = \frac{V_{j,k}}{V_{j,k+1}}$, as the quotient of consecutive values in the y direction. Index k ranges from 2 to $N_Y - 1$, being N_Y the number of nodes in the y direction (with $N_Y - 2$ unknowns). Notice that boundary values in the y direction in this set are $V_{j,1}$ and V_{j,N_Y}. We could guess the values of the y boundary if we got a value for $c_{y_{j,1}}$ and $c_{y_{j,N_Y-1}}$. Of course this set of values $c_{y_{j,k}}$ depend on the value function. In order to obtain these quantities at the boundaries we use a linear extrapolation, saying $c_{y_{j,1}} = 2 \cdot c_{y_{j,2}} - c_{y_{j,3}}$ and $c_{y_{j,N_Y-1}} = 2 \cdot c_{y_{j,N_Y-2}} - c_{y_{j,N_Y-3}}$. Depending on the solution shape at the boundary we must evaluate the convenience of using other higher order extrapolation in order to obtain better accuracy. Thus we can say $V_{j,1} = V_{j,2} \cdot c_{y_{j,1}}$ and $V_{j,N_Y} = V_{j,N_Y-1}/c_{y_{j,N_Y-1}}$. We calculate all transparent boundaries in a similar way, using the extrapolated values of the convenient quotient for each case. These transparent boundary values for the external electric potential $V_{j,k}$ are used in Eq. (3). Notice that not all the values for the quotient vector $c_{x_{j,k}}$ (or analogously $c_{y_{j,k}}$) are employed.

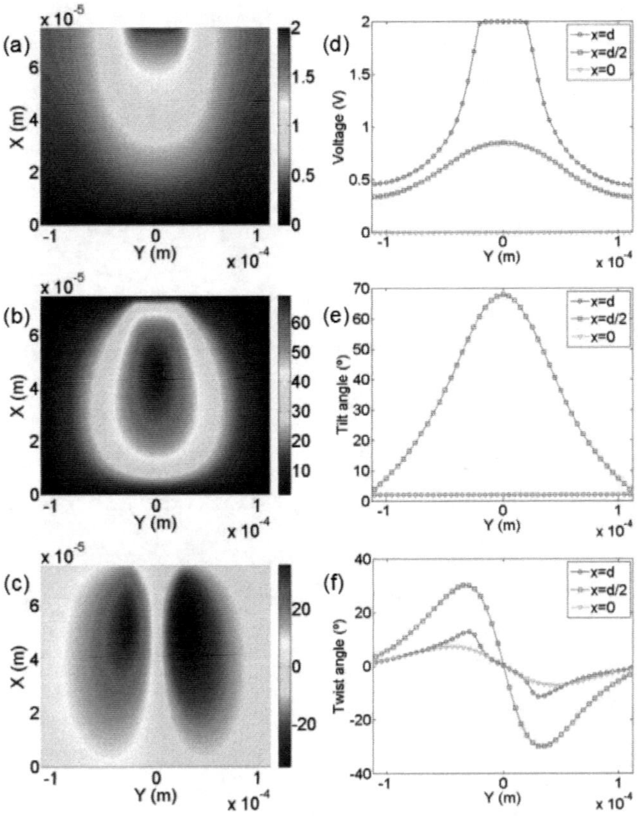

Figure 3. (a), (b) and (c) Show potential, tilt and twist angle distribution for $V = 2$ V respectively. (d), (e) and (f) Show three different sections of (a), (b) and (c) respectively. Blue circles, red squares and green triangles represent cuts of the three dimensional representation on the left around $x = d$, $x = d/2$ and $x = 0$ respectively. Notice that some curve might be overlapping the axes or other curve.

5. Results

5.1. One single electrode

The single electrode device represented in Fig. 1 has been used experimentally in order to guide nematicons in planar devices [30]. Fig. 3 shows the solution for the electric potential, the tilt angle and twist angle for $V = 2$ V. There it can be seen the effect of the transparent boundary condition. Although optical power affects the molecular orientation (as it can be seen in equations 10 and 11, we perform simulations for different voltages with no optical power, so that it doesn't appear any other electric excitacion than the external field. It can be seen in Fig. 4 the effect of the voltage on both the tilt and the twist angle. Both angles show different behavior: tilt angle changes more abruptly that twist angle. Changes begin at voltage $V = 1$ V since that supposes the minimal amount of energy necessary to start moving the LC molecules, what is known as the Friedericksz threshold. Models which don't take into account twist angle shouldn't be appropriate for describing voltages above one volt since from then on twist angle is no longer negligible. To our knowledge this is the first time that transparent boundary conditions are applied to these equations.

This one electrode device has already been studied to show the lateral propagation of a soliton

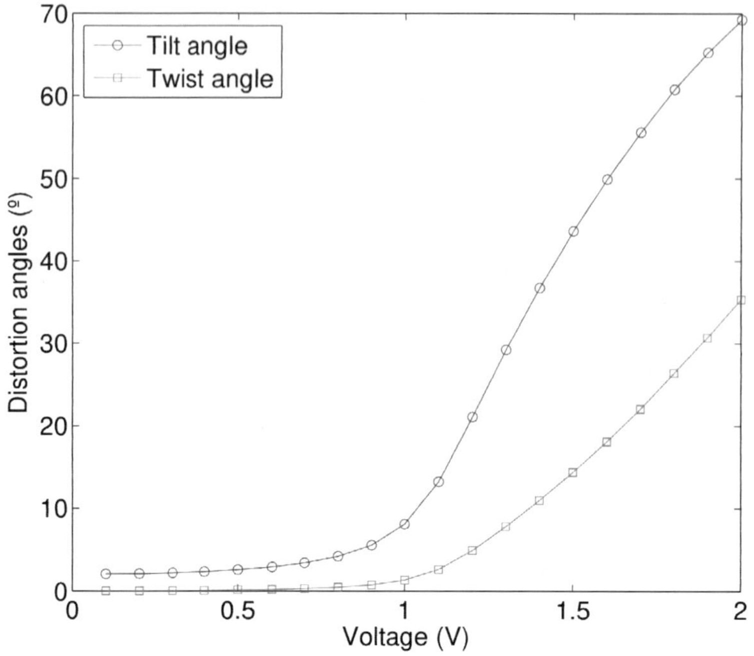

Figure 4. Voltage dependence over the peak values for the tilt and twist angles. These are the results of 200 simulations raging from $V = 0.01$ V to $V = 2$ V.

inside the NLCC. However it is said that it is necessary to extend artificially the computational domain in order to solver the electric field distribution inside the cell since nothing is known from the solution at the boundary [30]. The idea is setting a fixed value for the solution in an artifical boundary far enough from the real one so that one can say that it doesn't affect the solution inside the original numerical domain. Our TBC permits us to avoid this artifical extension of the 2D numerical domain, lowering this way the computational needs to solve the whole coupled problem.

5.2. Two different electrodes with gap in between

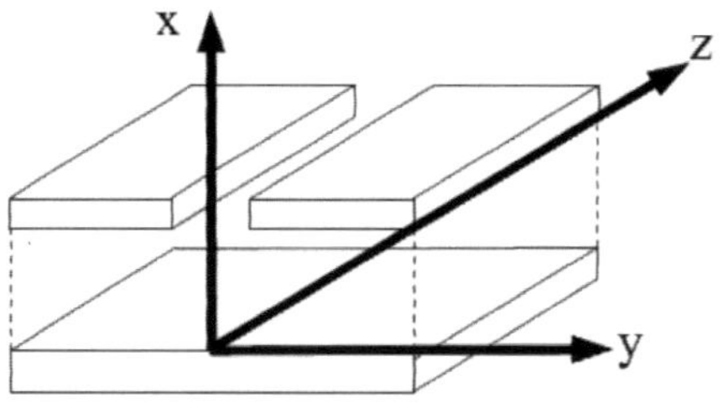

Figure 5. NLC Device with two different electrodes in the upper area.

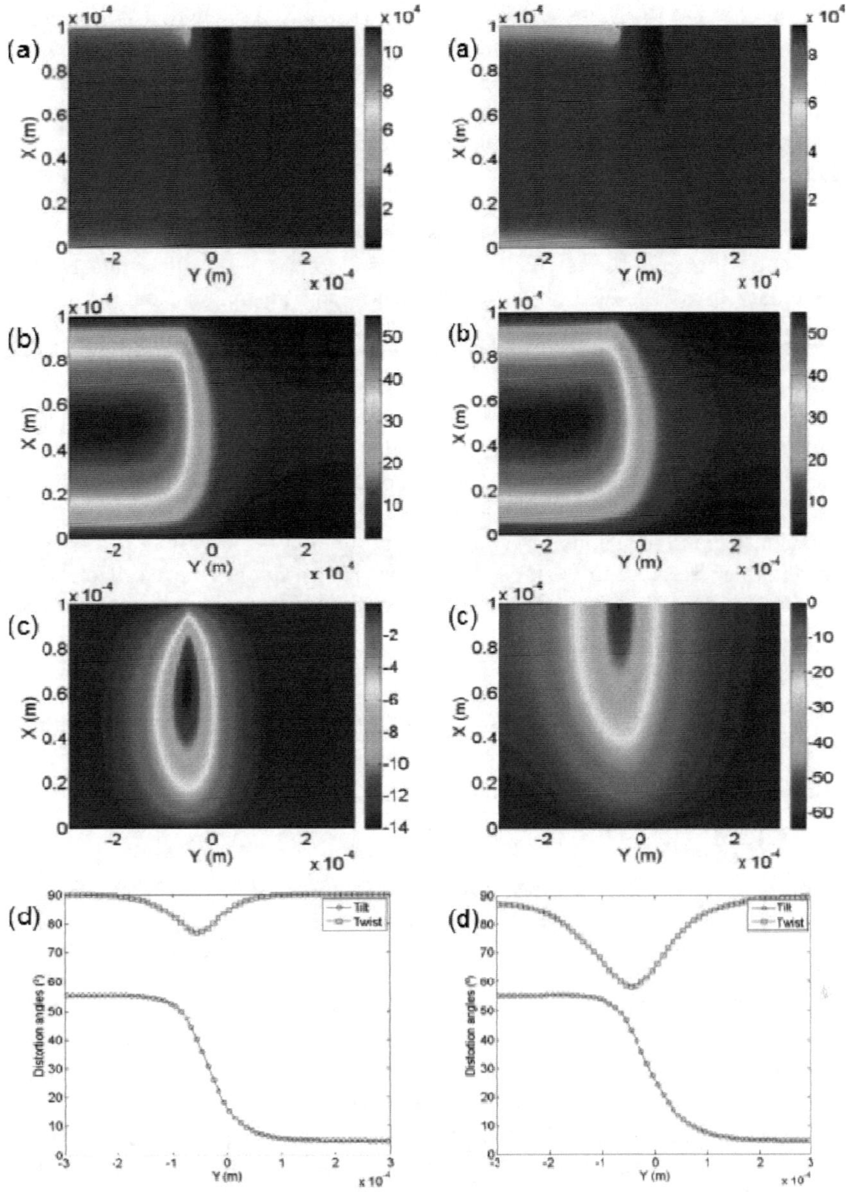

Figure 6. Solution for a two electrode device with Dirichlet boundary conditions at x boundaries (on the left) and using transparent boundary conditions at x boundaries for twist angle (on the right). Voltages are $V_1 = 1.5$ V and $V_2 = 0.7$ V. (a) Electric field, (b) tilt angle and (c) twist angle. Plot (d) show cross sections of both angles, tilt and twist, at $x = d/2$.

NLC Devices for lateral propagation and two electrodes with a gap in between have been studied for demonstrating tunable refraction and reflection of a soliton inside the nematics [31]. Fig. 5 shows which is the device we are referring to. Hard anchoring boundary conditions have been employed for the distortion angles, tilt and twist, in the x boundary since these values are fixed experimentally using a rubbing technique of the glass plates. Tilt angle is always fixed at the boundary since it is very difficult to allow a tilt movement of the molecules in a plane perpendicular to the glass plate there. However other boundary conditions are also realistic for the twist angle. Particularly, the *hybrid* case where the $x = 0$ boundary is defined by a hard anchoring condition and the $x = d$ boundary is transparent, allowing the molecules to twist at the boundary $x = d$, is also realistic. We study this case as a clear example of the use of our transparent boundary conditions. Notice that also transparent boundary conditions are employed for all three problems (V(x,y), $\theta(x, y)$ and $\varphi(x, y)$) in the y boundaries, allowing thus to study a single device of a finite extent.

In the device described in [31] the electrodes are used to create an effective index distribution so that it affects the nematicon trajectory in different ways. We have performed the simulations of such device in two different scenarios: total anchoring condition in the x boundary, and the hybrid case mentioned before. The results are illustrated in Fig. 6, where we show the difference among both configurations. No changes in tilt angle distribution are observed since the calculation remains the same in both cases. However, twist angle distributions differ because of the effect of the transparent boundary condition applied in $x = d$. In fact, the amplitude of the twist angle in the latter case is bigger since it is not restricted by the boundary. This also affects the electric field distribution which reaches lower values in that boundary.

The application of a TBC to NLC devices allow us to study new configurations where weak anchoring conditions are relevant. For example in-plane movements of the molecule (those corresponding to the twist deformation) at the boundary are common in twisted nematic liquid crystal cells which may be used as displays, for example.

6. Conclusions

We have developed a realistic two dimensional numerical model for the coupled problem of finding the electric field distribution in a planar NLC cell considering the effects of both the tilt and twist director deformations. Oseen-Frank theory for the director distribution has been employed. To our knowledge, transparent boundary conditions are applied for each differential equation for the first time, thus avoiding time-consuming codes where artificial extension of the numerical domain were necessary to simulate devices with patterned electrodes correctly. The use of TBC in the y direction allow us to simulate one single electrode device properly. In the x direction TBC are only used for the twist deformation in the single electrode device. It reaches non negligible values close to the upper boundary where the patterned electrode is located. Other devices have been simulated to show the robustness of our code, for example a two electrode device with different voltages has been calculated. A TBC for the twist deformation at all four boundaries has been used to show the differences between the code already solved in the literature and our code, which enlightens a bit more the weak anchoring situation (for the twist angle) in those kind of devices.

Acknowledgments

A.F. thanks Ministerio de Ciencia e Innovación of Spain by its financial support (TIN2006-12890 and TEC2010-1532). F.R.V. thanks Ministerio de Ciencia e Innovación (MICINN) of Spain by its financial support (TIN2009-12359 and MTM2010-19969). P.F.C. thanks MICINN of Spain by its financial support (TIN2009-12359).

References

[1] Peccianti M, De Rossi A, Assanto G, De Luca A, Umeton C and Khoo I C 2000 *Applied Physics Letters* **77**
[2] Peccianti M and Assanto G 2001 *Opt. Lett.* **26** 1690–1692
[3] Peccianti M, Conti C, Assanto G, De Luca A and Umeton C 2002 *Applied Physics Letters* **81**
[4] Fratalocchi A, Assanto G, Brzdkiewicz K A and Karpierz M A 2005 *Applied Physics Letters* **86** 051112
[5] Fratalocchi A, Assanto G, Brzdakiewicz K A and Karpierz M A 2005 *Opt. Lett.* **30** 174–176
[6] Beeckman J, Neyts K and Haelterman M 2006 *Journal of Optics A: Pure and Applied Optics* **8** 214
[7] Peccianti M, Dyadyusha A, Kaczmarek M and Assanto G 2008 *Phys. Rev. Lett.* **101**(15) 153902
[8] Kivshar Y 2006 *Nature Phys.* **2** 729
[9] Alexe-Ionescu A L, Barberi R, Barbero G and Giocondo M 1994 *Phys. Rev. E* **49**(6) 5378–5388
[10] Sodré N, Zola R S, Lenzi E K and Evangelista L R 2006 *Rev. Mex. Rev. Mex. de Física S* **52** 80–84
[11] Jérôme B 2008 *Handbook of Liquid Crystals* (Wiley-VCH Verlag GmbH) pp 535–548
[12] Allen M P and Frenkel D 1988 *Phys. Rev. A* **37**(5) 1813–1816
[13] Wang Q, He S, Yu F and Huang N 2001 *Optical Engineering* **40** 2552–2557
[14] Khoo I C 2006 *Liquid Crystals: Physical Properties and Nonlinear Optical Phenomena* (Cambridge University Press)
[15] Press W H, Teukolsky S A, Vetterling W T and Flannery B P 2007 *Numerical Recipes: The Art of Scientific Computing* (John Wiley & Sons, Inc.)
[16] Thomas J W 1995 *Numerical Partial Differential Equations* (Springer, New York)
[17] Engquist B and Majda A 1977 *Math. Comp* **31** 629–651
[18] Grote M and Keller J 1995 *SIAM Journal on Applied Mathematics* **55** 280–297
[19] Hagstrom T, Warburton T and Givoli D 2010 *Journal of Computational and Applied Mathematics* **234**
[20] Baskakov V and Popov A 1991 *Wave Motion* **14** 123 – 128
[21] Lubich C and Schädle A 2002 *SIAM Journal on Scientific Computing* **24** 161–182
[22] Kuska J P 1992 *Phys. Rev. B* **46**(8) 5000–5003
[23] Menza L D 1996 *Applied Mathematics Letters* **9** 55 – 59
[24] Sch A 2002 *Wave Motion* **35** 181 – 188
[25] Arnold A, Ehrhardt M and Sofronov I 2003 *Commun. Math. Sci.* **1** 501–556
[26] Lindquist J M, Neta B and Giraldo F X 2012 *Applied Mathematics and Computation* **218** 6666 – 6676
[27] Sun Z Z, Wu X, Zhang J and Wang D 2012 *Applied Mathematics and Computation* **218** 5187 – 5201
[28] Hadley G R 1991 *Opt. Lett.* **16** 624–626
[29] Hadley G 1992 *Quantum Electronics, IEEE Journal of* **28** 363–370
[30] Beeckman J, Chaubiska K and Neyts K 2006 *Ferroelectrics* **344** 225–231
[31] Peccianti M, Dyadyusha A, Kaczmarek M and Assanto G 2006 *Nat Phys* **2** 737–742

Catadioptric lenses in Visible Light Communications

J Garcia-Marquez[1,3], J C Valencia[2], H Perez[3], and S Topsu[3]
[1]*Laboratoire Commun de Métrologie LNE/CNAM, 1 rue Gaston Boissier, 75724, Paris, France*
[2] *Centro de Investigaciones en Optica, Loma del Bosque 115, 37150, León, Mexico*
[3] *OLEDCOMM-Lize, 10-12 Avenue de l'Europe 78140, Vélizy, France*

E-mail: jorge.garcia_marquez@cnam.fr

Abstract. Since few years ago, visible light communications (VLC) have experience an accelerated interest from a research point of view. The beginning of this decade has seen many improvements in VLC at an electronic level. High rates of transmission at low bit error ratios (BER) have been reported. A few numbers of start-ups have initiated activities to offer a variety of applications ranging from indoor geo-localization to internet, but in spite of these advancements, some other problems arise. Long-range transmissions mean a high BER which reduce the number of applications. In this sense, new redesigned optical collectors or in some cases, optical reflectors must be considered to ensure a low BER at higher distance transmissions. Here we also expose a preliminary design of a catadioptric and monolithical lens for a LI-FI receiver with two rotationally symmetrical main piecewise surfaces z_a and z_b. These surfaces are represented in a system of cylindrical coordinates with an anterior surface z_a with a central and refractive sector surrounded by a peripheral reflective sector and a back piecewise surface z_b with a central refractive sector and a reflective sector, both characterized as ideal for capturing light within large acceptance angles.

1. Introduction

Visible Light Communications (VLC) or LiFi (for Light Fidelity) is a relatively new technology that will complement HiFi communications and is intended to be used in places where HiFi interferes with EM sensible equipment. For instance we can name airplanes, hospitals, and places that will remain as radio frequency (RF) free communication environments. This technology can, in a certain measure, relive pressure on the saturated RF broadband and can also be used in places where communications must be secured. VLC can benefit from the huge network of LED based lamps worldwide as LED lamps are found in street illumination, buildings, homes, street fires, airplanes and vehicles. Applications ranges from downlink –where the emitter lamp broadcast information and the receiver only can download this information– to bidirectional communications where both the lamps and mobile devices are equipped with transmitter (Tx) and receiver (Rx) units to uplink and downlink purposes. Examples of the first are train stations or museums where light can transmit information while examples of the last are the communication between the light of two vehicles or the communication between cars and road markers or street lights; internet or machine to machine (M2M) communications.

VLC is passing from a development stage to a commercial stage; the number of publications can assess its increasing relevance. Nevertheless, the optical issues have not been taken into consideration yet. As a consequence the optical elements used to spread, direct of concentrate the light from LEDs or the light collectors in photodiodes should be redesigned for they to become useful not only for illumination but also for communications. The new optic elements will accomplished different tasks for different applications, *i.e.* for some configurations the light must cover a broad cone while for others a sharp cone is the better solution. In the first case, we can think for example, in a large space in a metro or train station where a powerful LED based lamp illuminates the platform as well as at the same moment it spread information concerning traffic, arrivals or ambience music. In the second case, lamps in a museum can illuminate an artwork or a masterpiece as well as spread useful concerning information on any mobile device. Here, the illumination cone will meet design specifications provided by museum specialists. From the receiver point of view, the same questions can be made. Probably the most important question is related to the use and applications that this technology will serve. Is it better to design multipurpose lenses, surely rendering expensive this technology, or to design specific purpose lenses that will reduce production costs? The same question is valid for reflectors in transmitters. It can be seen that a designer has several restrictions to consider and they will meet low costs, illumination purposes and data communications transfer.

From the electronics point of view, a transmission rate exceeds 1 Gb/s [1,2] at an ambience illumination up to 1200 lux. Typical distance between the transmitter (Tx) and the receiver (Rx) is 1 m, nevertheless distances up to 3 m have been reported [1]. Bi-directional communication with RGB LEDs for downlink and phosphor based LEDs for uplink, of some hundreds of Mb/s have recently been reported [3]. Different schemes have been also proposed to transmit a downstream at wide field-of-view [4]. Multi-Gb/s parallel data transmission using CMOS-controlled micro-LEDs will surely be a reality in the near future [5]. High transmission rates, broader field-of-transmission, longer transmission distances, robustness against ambience illumination are the main challenges of VLC. On-off keying modulation (OOK) is, by far, the most simple modulation scheme used in VLC. It consists in turning on and off the LED depending on the bit value where off is the logic state 0 and on the state 1. The IEEE 802.17.7 standard offers three physical (PHY) modulation schemes for VLC [6]. According to it, the PHY II operating mode in OOK modulation scheme has a data rate of 96 Mb/s.

In this article, we will refer to this modulation scheme to show a bi-directional uplink/downlink system to transmit Ethernet at 96 Mb/s. The maximum measured distance is 3 m when the environmental luminous intensity measured on the test table equals 3,000 lux (2,400 lux on the detector). LEDs used in the transmission are red (650 nm peak emission wavelength). Design considerations have taken into account this electronics choice.

2. Electronics

An Ethernet network is connected via an Ethernet-to-VLC transceiver (figure 1) to the LED driver for transmission, switching the LED at a cut-off frequency of 70 MHz.

2.1. Receiver

In reception mode a photodetector if fully covered by the LED beam collected by a concentrator lens. This receiver module consists in a silicon PIN photo-diode mounted with a silicon bipolar trans-impedance preamplifier integrated circuit (IC) in an optical subassembly (OSA). This OSA is mated to a custom silicon bipolar circuit that provides post-amplification and quantization. The module utilizes an InP PIN photodiode in the same configuration. The post-amplifier also includes a signal detection circuit that provides a positive emitter-coupled logic (PECL) output upon detection of a usable input optical signal level.

2.2. Transmitter

The transmitter section of the system consists of seven 650 nm high speed LEDs in an optical subassembly (OSA) for the uplink, and a lamp made of seven 650 nm high speed LEDs for the downlink, which mates to the lens. The OSA is driven by custom, silicon bipolar IC which converts differential PECL logic signals (ECL referenced to a +5 Volt supply) into an analogue LED diode drive current.

The LED Driver is a high speed LED Driver/current switch specifically targeted for use Fibre Channel 266 Mb/s optical transmitters. The device incorporates open collector outputs with a capability of driving peak currents of 100 mA. Since the output current switching circuitry simply switches current between the complementary outputs, the dynamic switching demands on the system power supply are greatly reduced. In addition, because the design is pure bipolar, the device current drain is insensitive to the data pattern and frequency of operation. In addition to this, to allow for open loop compensation for the LED's negative optical output power tracking over temperature, a circuit is included to provide an adjustable positive temperature tracking coefficient to the LED drive current. The driver is intended either to be integrated into high performance fiber optic modules or to be used stand-alone to drive a packaged optical LED device. The wide frequency response of the device allows it to be used to support 100 Mb/s that permits us to operate in PHY II.

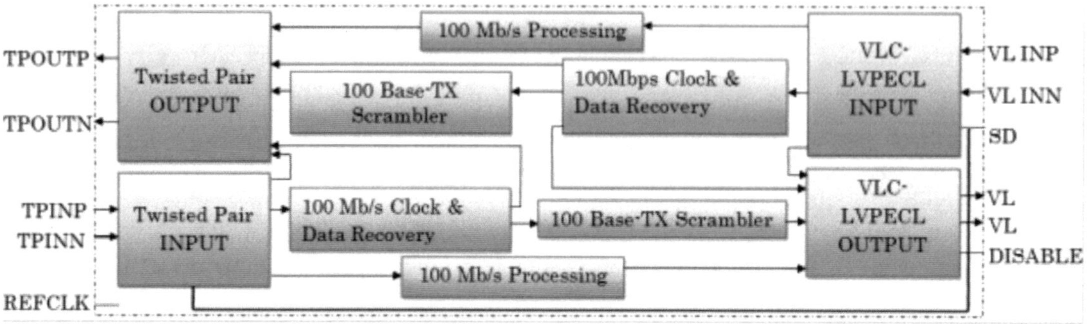

Figure 1. Ethernet to VLC transceiver.

3. Optical Parameters

There are several restrictions when designing optical systems for VLC. The most important ones are related with the cost of fabrication, the size, the weight and the particular application intended for. Here we concentrate our efforts in the design of a collector lens that permits the photodiode the correct interpretation of the transmitted signal. The light transmitted by a LED lamp carries the modulated signal; this light beam has a distribution typical from LED illumination lamps. As a consequence, the signal will be degraded rapidly with distance and with angle and this degradation will eventually induce errors in the receptor expressed as bit losses. Even if BER is a central concern for electronic engineers, optics will offer a low-cost solution to its reduction. Our proposal [7] consists in collecting the light reflected by walls, windows or objects located around the receiver by a grand acceptance angle collector, which should reduce BER. It means that the light and signal emitted will better arrive onto the photodetector.

The optical system will not display an image but uniformly distribute the intensity at the entrance pupil; it will have a huge acceptance angle α and low internal (or external) reflection loses. There are several possible solutions to obtain a low cost optical system. Yet another lens parameter consists in coupling the lens to a new generation mobile device such as a mobile telephone or ultra-compact computer devices. Ergonomic and functionality condition the lens to have a flat first surface and small dimensions even at the trade of reducing the intensity collection surface.

The optical system is only intended for capturing light and not to form an image; it has to be similar to a Galilean or Keplerian telescope, with or without a field lens, with an inverted configuration –

absolute angular augmentation or magnification |M| less than the unit or absolute reduction m = 1/|M|. It can or cannot invert the image, and it will have preferably a high acceptance degree α –the angular semi field of a field-angular lens. A uniform distribution of intensity in the entrance that permits the system to obtain a uniform intensity distribution at the exit is desired; a system that minimizes internal reflexions or refractions is then preferred. The coupled diaphragm is set at the first surface. As there are lots of possible solutions, it will always be possible to find one with good specifications and not expensive.

There are many examples of optical designs using condenser lenses with highly refractive power that can be either spherical or aspherical; usually they are hemispheres or spheres. Since our main purpose is to design an optical system that can be easily linked with portable devices, like cellular phones and portable computers, the following restrictions can be specified: For ergonomic and functional reasons, the first surface must be flat, the total length of the system and the volume must be minimal and finally, its production cost very low.

Proposal

Based on several concepts presented previously in [7-8], we show here a design of a catadioptric and monolithic optical system, assuming the object is at infinity, which form an non-magnified and inverted image, with multiple applications like: magnifiers, illumination systems, surveillance systems, solar collectors without heliostat and light detectors, and antennae, among many more.

The catadrioptic-monolithic system is composed by a first lenticular flat parabolic surface of revolution in a cylindrical coordinate system (x, z) having itself two sections: The first one is flat, occupy the center of the surface and is perpendicular to the optical axis z and has a diameter d_a. The second is peripheral (or marginal) and corresponds to an annular section of a revolution parabola around the optic axis, with a positive focal distance f_a. Thus, the first two sections of this surface –the central-flat and a the peripheral-convex-parabola– have a representation in the meridional plane that corresponds to

$$z_a = \begin{cases} 0, & \forall x \backslash x^2 \le \dfrac{d_a^2}{4} \quad \text{refractive} \rightarrow \quad n_a = n > 1. \\ \dfrac{x^2}{4f_a} - t_a, & \forall x \backslash x^2 > \dfrac{d_a^2}{4} \quad \text{reflective} \rightarrow \quad n_a = -1. \end{cases} , \tag{1}$$

the vertex of the parabola is placed at an axial distance t_a from the coordinate's origin. The flat central sector corresponds to a refractive interface towards a medium with a relative refraction index $n_a = n$ greater than the unit, and the parabolic-peripheral sector corresponds to a reflective interface, that is equivalent to $n_a = -1$, notation usually used for reflective optical systems.

As an approximation to the final design, we assume a total central thickness t, where a second surface is created, lenticular as well, flat-parabolic, concentric and inverted with respect to the first surface, and can be specified within the same coordinated system with a two-section function: The first is located in the central region, is flat and is perpendicular to the optical axis z at the point $(0, t)$ and has a diameter d_b; the second is peripheral and corresponds to an annular-section of another revolution-parabola with a negative focal distance f_b; thus, the second surface has a representation on the meridional plane corresponding to

$$z_b = \begin{cases} t, & \forall x \setminus x^2 \leq \dfrac{d_b^2}{4} \quad \text{refractive} \rightarrow \quad n_b = \dfrac{1}{n} < 1. \\ t + t_b + \dfrac{x^2}{4f_b}, & \forall x \setminus x^2 > \dfrac{d_b^2}{4} \quad \text{reflective} \rightarrow \quad n_b = -1. \end{cases} , \tag{2}$$

The parabolic surface is placed at an axial distance t_b from the point (0, t) to the point (0, $t + t_b$). The flat-central sector corresponds to a refractive interface towards a medium with a relative refractive index equivalent to $n_b = -1$.

Since both parabolas are opposite they must intercept at a point in space, limiting the frontier of the mirror-lens that makes it possible to form a monolithic optical system; thus, the maximum distance d of the monolithic bi-convex-bi-plane system can be determined by equaling sections in Eqs (1) and (2) evaluated at the semi diameter $d/2$:

$$\frac{d^2}{16f_a} - t_a = t + t_b + \frac{d^2}{16f_b}, \tag{3}$$

with a solution for d

$$d = 4\sqrt{\frac{f_a f_b (t_a + t + t_b)}{f_b - f_a}}. \tag{4}$$

By assuming that the focus of the parabola of the second surface is found in the coordinate origin of the catadioptric-monolithic system (refractive surface of entry), the next condition must be satisfied

$$-f_b = t + t_b, \tag{5}$$

given that f_b is negative. Moreover, by making the parabolic focus of the first surface coincide with the point (0, t), the following condition must also be accomplished.

$$f_a = t_a + t \tag{6}$$

The sagittae of the parabola can be calculated as a function of the cord diameters that corresponds to the diameter of the refractive windows, with

$$t_a = \frac{d_a^2}{16f_a} \tag{7}$$

and

$$t_b = -\frac{d_b^2}{16f_b}. \tag{8}$$

Substituting Eqs. (7 and 8) in Eqs. (5 and 6), and solving the quadratic equations for the valid focal distances it is easy to obtain

$$f_a = \frac{2t + \sqrt{d_a^2 + 4t^2}}{4}, \quad (9)$$

and

$$f_b = -\frac{2t + \sqrt{d_b^2 + 4t^2}}{4}, \quad (10)$$

given that $f_a > t/2$ and $f_b < t/2$.

Thus, according to standard ISO [9], the parabolas can be represented using the canonical Schwarzschild's formula, with a conical constant $K = -1$ and curvatures in the vertex $c_a = 1/(2f_a)$ and $c_b = 1/(2f_b)$.

Substituting Eqs. (9-10) in Eqs. (7-8) the sagittae can be represented as a function of de diameters of the refractive windows diameter

$$t_a = \frac{d_a^2}{4\left(2t + \sqrt{d_a^2 + 4t^2}\right)} = \frac{-2t + \sqrt{d_a^2 + 4t^2}}{4}, \quad (11)$$

and by symmetry

$$t_b = \frac{d_b^2}{4\left(2t + \sqrt{d_b^2 + 4t^2}\right)} = \frac{-2t + \sqrt{d_b^2 + 4t^2}}{4}. \quad (12)$$

The representation with the piecewise function of the two surfaces given by equations (1-2) can also be expressed as a function of the windows diameter and the central thickness. Substituting Eqs. (11-12) in Eqs. (1-2), rationalizing and simplifying, to obtain

$$z_a = \begin{cases} 0, & \forall x \backslash \ x^2 \le \dfrac{d_a^2}{4} \quad \text{refractive} \rightarrow \ n_a = n > 1. \\ \left(2t - \sqrt{d_a^2 + 4t^2}\right)\left(\dfrac{1}{4} - \dfrac{x^2}{d_a^2}\right), & \forall x \backslash \ \dfrac{d_a^2}{4} < x^2 < \dfrac{d^2}{4} \quad \text{reflective} \rightarrow \ n_a = -1. \end{cases} \quad (13)$$

and

$$z_b = \begin{cases} t, & \forall x \backslash \ x^2 \le \dfrac{d_b^2}{4} \quad \text{refractive} \rightarrow \ n_b = \dfrac{1}{n} < 1. \\ t - \left(2t - \sqrt{d_b^2 + 4t^2}\right)\left(\dfrac{1}{4} - \dfrac{x^2}{d_b^2}\right), & \forall x \backslash \ \dfrac{d_b^2}{4} < x^2 < \dfrac{d^2}{4} \quad \text{reflective} \rightarrow \ n_b = -1. \end{cases} , \quad (14)$$

Figure 2 shows an example of this design, showing the preset values for a productive process.

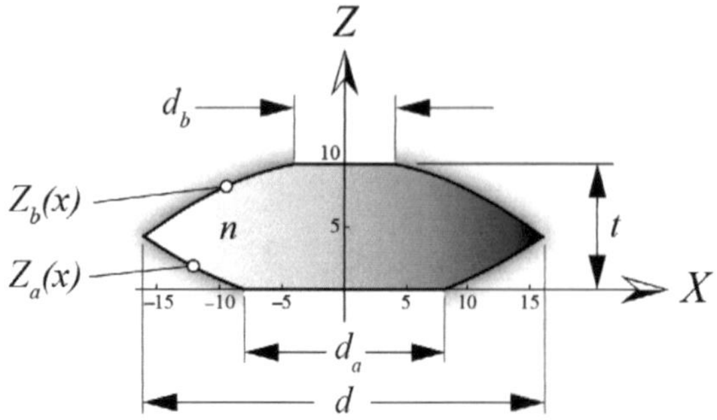

Figure 2. Here we show a catadioptric monolithic bi-flat bi-parabolic lens, with the parabolic sectors to be metalized with a thin layer of reflective metal, usually made of silver, aluminum, gold or platinum. The entrance window diameter corresponds to the diaphragm's diameter coupled to the first flat surface d_a = 16 mm, the exit window diameter d_b = 8 mm and the central thickness t = 10 mm. The total diameter is d = 31.0179 mm.

Figure 3 shows the ray tracing to show that the lens has image forming properties that are not dependent of the refractive index of the lens, and that it presents zero spherical aberration, minimal comma at the edge of the field, and low field curvature. The object is assumed to be flat and the image plane coincides with the entrance and exit windows respectively.

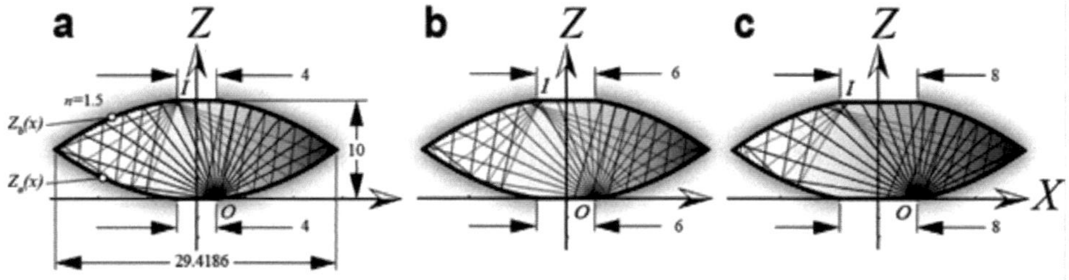

Figure 3. The tracing of rays in three catadioptric monolithical bi-flat bi-parabolic lenses is shown for the following parameters: Equal diameters in the entrance and exit windows $d_a = d_b$ = 4, 6 and 8 mm for lenses (Figs. 3a-3c respectively), and central thickness t = 10 mm. the total calculated diameter is d = 31.018 mm. The trace of rays that depart from an object point O is shown on the border of the bottom window's field, to try to form an image I in the upper window. One can also observe that the meridional comma aberration grows when the proportion $d_a / t = d_b / t$ increases. For rays to emerge rapidly from the top window, it is convenient to cut off the upper plane that corresponds to the lower parabolic focal plane with defocus; thus, diminishing the noise in the detector and avoiding lens heating. All dimensions are in millimetres. No magnification is observed, but there is an apparent magnification increase due to the fact that the detector seems to be closer to an observer.

Thus, analytically, it can be inferred that:

1. The proposed catadioptric design with lenticular flat-parabolic surfaces forms an inverted image with almost the same size, regardless of its windows' diameter ratio d_b/d_a.
2. That radiation losses by reflected rays decrease to a minimum when there's no magnification.

Journal of Physics: Conference Series **605** (2015) 012029 doi:10.1088/1742-6596/605/1/012029

3. That radiation losses by reflected rays decrease when thickness increases.
4. A thick lens is necessary to avoid lens heating and noise in the detector.
5. That the image does not depend on the lens refractive index.
6. Aberrations are small.

In figure 4, we show a solution that can be used in signal detections or as a LiFi collector assuming, an object far away or at infinity. Another advantageous property of this proposed system can be observed: It drive collimated beams towards the exit window over the entry window with extreme acceptance angles. Here only the frontal (bottom) part of the lens is visible in mobile apparatus. The visible part of the lens is flat and can be touched. Great acceptance angles will permit the collector to trap light reflected by reflecting surfaces in the vicinity of the photodetector rendering, in this way, the receiver more robust and decreasing the BER at the same time.

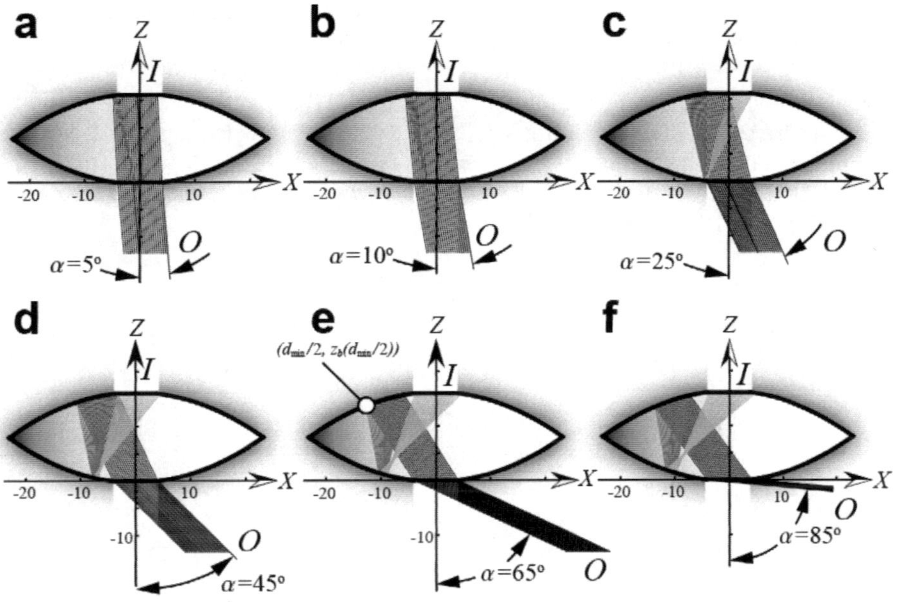

Figure 4. Wide acceptance angle receiver-lens. Here we show the tracing of rays up to the second reflection in six catadioptric monolithic bi-flat-parabolic lenses with an external object in infinity, with different acceptance angles α = 5, 10, 25, 45, 65 and 85 degrees; 4a-4f, respectively. Incoming rays enter from bottom to top. Prescribed parameters for all the lenses: detection zone diameter d_b = 8 mm, entrance window diameter d_a = 8 mm, central thickness t = 16 mm and refractive index n = 1.7. The total preliminary diameter was calculated with d = 46.298 mm. The tracing of rays for a collimated beam that comes from an object O in infinity (upwards) try to form an image *I* in the upper window. The tracing of emerging rays is suppressed because we assume a detector in direct contact with the exit window. A small loss of intensity is observed when $\alpha > 25°$.

4. Conclusions

In this article we have shown the perspectives of applications of VLC systems and the electronics for a state-of-the-art Ethernet LED-based commercial system known as VLC or LiFi. Our system enters now in production and is robust in indoor ambience. We have also introduced to a new incoming optical design technology that will be charged to reduce BER in VLC systems as well as given some optical receiver wide acceptance angle examples to reduce it. The lens here briefly describes is a part of a complete family of new great acceptance angle, compact and low fabrication cost collecting lenses to be used in photoreceivers for LiFi application. A similar design approach can be also used to design optical reflectors for several different applications of LiFi technology. We hope this article will motivate optical designers as well as specialists in photometry and radiometry to contribute to this emerging technology.

5. References

[1] Le Minh H, Brien D O, Faulkner G, Bouchet O, Wolf M, Grobe L, and Li J 2010 *IEEE Photon. Technol. Lett.* **22** 1598.
[2] Azhar A H, Tran T, and Brien D O 2013 *IEEE Photonics Technol. Lett.* **25** 171.
[3] Wang Y, Wang Y, Chi N, Yu J, and Shang H 2013 Opt. Express **21** 1203.
[4] O'Brien D, Turnbull R, Le Minh H, Faulkner G, Bouchet O, Porcon P, El Tabach M, Gueutier E, Wolf M, Grobe L, and Li J 2012 *J. Light. Technol.* **30** 2181.
[5] Mckendry J J D, Massoubre D, Zhang S, Rae B R, Green R P, Gu E, Henderson R K, Kelly A E, and Dawson M D 2012 J. Lightwave Technol. **30** 61.
[6] Rajagopal S, Roberts R D, and Lim S-K, 2012 *IEEE Commun. Mag.* **50** 72.
[7] Valencia Estrada J C, García-Márquez J, Malacara D. 2014 Lentes catadióptricas y monolíticas y método para producirlas Patent Instituto Nacional de la Propiedad Industrial, Mexico (2014).
[8] Valencia-Estrada J C, Bedoya-Calle A H and Malacara D 2013 *J. Opt. Soc. Am.* A **30** 1814.
[9] ISO 10110-12:2007 Optics and photonics–"*Preparation of drawings for optical elements and systems Part 12: Aspheric surfaces.*"

Measurement of Length and Position with Frequency Combs

C Weimann[1,2*], F Hoeller[3], Y Schleitzer[3], C A Diez[3], B Spruck[3], W Freude[1,2], G Boeck[4], C Koos[1,2*]

[1]Institute of Photonics and Quantum Electronics, Karlsruhe Institute of Technology (KIT), Karlsruhe, Germany
[2]Institute of Microstructure Technology, KIT, Karlsruhe, Germany
[3]Corporate Research and Technology, Carl Zeiss AG, Oberkochen, Germany
[4]Microwave Engineering Lab., Berlin Institute of Technology, Berlin, Germany

*Email: claudius.weimann@kit.edu, christian.koos@kit.edu

Abstract. We show two different absolute distance measurement methods with micrometer accuracy based on frequency combs, and we discuss possible applications. Using a mode-locked laser and MEMS-based tracking optics, we measure the 3D position of a retroreflector within 10 ms and with a 24 μm volumetric accuracy. We also investigate modulator-based combs and show that they enable highly sensitive surface topography measurements with microsecond acquisition times and micrometer precision. Moreover, the potential for photonic integration of frequency comb sources is explored.

1. Introduction

Dimensional metrology such as measurement of distance and position in three-dimensional (3D) space are of great importance in many scientific fields as well as in industrial applications such as inline inspection, robotics and machinery assembly. Optical 3D position measurement systems can be used for large-scale or inline measurements, ideally in combination with an optical sensor head to enable non-contact, fast and precise investigation of object surfaces. In this paper, we will present the use of optical frequency combs to fulfil these measurement tasks, first discussing the optical measurement of 3D coordinates and then the realization of an optical surface sensor.

Different approaches for an optical position measurement have been taken in the past, such as laser trackers [1, 2], photogrammetry [3, 4], camera or diode-based triangulation [3, 5, 6] and multilateration [7–11]. Two main parameters for such systems are accuracy and measurement range. This parameter space is sketched in Figure 1 for the different measurement principles employed. In terms of accuracy, only schemes based on laser trackers, measuring two angles and one distance, and trilateration or multilateration schemes, measuring at least three distance or distance differences, can potentially compete with the classical coordinate measurement machines (CMM) based on serial kinematics, where a high-precision actuator is used for each degree of freedom. However, while the accuracy of laser trackers is mainly limited by the angular accuracy, multilateration and trilateration schemes can achieve a measurement uncertainty similar to that of the underlying length measurement system, depending on the configuration of the measurement geometry [8].

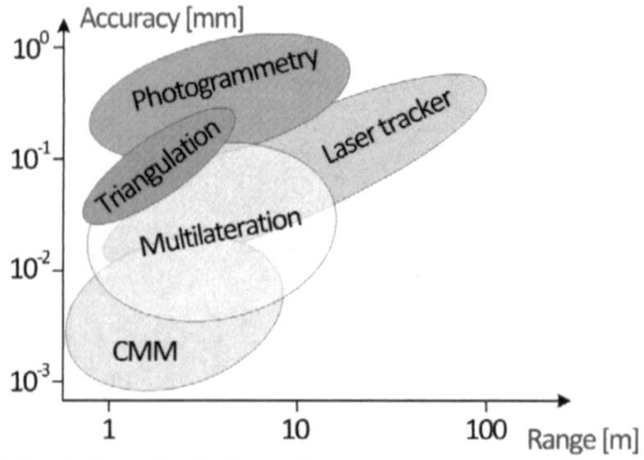

Figure 1. Accuracy and range of classical mechanical coordinate measurement machines (CMM) in comparison to different optical measurement principles, adapted from [3].

While many of the works on multilateration employ single-wavelength heterodyne laser interferometers with sub-µm distance measurement uncertainties [9, 10, 12, 13], the practical applicability of this approach is limited. The interferometric distance measurement has to be incremental, and no absolute distances can be measured. This forbids any interception of the line-of-sight to the retro-reflector target. A measurement principle capable of measuring absolute distances is of key importance for applications in an industrial environment. In this context, optical frequency combs as a metrological tool for precise and absolute distance measurements have received a lot of attention in the last years [14–20]. However, these works concentrate on distance measurements, and no 3D position measurements have been demonstrated so far.

In this paper, we demonstrate 3D position measurements based on trilateration using optical frequency combs. The distances to the retroreflector are absolutely measured using the 240th and 241th harmonic inter-mode beats of the frequency comb formed by a femtosecond pulsed laser with 100 MHz repetition rate. Comparing length measurement results of our system to those obtained from a commercial interferometer, we achieve a length measurement accuracy of 2.2 µm with an integration time of 10 ms. For the used triangular arrangement of three measurement modules spaced by 200 mm this results in a volumetric accuracy of 24 µm. Target tracking is based on MEMS mirrors and four-quadrant diodes. Furthermore we present developments towards a high-sensitivity surface topography sensor employing modulator-based frequency combs with the potential for photonic integration. We demonstrate distance measurements to natural scattering surfaces with acquisition times of 8.3 µs and accuracies better than 10 µm. These data are maintained over a dynamic range of more than 37 dB. Modulator-based frequency comb sources are well suited for photonic integration, which is demonstrated using the silicon-organic hybrid (SOH) integration platform.

2. Distance measurement with inter-mode beats of a frequency comb

The distance measurement scheme employed here is based on the phase evaluation of inter-mode beats of a frequency comb, as first published by Kaoru Minoshima in 2000 [18]. Direct detection of light from a mode-locked laser (MLL) with a high-bandwidth photodiode leads to an electrical signal comprising a multitude of harmonic beat frequencies at integer multiples of the repetition rate. The stability and accuracy of these higher-order harmonics is directly linked to the pulse repetition rate, which can be measured and compared to a frequency standard. Thus, the broad spectrum of a femtosecond pulse laser offers a convenient access to accurate high-frequency signals.

In this work we use a MLL with a repetition rate of 100 MHz and a 3 dB bandwidth of 10 nm centered at a wavelength of 1556 nm. The experimental setup is depicted in Figure 2. Light from the MLL is split using a fiber coupler. One part is incident on a reference photodiode (PD) with a bandwidth of 40 GHz, the other part is directed to a free-space path towards a retroreflector. The reflected light is guided via a circulator to a measurement PD with identical specifications as the reference PD. A heterodyne scheme with two intermediate frequency stages is used to evaluate the phases of the 240th and 241th harmonic at 24 GHz and 24.1 GHz simultaneously. After pre-amplification and bandpass filtering, the lines of interest are mixed with a 24.04 GHz local oscillator (LO), which is phase-locked to the repetition rate of the MLL. This generates the first intermediate frequencies at 40 MHz and 60 MHz, which maintain the phase relations of the original signals. To evaluate the phases of both lines in reference and measurement paths, the signals are split and the individual lines are filtered by additional bandpass filters, which suppress the corresponding unwanted frequency. Additional LO with frequencies of 39.875 MHz and 39.9 MHz mix the signals to second intermediate frequencies at 100 kHz and 125 kHz. These signals are then analog-to-digital converted, and the phase differences are evaluated in the digital domain.

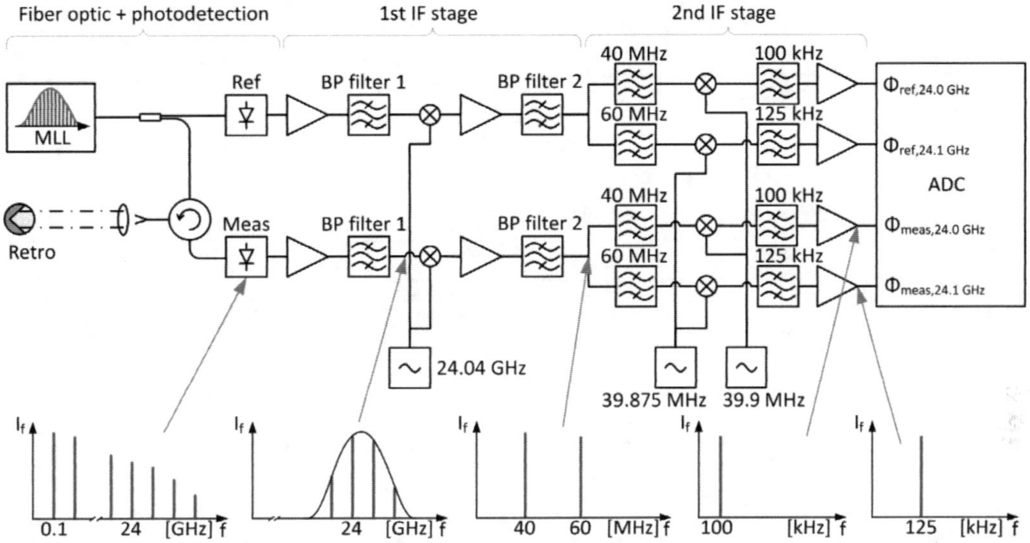

Figure 2. Schematic overview of distance measurement unit and electrical signal spectra (below). Light from a mode-locked laser (MLL) is split in a fiber coupler. One part is detected directly by a reference photodiode (Ref), the other part propagates to the retroreflector (Retro) before it is incident on a measurement photodiode (Meas). The generated electrical beat signals at multiples of the MLL repetition rate are amplified, the 240th and 241th harmonics are band-pass filtered and mixed to first intermediate frequencies of 40 MHz and 60 MHz with a local oscillator (LO) at 24.04 GHz. Further band-pass filters isolate each line, before second intermediate frequencies at 100 kHz and 125 kHz are generated by mixing with a second LO. These signals are analog-to-digital converted (ADC), and the phase differences of reference and measurement signals are evaluated in the digital domain.

The phases $\varphi_{240\text{th}}$ and $\varphi_{241\text{st}}$ of the signals at $f_{240\text{th}} = 24.0\,\text{GHz}$ and $f_{241\text{st}} = 24.1\,\text{GHz}$ are measured relative to the reference signal. The optical path length difference $2D$ can be directly calculated from these phases, although the 2π periodicity of the phase limits the unambiguity range of the measurement:

$$\varphi_{240\text{th}} = \left(2\pi \frac{2Df_{240\text{th}}}{c}\right) \bmod 2\pi = \left(2\pi \frac{2D}{\lambda_{240\text{th}}}\right) \bmod 2\pi \tag{1.1}$$

The free-space wavelength of the evaluated beat signal is $\lambda_{240\text{th}} = 12.5\,\text{mm}$. To extend the absolute measurement range of this fine-scale measurement, the phase difference $\varphi_{241\text{st}} - \varphi_{240\text{th}}$ is evaluated as well:

$$\varphi_{241\text{st}} - \varphi_{240\text{th}} = \left(2\pi \frac{2D\left(f_{241\text{st}} - f_{240\text{th}}\right)}{c}\right) \bmod 2\pi = \left(2\pi \frac{2D}{\Lambda}\right) \bmod 2\pi \tag{1.2}$$

This large-scale measurement with a wavelength $\Lambda = \frac{c}{f_{241th} - f_{240th}} = 3\,\mathrm{m}$ extends the absolute measurement range significantly. An absolute distance measurement can be made even if the line-of-sight to the retroreflector has been intercepted temporarily.

The measurement results can be seen in Figure 3. The repeatability of the measurement and the influence of the measurement rate were evaluated by repeatedly measuring the distance to a fixed target for different integration times per measurement. The standard deviation is as low as 0.5 µm for an integration time of 50 ms, and fast measurements with update rates of 10 kHz are possible while standard deviations remain in the single-µm regime. The absolute accuracy of the distance measurement was evaluated by comparing to a commercial heterodyne laser interferometer with a specified resolution of 20 pm (Sios MI 5000). The target was incrementally moved along a line with a length of 1 m while recording the relative displacement with the interferometer and simultaneously measuring the distance with our system. The difference in measurement values is plotted in Figure 3(b). The standard deviation of the measurement differences amounts to 2.2 µm. We attribute the observed systematic deviations to residual phase nonlinearities.

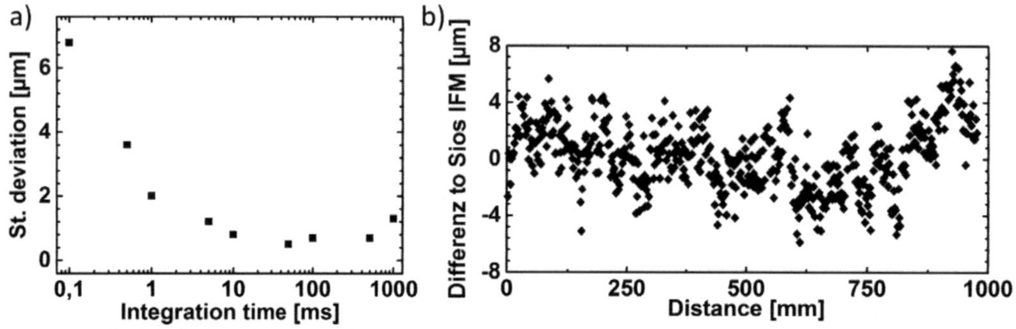

Figure 3. a) Standard deviation of repeated distance measurements to a fixed target as a function of integration time. For integration times of 50 ms, the standard deviation is as low as 0.5 µm. b) Differences of measurement results with 10 ms integration time compared to a commercial HeNe laser interferometer (Sios IFM) for a distance of up to 1 m. The standard deviation of the measurement differences amounts to 2.2 µm.

The optical part of the distance measurement setup relies mainly on comparatively cheap telecommunication-grade fiber-optic components. With the use of fiber couplers and splitters, the light of the single MLL with an optical power of 16 dBm can be easily distributed to three identical receiver units. Combined with MEMS-based tracking optics, this opens the way to measure the 3D-position of a retroreflector by measuring three distances in a trilateration-based approach.

3. µGPS position measurement by trilateration

Many approaches for 3D localization reported in literature use a minimum of four stations and relative distance measurements, e.g. by using single-wavelength interferometry. The redundant number of stations enables a self-calibration with respect to the relative spatial relationship of the stations [9–11]. Such a configuration also allows to determine the initial offset distance to the retroreflector, which is especially important for schemes employing relative distance measurements. However, an interception of the line-of-sight requires a recalibration to again determine the current offset distances to the retroreflector, which have to be added to the measured relative distances. In contrast to these works, we employ a trilateration scheme using only three absolute distance measurements. Our system is calibrated using a similar algorithm as employed in the Global Positioning System (GPS), measuring the distances to a number of known retroreflector positions [21].

Our system comprises only three distance measurement modules with the capability of tracking a retroreflector which are combined in a triangular configuration as depicted in Figure 4(b). The spacing of the modules was chosen to be 200°mm along the sides of the triangle to realize a relatively compact measurement unit. Larger module spacing would result in higher measurement precision at the cost of

Journal of Physics: Conference Series **605** (2015) 012030 doi:10.1088/1742-6596/605/1/012030

a bulkier experimental setup. To calibrate the distance measurement offsets $l_{0,j}$ as well as the relative coordinates (x_j, y_j, z_j) of the pivot points of each module j, the distances $d_{i,j}$ to a number i of retroreflector positions with known coordinates (x_i, y_i, z_i) are measured for each module. To obtain a good precision, in practice at least 10 different known retroreflector positions are used to obtain pivot point coordinates and the distance measurement offsets. This corresponds to a least-square solution of an over-determined set of equation The wanted parameters are obtained by a minimization:

$$S_j = \sum_i \left| \sqrt{(x_i - x_j)^2 + (y_i - y_j)^2 + (z_i - z_j)^2} + l_{0,j} - d_{i,j} \right|^2$$
$$S_j \to \min \qquad (1.3)$$

Experimentally, this is realized by attaching the reflector to a high-precision CMM (Zeiss Eclipse). Once the system is calibrated, three distance measurements are enough to determine the 3D coordinates of the retroreflector, as a possible second solution can be omitted by the known physical layout of the system configuration [22,23].

The tracking of a retroreflecting target while it is moved through space was realized with a MEMS mirror and a four-quadrant (4Q) photodiode. The optical setup is sketched in Figure 4(a). The beam from the fiber collimator is steered by a MEMS-actuated mirror with a mechanical angular range of ±5.5°. To realize a larger tracking volume, angle enlargement optics are used to realize an optical beam deflection of ±30° around a pivot point. The target-tracking feedback loop processes the output of a 4Q photodiode, which is illuminated by using a beam splitter in the optical path.

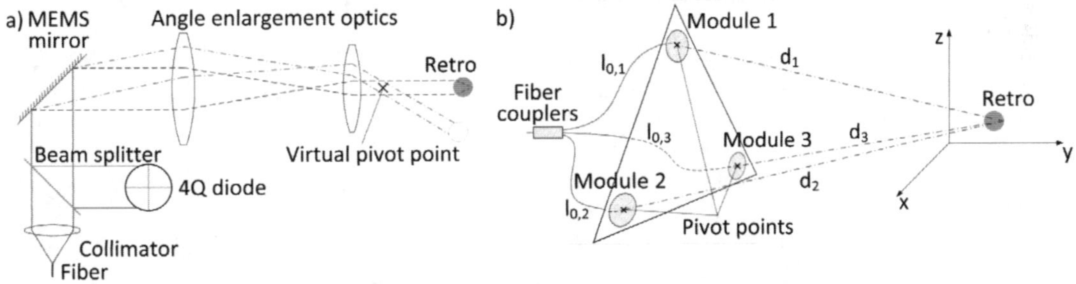

Figure 4. a) Schematic of optical setup for one module. The target is tracked with a four-quadrant photodiode (4Q) and a MEMS mirror. To realize a scanning range of ±30°, optical elements enhance the ±5.5° angle of the MEMS mirror. The beam is mainly deflected around a single pivot point. b) Schematic overview of the GPS-like coordinate measurement. Three fully functional distance measurement modules are arranged in a triangular geometry. They are fed by the same laser source and track a single retroreflector. Coordinate measurements are possible after a calibration, determining the relative coordinates of pivot points and distance offsets $l_{0,i}$, given by the length of the fibers that connect each of the modules.

To characterize the system's accuracy, the same high-precision CMM used for calibration is used to position the target at different locations within a field of 350 × 450 mm^2 roughly 1 m away from the modules. The 3D coordinates are calculated based on the three measured distances d_j acquired with 10 ms integration time. The results are shown in Figure 5. The deviations of the measured coordinates to the set coordinates are plotted as a histogram for X,Y and Z. The measurement geometry and the definition of the axes are the same as defined in Figure 4(b). It is not unexpected that the deviation along the Y axis is the smallest with a standard deviation of 4.4 µm, as the compact module configuration leads to a geometry tending to collinearity for the measurement volume. The measurement standard deviation along X and Z axes amounts to 16…17 µm, leading to a volumetric accuracy of $\sqrt{(16.1\mu\text{m})^2 + (4.4\mu\text{m})^2 + (17.5\mu\text{m})^2} = 24.1\mu\text{m}$. These values fit well to a simulation of the measurement geometry with an estimated distance measurement error of 2.2 µm. The simulation leads to a standard deviation in Y direction of 3.4 µm and to standard deviations of 15.0 µm in X, Z

Journal of Physics: Conference Series **605** (2015) 012030 doi:10.1088/1742-6596/605/1/012030

directions. If different geometrical configurations are used, e.g., by placing the modules farther apart, also better accuracies can be achieved [8].

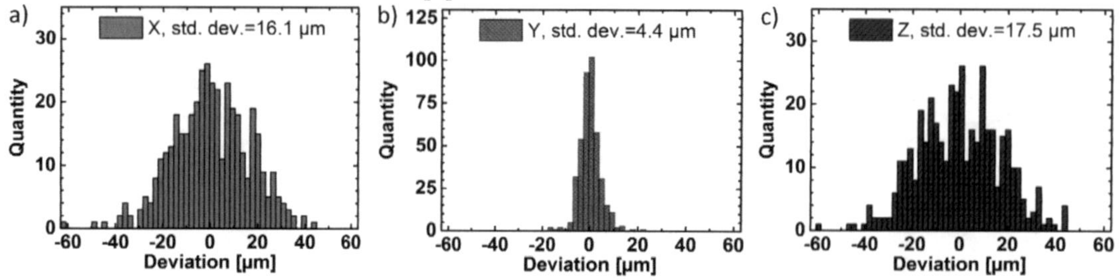

Figure 5. Deviations of optically measured coordinates to set coordinates of a CMM. The measurement geometry and the coordinate axes correspond to those depicted in Figure 4. Values along Y exhibit a smaller variance with a standard deviation of 4.4 μm, while standard deviations along X and Z amount to ~17 μm, resulting in a volumetric error of 24 μm, close to expected values for the chosen configuration with a 200 mm spacing between modules.

4. Modulator-based frequency combs for multi-heterodyne surface topography measurements

Frequency combs generated by mode-locked femtosecond lasers (MLL) typically provide a broadband spectrum with bandwidths of tens of nanometers. Combined with direct detection of the inter-mode beat signals as demonstrated in the sections above, accurate absolute distance measurements are possible. At the expense of increased complexity in the optical domain, the high-frequency electronics for photodetection and phase evaluation can be replaced by slower components with bandwidths of only hundreds of MHz, by employing a second MLL with slightly different line spacing for multi-heterodyne detection [14]. However, considerable technical effort is required to ensure stable drift-free operation of two femtosecond lasers in an industrial environment.

Here we demonstrate distance measurements using continuous-wave (cw) lasers and electro-optic modulators for generation of two frequency combs [24]. In this scheme, the line spacing of the frequency combs is determined by the frequency of the electrical drive signals, which can be precisely adjusted by highly stable RF synthesizers. In contrast to schemes generating only two modulation sidebands for synthetic wavelength interferometry or frequency sweeping methods [25, 26], we generate spectrally flat frequency combs that comprise a multitude of lines with equidistant 40 GHz spacing. The sensitivity of the presented distance measurement method is investigated for different levels of return loss: For optimum return power levels of −13 dBm, we obtain measurement errors of less than 3 μm in an acquisition time of only 8.3 μs, and a measurement error of less than 10 μm can be maintained over a dynamic range of more than 37 dB. The high dynamic range allows a proof-of-principle demonstration of the measurement scheme as a sensitive and fast surface topography sensor. The height of a step in a metal surface is measured and the results are compared to a tactile measurement.

Modulator-based frequency comb sources lend themselves to optical integration, thereby increasing robustness while decreasing size and cost. We demonstrate the realization of nanophotonic frequency comb sources using silicon-organic hybrid (SOH) integration [27]. These sources can be co-integrated with other optical elements available on the silicon-on-insulator SOI platform such as waveguides, splitters and photodiodes, thereby forming a miniaturized distance sensor with a footprint in the mm^2-range.

4.1. Experimental setup and measurement principle

The experimental setup of our modulator-based distance measurement system is shown in Figure 6. A pair of Mach-Zehnder modulators (MZM) with carefully adjusted sinusoidal drive signals is used to generate two flat-top frequency combs [28] featuring 7 lines within a power range of 1.25 dB (9 lines within 5.45 dB), see Insets 1 and 2 of Figure 6. The line spacings of the two combs differ by 50 MHz, leading to discrete beat notes when the optical signals are superimposed on a balanced detector (BD) for multi-heterodyne detection, see Inset 3 of Figure 6. To extract the phases of the individual beat notes, both detector signals are analogue-to-digital converted (A/D, sample rate 3.6 GS/s) and a fast Fourier transform is performed. A portion of the first frequency comb is coupled into the free-space measurement arm, and interferes with a portion of the second comb on BD 1. The phases Φ_m of the detected beat notes are directly linked to the phase shifts that the optical waves accumulate along the measurement arm. As a reference, a second balanced detector (BD 2) is used to detect the beat note phases Ψ_m of the frequency combs before propagation through the measurement arm. After calibration of the initial phases, the phase differences $\Delta\varphi_m = \Phi_m - \Psi_m$ are given by $\Delta\varphi_m = (2\Delta L m\omega_r / c + \Delta\varphi_{0,\Delta L}) \bmod 2\pi$, where $2\Delta L$ is the optical path length difference, ω_r is the line spacing of the comb propagating the measurement arm and $\Delta\varphi_{0,\Delta L}$ is a constant phase offset. After unwrapping the phases obtained from the various beat notes, ΔL can be obtained by a linear fit of $\Delta\varphi_m$ as a function of m. This corresponds to a synthetic wavelength interferometry with a redundant number of wavelengths, making the scheme robust against fluctuations of the measurement signal. No knowledge of the absolute optical wavelengths is needed, as the synthetic wavelengths are solely defined by the spacing ω_r of the measurement comb lines, which is given with very high precision by the RF source used to drive the modulators. The largest synthetic wavelength is linked to the line spacing ω_r of the measurement comb and amounts to $\Lambda = 7.5$ mm, which leads to a range of non-ambiguity of $\Lambda/2 = 3.75$ mm. The measurement path is equipped with a lens system, leading to a Gaussian beam with a waist radius of $w_0 = 13$ µm and a confocal parameter of $b = 340$ µm on the sample side. Less than −8 dBm of output power at around 1550 nm are emitted towards the sample.

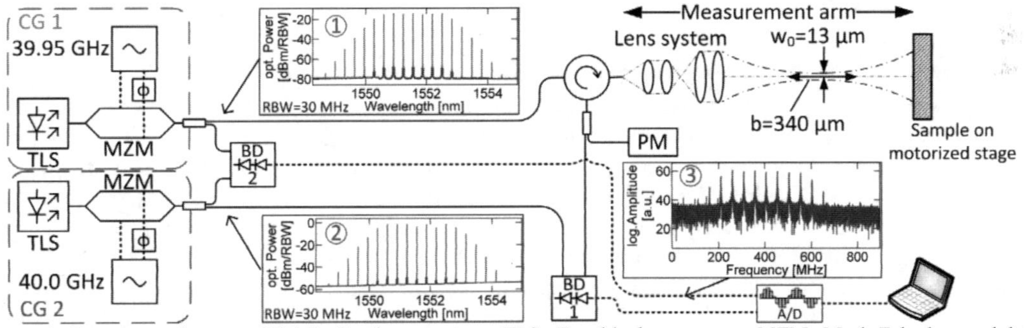

Figure 6. Experimental setup. CG 1,2: Comb generators, TLS: Tunable laser source, MZM: Mach-Zehnder modulator, BD 1,2: Balanced detectors, PM: Power meter, A/D: Analog-to-digital converter; Dashed lines: Electrical connections, Solid lines: Optical fiber. Inset 1+2: Optical spectrum. Inset 3: Electrical spectrum

4.2. Experimental demonstration

To characterize the system, we measure the distance to a plane mirror on a motorized linear translation stage that features a linear encoder with 0.1 µm resolution. The use of the lens system leads to strongly varying power levels at BD 1 when the mirror is moved through its focus. This effect is used to characterize the sensitivity of the system. For measurements of natural scattering surfaces, the sample is kept in focus and the lens system captures scattered light. Each measurement is based on recorded photodiode signals with a duration of 8.3 µs, and the measurement is repeated ten times for each stage position. Figure 7(a) shows the measured distance as a function of stage position along with the corresponding measurement errors; Figure 7(b) shows the power levels along with the standard deviations that were calculated from the ten measurement results obtained at each position. As expected, measurement errors increase for decreasing power levels. Nevertheless, we find that the absolute errors remain smaller than 10 µm over virtually the full range with standard deviations below 4 µm, even though the returned optical power drops by more than 37 dB. Oscillations in the error are attributed to thermal fluctuations in the optical fibers during the experiments, which influence the optical path lengths. For optimum power levels between -13 dBm and –24 dBm, the error is smaller than 3 µm and sub-µm standard deviations are achieved.

To confirm the viability of the technique when used with naturally scattering surfaces, we measure a step milled into a steel surface, Fig. 2(c). Using a tactile coordinate measurement machine (Carl Zeiss O-Inspect 442), a step height of 464.4 µm was obtained. This is in excellent agreement with the height of 461.2 µm as measured with our system, given the roughness of the steel surface and the limited uniformity of the milled step.

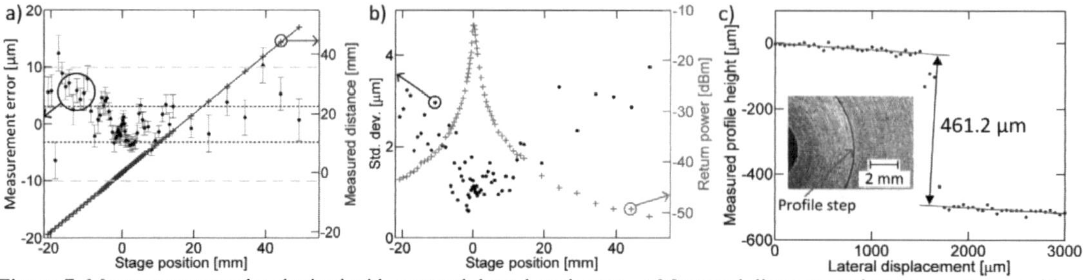

Figure 7. Measurement results obtained with our modulator-based setup. a) Measured distances and errors vs. stage position. Error bars denote the standard deviation from the mean for 10 subsequent measurements taken at each position. Red dotted line marks an error of ±3 µm, green dashed line marks an error of ±10 µm. b) Power coupled back to the measurement detector as a function of stage position along with the standard deviation of 10 subsequent measurements at each stage position (black). c) Measured values for a step milled into a steel surface. The step height as measured with our setup deviates by 3.2 µm from a tactile reference measurement. Inset: Microscopic picture of the surface. The step is indicated by an arrow.

5. Towards an integration on the silicon photonic platform

Silicon photonics offers tremendous potential for large-scale photonic-electronic integration through joint processing of photonic and electronic circuitry and by fabless fabrication. The silicon photonics platform offers a broad portfolio of optical components such as waveguides, phase shifters, splitters and photodetectors [29]. However, to realize a miniaturized distance sensor system on a chip following the principles discussed in the preceding section, it is necessary to realize miniaturized frequency comb sources on the silicon chip. This is a challenging task: The generation of broadband frequency combs requires a large modulation depth to achieve a large number of sidebands along with a high modulation frequency to obtain useful line spacings of tens of gigahertz. This is hard to achieve using conventional silicon photonic phase modulators that exploit free-carrier depletion or injection in p-n, p-i-n, or metal-oxide-semiconductor (MOS) structures [30]. To overcome the limitations of all-silicon devices, we use the concept of silicon-organic hybrid (SOH) integration [31, 32] which combines nanophotonic SOI waveguides with organic cladding materials to realize highly efficient broadband phase shifters [33, 34]. This hybrid integration leads to devices that exhibit voltage-length products down to $U_\pi L = 0.5$ V mm at DC, which allows high modulation depths at low voltages [35]. Using a dual-drive SOH Mach-Zehnder modulator (MZM), we demonstrate generation of a flat-top frequency comb comprising 7 lines with a spacing of 40 GHz and a spectral flatness of 2 dB that are perfectly suited for high-speed data transmission [27].

The SOH phase modulator [33, 35, 36] consists of an optical slot waveguide [37] electrically connected to metallic transmission lines through 60 nm thick doped silicon strips. A schematic cross section and a simulated mode profile of the optical slot waveguide are depicted in Figure 8(a). The modulation voltage applied to the transmission lines drops across the narrow slot, which has a typical width of 100 … 150 nm. This results in a high electric field that strongly overlaps with the optical mode, and this strong overlap leads to highly efficient electro-optic interaction and hence to a low voltage-length product $U_\pi L$. The silicon waveguides are fabricated on a SOI wafer with a 2 µm thick buried oxide and a 220 nm thick device layer using 193 nm deep-UV lithography. After processing the slot waveguides and the transmission lines, the structures are covered with the electro-optic chromophore DLD164 [35], which entirely fills the slot. The electro-optic cladding is applied by spin coating and poled at elevated temperatures with a DC voltage across the slotline electrodes. More details about the phase modulator can be found in [36]. Our comb generators consist of two phase modulators in a Mach-Zehnder interferometer. These phase modulators are driven independently in a dual-drive configuration and have a length of about 2 mm.

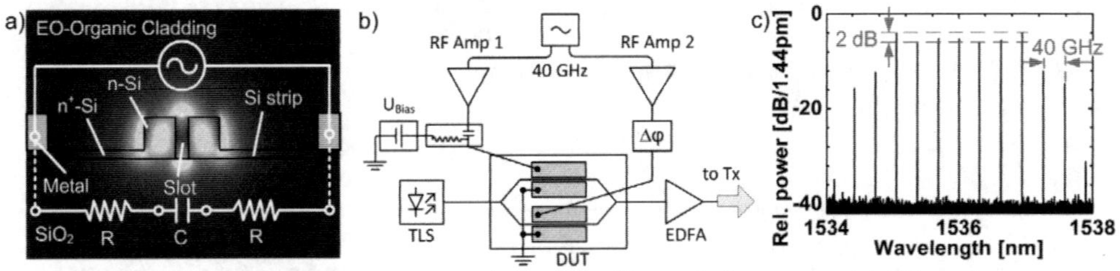

Figure 8. Dual-drive frequency comb generator and corresponding optical spectrum. (a) Schematic cross-section and simulated optical mode of a silicon-organic hybrid (SOH) slot-waveguide phase modulator. The two rails of the silicon slot waveguide are electrically connected to metal electrodes by 60 nm high n-doped silicon strips. The waveguide is covered by an electro-optic cladding material (DLD164), which entirely fills the slot. (b) Integrated dual-drive frequency comb generator: A tunable laser source (TLS) is coupled to a dual-drive SOH Mach-Zehnder modulator (MZM) via grating couplers. The arms are driven by two sinusoidal 40 GHz signals. Flat combs are obtained by carefully adjusting the signal powers and phases along with the bias voltage U_{Bias}. Fiber-chip coupling losses are compensated by an erbium-doped fiber amplifier (EDFA). (c) Flat-top spectrum obtained for electrical drive powers of 28 dBm and 23 dBm measured at the output of the amplifiers, showing 7 lines within 2 dB flatness. The lines are spaced by 40 GHz.

The experimental setup for characterizing the frequency combs is shown in Figure 8(b). A tunable laser source (TLS) is coupled to the chip via grating couplers [38], and the electrodes for the sinusoidal drive signals are contacted with RF probes. Using sinusoidal drive signals with carefully adjusted amplitudes and phases [28], a spectrally flat frequency comb can be generated. The voltage-length product of each phase modulator was measured at DC to be $U_{\pi}L = 2$ V mm. This allows an efficient generation of higher-order optical sidebands with electrical drive powers of approximately 28 dBm and 23 dBm in the interferometer arms. The resulting optical spectrum is shown in Figure 8(c). We achieve a spectral flatness of 7 lines within 2 dB for spacings 40 GHz. By comparing the measured optical spectrum to simulations, we estimate a peak-to-peak modulation depth in the two MZM arms of 3.6π and 2.7π, respectively. The estimated phase difference of about π matches nicely to theoretical predictions for achieving an optimal flat comb [28].

6. Conclusion

We demonstrate the use of frequency combs for trilateration-based 3D position measurements, and for sensitive surface topography measurements. A frequency comb generated by a MLL is used in combination with direct detection and electrical heterodyne down-mixing to measure the distances from three MEMS-based tracking units to a retroreflector target. The measurements allow a position determination of the target within a measurement time of 10 ms, and a volumetric accuracy of 24 µm using a compact triangular configuration of the measurement modules with 20 cm side length. In a second set of experiments, modulator-based frequency combs are generated from cw lasers and are used with a sensitive multi-heterodyne detection technique for measuring distances to natural scattering surfaces. Measurement errors are below 10 µm over a dynamic range of more than 37 dB. For optimum power levels and integration times of 8.3 µs, the absolute errors become smaller than 3 µm. The comb-generation principle lends itself to photonic integration, and frequency comb sources based on silicon-organic hybrid (SOH) electro-optic modulators are shown.

References

[1] Bridges B, 2012 Technology white paper: Laser Tracker. Available at https://faro.blob.core.windows.net/sitefinity/white-papers/04ref707-012-de---whitepaper_laser-tracker.pdf?sfvrsn=6, accessed on 03.09.2014

[2] Leica Geosystems 2012 Leica absolute interferometer. Available at http://metrology.leica-geosystems.com/downloads123/m1/metrology/general/white-tech-paper/Leica Absolute Interferometer_white paper_en.pdf, accessed on 03.09.2014

[3] Cuypers W, Van Gestel N, Voet A, Kruth J-P, Mingneau J and Bleys P 2009 *Optics and Lasers in Engineering* **47** 292.

[4] Stanley A and Klein M 2010 *Applied Mechanics and Materials* **24-25** 167.

[5] Milanović V, Siu N, Kasturi A, Radojičić M, Su Y, Schenk H and Piyawattanametha W *Proceedings of SPIE* vol. **7930** pp 79300U.

[6] Nikon 2010 K-Series Optical CMM solutions. Available at http://www.nikonmetrology.com/de_EU/content/download/11040/220148/version/4/file/Optical_CMM_EN.pdf, accessed on 03.09.2014

[7] Takatsuji T, Goto M, Kirita A, Kurosawa T and Tanimura Y 2000 *Meas. Sci. Technol.* **11** 477.

[8] Zhang D, Rolt S and Maropoulos P G 2005 *Meas. Sci. Technol.* **16** 2541.

[9] Ibaraki S, Takeuchi K, Yano T, Takatsuji T, Osawa S and Sato O 2011 *Journal of Mechanics Engineering & Automation* **2011** 313.

[10] Schwenke H and Warmann C 2011 10th IMEKO TC14 *Symposium on Laser Metrology for Precision Measurement and Inspection in Industry* Braunschweig, Germany, September 12-14.

[11] Takatsuji T, Goto M, Kurosawa T, Tanimura Y and Koseki Y 1998 *Meas. Sci. Technol.* **9** 38.

[12] Santolaria J, Majarena A C, Samper D, Brau A and Velázquez J 2014 *The Scientific World Journal* **2014** 1.

[13] Linares J-M, Chaves-Jacob J, Schwenke H, Longstaff A, Fletcher S, Flore J, Uhlmann E and Wintering J 2014 *Precision Engineering* **38** 578.

[14] Coddington I, Swann W C, Nenadovic L and Newbury N R 2009 *Nature Photonics* **3** 351.

[15] Cui M, Zeitouny M G, Bhattacharya N, van den Berg S A and Urbach H P 2011 *Opt. Express* **19** 6549.

[16] Hyun S, Kim Y-J, Kim Y, Jin J and Kim S-W 2009 *Meas. Sci. Technol.* **20** 95302.

[17] Kajima M and Minoshima K 2010 *Appl. Opt.* **49** 5844.

[18] Minoshima K and Matsumoto H 2000 *Appl. Opt.* **39** 5512-5517.

[19] Liu T-A, Newbury N R and Coddington I 2011 *Opt. Express* **19** 18501.

[20] Doloca N R, Meiners-Hagen K, Wedde M, Pollinger F and Abou-Zeid A 2010 *Measurement Science and Technology* **21** 115302.

[21] Bancroft S 1985 IEEE Trans. Aerosp. *Electron. Syst.* **21** 56.

[22] Lee J-K, Kim Y, Lee J-H and Kim S-C 2014 IEEE *Commun. Lett.* **18** 1591–1594.

[23] Zhuang H, Li B, Zvi S. Roth and Xie X 1992 *Robotics and Autonomous Systems* **9** 255.

[24] Weimann C, Meier D, Wolf S, Schleitzer Y, Totzeck M, Heinrich A, Hoeller F, Leuthold J, Freude W and Koos C 2013 *Conference on Lasers and Electro-Optics* pp CTu2I.3.

[25] Lay O P, Dubovitsky S, Peters R D, Burger J P, Ahn S-W, Steier W H, Fetterman H R and Chang Y 2003 *Opt. Lett.* **28** 890.

[26] Samuel Choi, Mitsufumi Yamamoto, Daisuke Moteki, Tatsutoshi Shioda, Yosuke Tanaka and Takashi Kurokawa 2006 *Opt. Lett.* **31** 1976.

[27] Weimann C, Schindler P C, Palmer R, Wolf S, Bekele D, Korn D, Pfeifle J, Koeber S, Schmogrow R, Alloatti L, Elder D, Yu H, Bogaerts W, Dalton L R, Freude W, Leuthold J and Koos C 2014 *Opt. Express* **22** 3629.

[28] Sakamoto T, Kawanishi T and Izutsu M 2007 *Opt. Lett.* **32** 1515.

[29] Biberman A and Bergman K 2012 *Reports on Progress in Physics* **75** 46402.

[30] Reed G T, Mashanovich G, Gardes F Y and Thomson D J 2010 *Nature Photonics* **4** 518.

[31] Baehr-Jones T W and Hochberg M J 2008 *J. Phys. Chem. C* **112** 8085.

[32] Brosi J-M, Koos C, Andreani L, Waldow M, Leuthold J and Freude W 2008 *Opt. Express* **16** 4177.
[33] Alloatti L, Korn D, Palmer R, Hillerkuss D, Li J, Barklund A, Dinu R, Wieland J, Fournier M, Fedeli J, Yu H, Bogaerts W, Dumon P, Baets R, Koos C, Freude W and Leuthold J 2011 *Opt. Express* **19** 11841.
[34] Koos C, Brosi J, Waldow M, Freude W and Leuthold J 2007 *Proc. European Conf. on Optical Communication,* paper P056
[35] Palmer R, Koeber S, Heni W, Elder D L, Korn D, Yu H, Alloatti L, Koenig S, Schindler P C, Bogaerts W, Pantouvaki M, Lepage G, Verheyen P, van Campenhout J, Absil P, Baets R, Dalton L R, Freude W, Leuthold J and Koos C 2013 *Proc. European Conf. on Optical Communication,* paper We.3.B.3.
[36] Palmer R, Alloatti L, Korn D, Schindler P C, Baier M, Bolten J, Wahlbrink T, Waldow M, Dinu R, Freude W, Koos C and Leuthold J 2013 *Photonics Technology Letters*, IEEE **25** 1226.
[37] Almeida V R, Xu Q, Barrios C A and Lipson M 2004 *Opt. Lett.* **29** 1209.
[38] Taillaert D, van Laere F, Ayre M, Bogaerts W, van Thourhout D, Bienstman P and Baets R 2006 *Jpn. J. Appl. Phys.* **45** 6071.

Journal of Physics: Conference Series **605** (2015) 012031 doi:10.1088/1742-6596/605/1/012031

A Novel Behavior of Pump Power in the Instability Induced Supercontinuum Generation of Saturable Nonlinear Media

K Nithyanandan and K Porsezian

Department of Physics, Pondicherry University, Puducherry, India.

E-mail: nithi.physics@gmail.com

Abstract. We investigate the modulational instability (MI) induced Supercontinuum generation (SCG) in exponential saturable nonlinearity. The pump power (P) is observed to behave in a unique way such that unlike the conventional Kerr case, the effective nonlinearity of saturable nonlinear system does not monotonously increases with an increase in power. The supercontinuum is observed at the shortest distance of propagation at power equal to the saturation power (P_s), whereas for all combinations of powers ($P < P_s$ or $P > P_s$) spectral broadening occurs at longer distance.

1. Introduction

One of the most advanced frontiers of nonlinear optics is the generation of intense ultrabroadband spectrum known to the scientific world as the supercontinuum generation (SCG). The potential applications of such as frequency metrology, optical coherence tomography, ultra-short pulse compression, sensor technique, fiber characterization etc., have made SCG as one of the dynamic and elegant nonlinear phenomena in nonlinear optics. Thus SCG in the modern days has been recognized as the "White-light laser" [1–3]. All the applications are powered by the exceptional properties such as high brightness, coherently pulsed nature, high degree of spatial coherence, broad bandwidth and the requirement of low pulse energies. The extensive interest stem not only on its application perspective but also on the scientific curiosity due to the rich physics. The invention of photonic crystal fiber (PCF) have revolutionized the concept of SCG, this is because of its exceptional feautres such as the requirement of less input intensity, engineerable zero dispersion wavelength by tailoring the waveguide dispersion, endlessly single mode operation, and elevated nonlinearity. The first demonstration of SCG in PCF that span two octaves from UV to the infrared was reported by Ranka *et al.*. This futuristic work sets the tone for new innovation in the context of SCG in PCF [4, 5].

The physics underlying SC generation is now generally well understood, the primary mechanism of generating broadband source is identified as soliton fission and modulational instability (MI) [1, 6]. Former is the well-known mechanism of generating high quality broadband spectrum driven by the soliton dynamics . Latter is the MI induced SCG (MI-SCG), where the initial dynamics are governed by the noise, which lead to the breakup of the pulses similar to the case of SF.

In recent times, there has been renowned interest in the concept of saturable nonlinearity (SNL), due to the inherent nature in some nonlinear materials. In reality, for any real glass material, there is an upper limit to the nonlinear refractive index that can be induced optically. Beyond a certain level of the pump power, higher-order nonlinear susceptibilities inevitably come into play and manifest themselves in qualitatively different ways depending on the type and nature of the dopants that are incorporated in the glass material. For instance, system such as semiconductor doped fiber [7, 8] and liquid core photonic crystal fiber [9, 10] the nonlinear response tends to saturate even at low power due to its enhanced nonlinear index. We consider such a system in the saturation window for our investigation and hence the inclusion of saturable nonlinearity is obvious to reveal the actual physics associated with such system. Nonetheless, high power operation not only saturates the nonlinear response of the medium but also elevates soliton order. This increase in soliton order favors the MI process over soliton fission, and thereby making MI as the prominent governing mechanism in the SCG in the SNL system. Lyra *et al*, reported the different functional form of the saturable nonlinearity, out of which exponential saturable nonlinearity (ESN) is observed to generate broadest spectrum at relatively shorter distance [11, 12]. In the similar lines to the earlier report [12], we describe in the manuscript about the unique role of power in the instability induced SCG.

2. Theoretical Model

The propagation of optical beam through a fiber with SNL is described by modified nonlinear Schrödinger equation (MNLSE). The MNLSE of the slowly varying envelope $U(z,t)$ along the z-axis in a retarded reference time centered at frequency ω_0 is given by [7, 8, 12]

$$\frac{\partial U}{\partial z} + \sum_{n=2}^{4} \beta_n \frac{i^{n-1}}{n!} \frac{\partial^n U}{\partial t^n} = i \frac{\gamma |U|^2}{1+\Gamma|U|^2} U, \tag{1}$$

where, z and t are the longitudinal coordinate and time in the moving reference frame, respectively. β_n is the dispersion coefficient resulting from the Taylor expansion of the propagation constant around the center frequency ω_0 and $\gamma = \frac{n_2\omega_0}{cA_{eff}}$ is the Kerr coefficient, where A_{eff} is the effective core area, n_2 is the nonlinear index coefficient. P_s is the power threshold corresponding to the saturation of the nonlinear response. Γ is the saturation parameter whose expression is given by $\Gamma = 1/P_s$.

3. MI induced supercontinuum generation

The propagation of optical beam in ESN governed by Eq. (1) is numerically studied using split step Fourier method. The initial envelope of the soliton at z = 0 is given by $U(0,t) = \sqrt{P_0} sech(t)$ of width 10 ps at λ_0=1.55μm. The numerical simulation is carried out with a resolution of 2^{20} grid points. The system parameter are as follows: The dispersion coefficients takes β_2 = -4.50$\times10^{-2}$ ps^2/m, β_3 = 6.710$\times10^{-4}$ ps^3/m, β_4 = -6.722$\times10^{-7}$ ps^4/m and the nonlinear coefficient can be written as γ = 2.18 W^{-1}m^{-1}[9, 10]. The high power operation in the saturation window of the nonlinear response and the choice of longer pulse width in ps time scale favors the MI process over soliton fission and thereby confirms the MI as the dominant mechanism in SCG of SNL system.

Following our earlier report [12], the dispersion relation corresponding to the MI gain in ESN can be written as

$$K = \beta_3 \frac{\Omega^3}{6} \pm \left[\left(\beta_2 \frac{\Omega^2}{2} + \beta_4 \frac{\Omega^4}{24} \right) \left(\beta_2 \frac{\Omega^2}{2} + \beta_4 \frac{\Omega^4}{24} + 2\gamma_1 P_0 \right) \right]^{1/2}, \quad \gamma_1 = \gamma\, e^{-\Gamma P_0} \tag{2}$$

In the case of anomalous dispersion regime, β_2 takes negative value and thereby serve the required phase matching for MI [6, 13]. Instability is found to be possible only within the threshold frequency known as the critical modulation frequency (CMF) and for frequencies $\Omega_{CMF} > \Omega$, K becomes complex and the spectral sideband grows exponentially down the fiber. It is worth noting that when the saturation parameter $\Gamma = 0$, both CMF and G_{max} revert back to the conventional Kerr type nonlinear response.

The MI/FWM process typically manifest as follows: (*i*) the primary stage is the emergence of spectral sidebands and the subsequent higher order sidebands (*ii*) the second stage is the evolution of ultra-broadband spectrum through the nonlinear phase matching. In time domain, the MI can be recognized as the growth of weak perturbation leading to the fast modulation of pulse envelope and results in the fine structure on the waveform.

4. Supercontinuum generation in saturable nonlinear media

This section features the insight analysis of the characteristics of ESN in the supercontinuum generation. To demonstrate the impact of pump power in the SCG, we have considered different input power at a constant saturation power. It is observed that the increase in input powers does not lead to monotonous enhancement of SCG. The enhancement we quote here is the ability of the system to achieve SCG at shorter distance of propagation as possible. The maximum enhancement of SCG is achieved only at operating the input power around the saturation threshold, for all other choice of input power the SCG inevitably gets suppressed. To illustrate, we discuss three combinations of powers as shown in the Fig. (1), one at the satura-

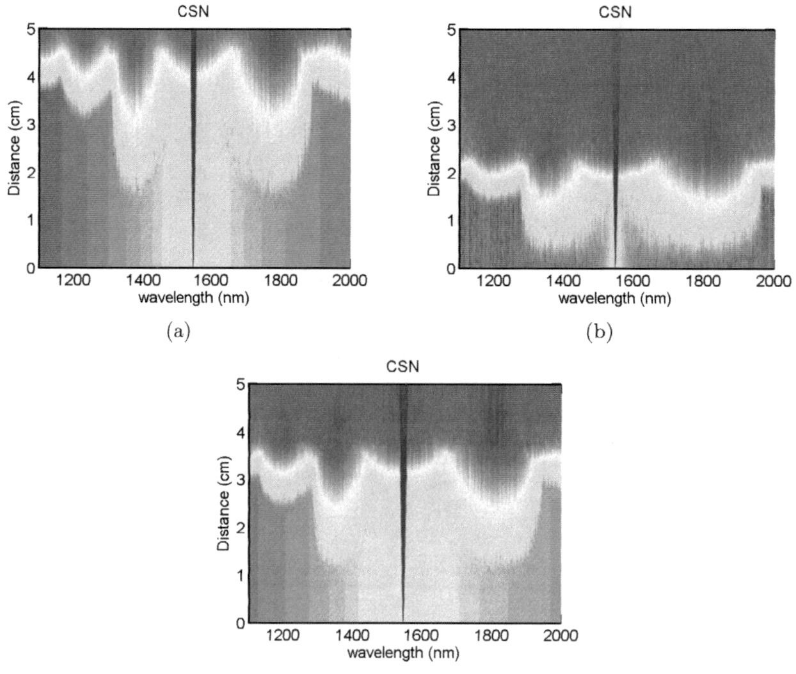

Figure 1: Spectral evolution of MI-SCG in SNL at P (=500 W (a), 1000 W (b) and 3000 W (c) and $P_s = 1500$ W for a propagation length $L = 5$ cm.

tion power $P = P_s$, others at below $P < P_s$ and above $P > P_s$ the saturation power. Without any loss of generality, we consider the saturation power as $P_s = 1500W$ and the operating input powers as $P(A \rightarrow 500, B \rightarrow 1500, C \rightarrow 3000)$W. Among the three cases, one can readily observe that at $P = P_s$, the sidebands evolve much earlier than the rest of the two cases. Precisely, the primary sideband appears at $z \approx 0.5$ cm and secondary sideband at $z \approx 1.6$ cm, whereas on the other side, for all other representative power levels the sidebands manages to evolve only at higher distance of propagation. Similarly, the corresponding temporal evolution is given in the Fig. (2). The pulse breaking occurs at shortest possible distance corresponding to $P = P_s$ *(i.e.)* at $z \approx 1.9$ cm as shown in Fig. (2).

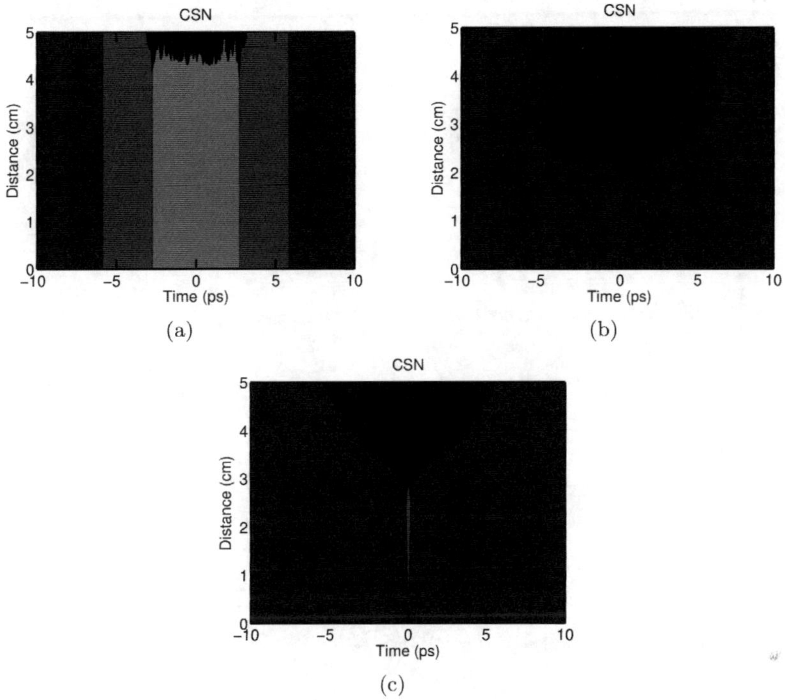

Figure 2: Temporal evolution of MI-SCG in SNL at P (=500 W (a), 1000 W (b) and 3000 W (c) and P_s = 1500 W for a propagation length L = 5 cm.

To give an insight, we consider the unsaturated Kerr (USN) type nonlinear response case with the same parameters as in the case of SNL. The spectral and temporal evolutions corresponding to the power $P(A \rightarrow 500, B \rightarrow 1500, C \rightarrow 3000)$W are given in Fig. (3) and Fig. (4). It is obvious from the figures that MI sidebands occur much earlier than the SNL counterpart. For instance, primary sideband occurs at $z \approx 0.3$cm and secondary sideband occurs at $z \approx 0.8$ cm for an input power of $P = 1500W$ (Refer Table. 1). Also, one can straightforwardly notice that increase in power certainly increases the effective nonlinearity and thereby enhances the SCG by achieving at shorter distance of propagation. Thus, our numerical study strongly illustrates that the incorporation of the nonlinear saturation of the medium certainly suppresses the MI.

On the other hand, saturable nonlinear system behaves quite differently, than the conventional Kerr type nonlinearity. The difference in the distance for the evolution of sidebands in

SNL is attributed to the fact that the power is not directly proportional to the effective nonlinearity ($\tilde{\gamma}$) as in the case of the conventional Kerr medium. It is observed that for power less than the saturation power, the $\tilde{\gamma}$ increases with an increase in power and for power greater than the saturation power, the $\tilde{\gamma}$ decreases with an increase in power. This change in the effective nonlinear coefficient changes the nonlinear length of the system. In principle, it is a proven fact that the MI occurs at the balance between the dispersion and the nonlinear length of the system. This change in the nonlinear length is in accordance with the required phase matching for the FWM process. For instance, the nonlinear length ($L_{nl} = \frac{1}{\gamma P_0}$) of the SNL at different power follows the order $Lnl_B < Lnl_C < Lnl_A$ and hence the distance of propagation required to achieve MI/FWM, or, in other words the ease of achieving PMC follows the order $PMC_B > PMC_C > PMC_A$.

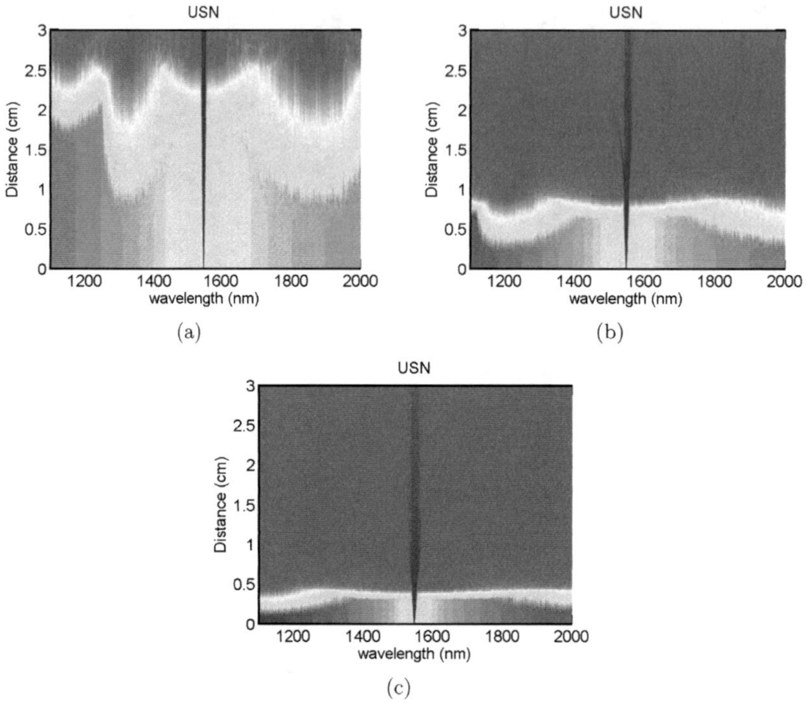

Figure 3: Spectral evolution of MI-SCG in USN at P (= 500 W (a), 1500 W (b) and 3000 W (c)) for a propagation length $L = 3$ cm.

Thus among the different combination of power, case B register the least value of L_{nl} thus enables the pulse breaking at shorter distance. This is due to the fact that operating the system around the saturation power records the highest effective nonlinearity, thereby enables to achieve PMC at relatively shorter distance which results in the maximum enhancement of SCG and the associated nonlinear effects.

5. Results and Discussion

Thus, our numerical study strongly illustrates that the incorporation of the SNL of the medium certainly suppresses the MI and the associated dynamical behavior. Our numerical analysis affirms the evolution of SCG from the MI, some of the valid reason to justify the same are as follows (i) no initial soliton contraction and propagation of higher order soliton as in SF, (ii) pulse breaking is purely due to the numerical noise included, (iv) extreme sensitivity to noise are some of the classical evidence of the MI induced SCG.

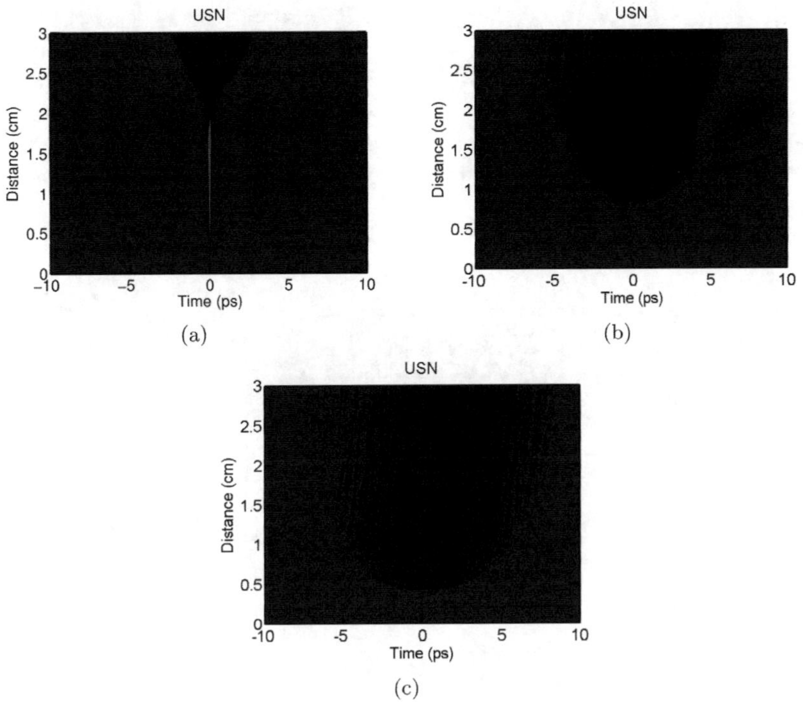

Figure 4: Temporal evolution of MI-SCG in USN at P (= 500 W (a), 1500 W (b) and 3000 W (c)) for a propagation length $L = 3$ cm.

To give a comprehensive idea of the impact of power in the SCG spectrum, we consider the evolution of SCG from MI for some representative cases of power at different propagation length. Fig. (5) infer that the pattern of evolution of the spectral sidebands in all cases remains the same. The saturation power is fixed as $P_s = 1500W$ and the input power corresponds to $P(A \rightarrow 500, B \rightarrow 1500, C \rightarrow 3000)W$. It is evident from the figures that among the three representative powers, curve B corresponds to $P = P_s$ records the maximum gain (arb.units). Despite the high value of power curve C can only find less gain than curve B. Thus operating at saturation power enables one to achieve SCG at shorter distance than for all combinations of powers.

For numerical appreciation about the role of pump power in the SCG, we tabulated the distance corresponding to the pulse breaking and the evolution of sidebands in Table.1. For the better understanding Kerr type unsaturated nonlinear response is also considered alongside the saturable nonlinearity. One can straightforwardly infer from the table that for the case of Kerr response the distance corresponding to the evolution of sidebands and pulse breaking occurs at shorter distances for higher powers. Whereas in the SNL system both the pulse breaking and the spectral sideband evolution is no longer proportional to the power and register least at $P = P_s$. Interestingly one could also observe that power corresponding to the particular distance indeed

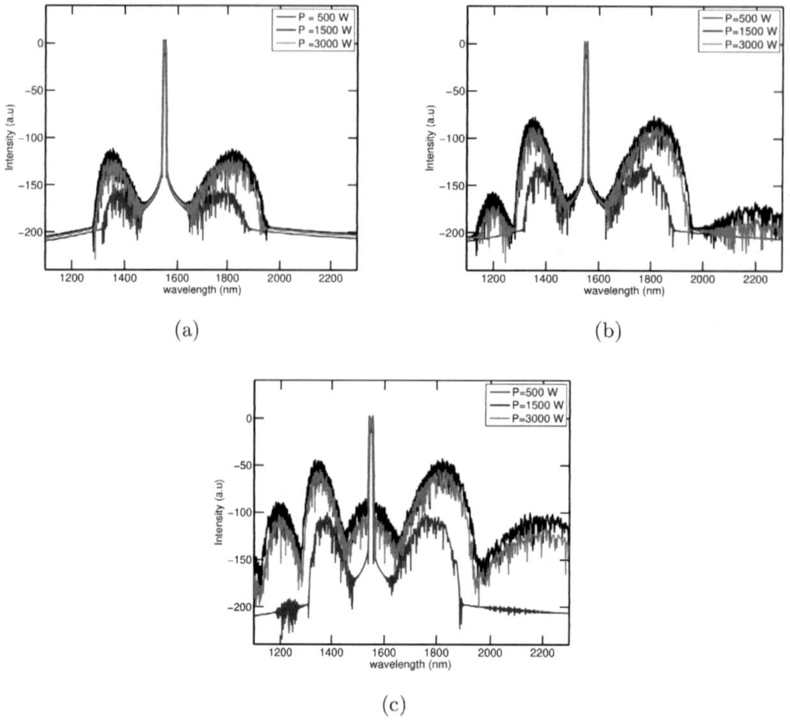

Figure 5: Sideband evolution of SNL for power P(=500, 1500, 3000)W at a distance of 2 cm (a), 2.5 cm (b) and 3 cm (c).

occurs in pairs *(i.e.)* there exist two value of power at which the distance corresponding to the pulse breaking and spectral evolution register same value. This we reckon as the two state behavior in the context of MI, which can be readily observed in SCG, as well. This observation is a clear manifestation of evolution of SCG from MI.

6. CONCLUSION

In summary, we theoretically investigated SCG in the SNL system using MNLSE. We have reported that the behavior of pump power in the SNL system is indeed unique such that, the maximum enhancement of SCG (SCG at shortest distance) is observed only at $P = P_s$ and

Table 1: Summary of MI-SCG - Evolutions of sidebands (in cm) with power (W)

Power (W)	Primary Sideband(cm)		Secondary Sideband(cm)		Pulse breaking(cm)	
	USN	SNL	USN	SNL	USN	SNL
500	0.9	1.7	1.7	3.2	2.3	4.2
1000	0.7	1.0	1.1	2.2	1.4	2.9
1500	0.3	***0.5***	0.8	***1.5***	0.9	***1.9***
3000	0.2	1.3	0.4	2.6	0.4	3.3

for all combinations of powers the SCG gets suppressed. To give a comprehensive idea about SCG in SNL system, Kerr type nonlinear response is also considered alongside the SNL, similar to the earlier published reports that the SNL inevitably suppresses the SCG. However, our present results deviate from the rest, such that we identified the active regime in SNL system, where maximum enhancement of the SCG is possible. Also, the two state behavior we reported enables one to manipulate the desired SCG characteristics in a relatively simple means by merely changing the input power. Thus in the report, we bring to the light on the unusual role of pump power in the SCG process in the SNL system. We understood that there is no experimental report till now in the context of SCG in SNL system, and we believe that our presented theoretically reported results will set benchmark for new experiments aiming to probe the various nonlinear effects in the SNL system.

Acknowledments
KP thanks DST, DAE-BRNS, CSIR, and UGC, Government of India, for the financial support through major projects. K. Nithyanandan thanks CSIR, Government of India for providing financial support by awarding Senior Research Fellowship (SRF).

References
[1] Dudley J M, Genty G and Coen S 2006 *Rev. Mod. Phys.* **78** 1135–1184
[2] Zheltikov A (2006) *Physics-Uspekhi* **49** 605
[3] Alfano R R 2006 *The Supercontinuum Laser Source* (Berlin:Springer)
[4] RS Windeler, J and AJ Stentz (2000) *Opt. Lett. Lett.* **25** 25
[5] Husakou A V and Herrmann J 2002 *J. Opt. Soc. Am. B* **19** 2171–2182
[6] Agrawal G 2012 *Nonlinear fiber optics* (San Diego, 5th edition: Academic Press)
[7] Dinda P T and Porsezian K 2010 *J. Opt. Soc. Am. B* **27** 1143–1152
[8] Hickmann J M, Cavalcanti S B, Borges N M, Gouveia E A and Gouveia-Neto A S 1993 *Opt. Lett.* **18** 182–184
[9] Ganeev R, Ryasnyansky A, Baba M, Suzuki M, Ishizawa N, Turu M, Sakakibara S and Kuroda H 2004 *Applied Physics B* **78** 433–438 ISSN 0946-2171
[10] Vieweg M, Gissibl T, Pricking S, Kuhlmey B T, Wu D C, Eggleton B J and Giessen H 2010 *Opt. Express* **18** 25232–25240
[11] Lyra M L and Gouveia-Neto A 1994 *Opt. Communication* **108** 117–120
[12] K Nithyanandan Vasantha Jayakantha Raja R P K and Uthayakumar T 2013 *Opt. Fib. Tech.* **19** 348
[13] Porsezian K, Nithyanandan K, Raja R V J and Shukla P K 2012 *J. Opt. Soc. Am. B* **29** 2803–2813

Journal of Physics: Conference Series **605** (2015) 012032 doi:10.1088/1742-6596/605/1/012032

Influence of the functional form of nonlinearity in the Modulational Instability spectra of relaxing saturable nonlinear system

K Nithyanandan and K Porsezian

Department of Physics, Pondicherry University, Puducherry, India.

E-mail: nithi.physics@gmail.com

Abstract. We investigate the modulational instability (MI) of the optical beam propagating in the relaxing saturable nonlinear system. We identify and discuss the salient features of various functional forms of saturable nonlinear responses such as exponential, conventional and coupled type on the MI spectrum. Using Debye relaxation model, the relaxation of nonlinear response is effectively included along with the saturable nonlinear response (SNL). Using linear stability analysis, an explicit dispersion relation is determined for considering different functional forms of SNL. Firstly, we analyze the impact of SNL on the MI spectrum and found that the MI gain and bandwidth is maximum for exponential nonlinearity in comparison to other types of SNL's. Latter the relaxation of the nonlinearity is included, the inclusion of the finite value of the response time extends the range of the unstable frequencies literally down to infinite frequencies. In the regime of slow response, the MI inevitably suppressed regardless of the sign of the dispersion coefficient. To give insight into the MI phenomena, the maximum MI gain and the optimum modulation frequency is drawn as a function of the delay. Thus the MI dynamics in the system of relaxing saturable nonlinear media is emphasized and the significance of various functional forms of SNL are highlighted.

1. Introduction

In the classical optics, the intensity dependent process in the optical system seeds the emergence to the field of nonlinear optics [1], which deserves a considerable attention for more than five decades in terms of both fundamental and application perspectives. One such exciting prospect that earns serious scientific interest nearly for three decades is the so-called modulational instability (MI) [1]. MI is an ubiquitous good old phenomena studied in diverse fields such as fluid mechanics[2], plasma [3], nonlinear optics [1, 4–6] etc., The dynamics of MI is governed by the nonlinear Schrödinger equation (NLSE), which admits soliton or solitary wave through a conservative interaction between anomalous group velocity dispersion (GVD) and self focusing Kerr nonlinearity [4–6]. The potential applications of MI in various disciplines such as optical communication and signal processing systems, ultrafast pulse generation, supercontinuum generation, new laser sources, optical amplification of weak signal, material absorption and loss compensation, all-optical switching etc., are some of the examples of MI, which provides a continuing interest. [7–12].

In the context of optical fiber, MI admits natural means of generating ultrashort pulses was theoretically proposed by Hasegawa in 1984 [4]. Latter, the experimental demonstration was given by Tai *et al* [5] in 1986. MI and four wave mixing are recognized as the time/frequency analog of the same physics, such that the two degenerate pump beams under nonlinear interaction evolves into a pair of parametric sidebands that are frequency downshifted (Stokes) and upshifted (anti-Stokes) relative to the pump [1, 7, 13]. The parametric sideband grow exponential with a gain given by G $= [(\gamma P_0)^2 - (\rho/2)^2]^{1/2}$, where $\rho = 2\gamma P_0 + 2\sum_{n=1}^{\infty}[\beta_{2n}/(2n)!]\Omega^{2n}$ is the phase mismatching condition, and Ω is the angular frequency shift from the pump and β_{2n} are the even order dispersion coefficients of the propagation mode. The maximum gain $G_{max} = \gamma P_0$ corresponds to $\rho = 0$.

The perception of analyzing MI takes different dimension at different stages depending upon the nonlinear contribution of the refractive index. As a consequence, system with higher order nonlinearity like quintic and non-Kerr nonlinearity, saturable nonlinearity [14–18], relaxing nonlinearity [19–21] and relaxing saturable nonlinearity [22] are subject of high interest due to its potential applications.The resonant materials such as semiconductor-doped glass fibers (SDF) are the modern examples of scientific curiosity. In SDF, in addition to the fast response due to electronic contribution, thermal mechanisms may lead to additional contribution to the nonlinear response whose time scale ranges from tens of picoseconds to hundreds of nanoseconds [12]. Such system are identified as the relaxing nonlinear system due to the delayed nature of its nonlinear response. In the ultrashort pulse regime, besides the slow response, due to the high nonlinearity of SDF, at high intensities there is an upper limit for the optically induced refractive index variation, beyond which the nonlinear response of the medium saturates and such type of material is known to be saturable nonlinear media [14–18, 23]. Thus index relaxation and saturation are integral part of the high index material in the ultrashort regime, and hence both need to be considered together for any dynamical study [22].

The role of delayed Raman response on the MI spectrum was illustrated in [24]. Recently, the impact of finite response time in the MI spectrum was explained by Liu *et al.* using a simple model called as Debye relaxational model [19]. The authors identified that for any finite relaxation time the range of unstable frequencies extends literally upto infinite frequencies [19]. Similar analysis can be seen in [22], where the authors reported the role of relaxation in the saturable nonlinear media. The existence of two unstable bands in the anomalous GVD regime, namely, instantaneous and Raman band was reported in ref. [22]. In [22], what the authors have considered as saturable response is known to be the conventional type (CSN). But the earlier report of Lyra *et al.* reveals the existence of two other types of functional forms of SNL, which we named as exponential (ESN) and third one known as coupled type saturable nonlinearity (TSN). Thus, in this article we intent to investigate the influence of different functional form of SNL on the MI.

The organization of the paper is as follows: Following the detailed introduction in the Section I. Section II features the theoretical model, followed by the linear stability analysis in Section III. Section IV is meant for the detailed MI analysis under different sings of the dispersion coefficients. Section V concludes the paper with a detailed summary of the results.

2. Theoretical model

The optical beam propagation in SNL is governed by modified nonlinear Schrödinger equation (MNLSE). The MNLSE for the slowly varying envelope $E(z, t)$ along the z-axis in a retarded reference time is given by [19, 22, 25]

$$i\frac{\partial E}{\partial z} = \frac{\beta_2}{2}\frac{\partial^2 E}{\partial t^2} - \gamma\frac{f(\Gamma|E|^2)}{\Gamma}E, \tag{1}$$

Journal of Physics: Conference Series **605** (2015) 012032 doi:10.1088/1742-6596/605/1/012032

where, t and z are the time and longitudinal coordinate in the moving reference frame, respectively. β_2 is the group velocity dispersion coefficient and $\gamma = \frac{n_2\omega_0}{cA_{eff}}$ is the Kerr parameter, A_{eff} is the effective core area, n_2 is the nonlinear index coefficient.

The term on the extreme right accounts for the saturation of the nonlinear response. The commonly studied saturable nonlinear response in the literature are as follows
Type -I is the CSN of the form [26, 27]

$$f_1(\Gamma|E|^2) = \frac{\Gamma|E|^2}{1+\Gamma|E|^2}, \tag{2a}$$

Type -II is the ESN takes the from [28, 29]

$$f_2(\Gamma|E|^2) = 1 - \exp(-\Gamma|E|^2) \tag{2b}$$

Type - III is the TSN of the form [30]

$$f_3(\Gamma|E|^2) = \frac{\Gamma|E|^2(2+\Gamma|E|^2/2)}{2(1+\Gamma|E|^2/2)^2} \tag{2c}$$

where, $\Gamma = 1/P_s$ is the saturation parameter and P_s is the saturation power. It should be noted that for $\Gamma \to 0$, the nonlinear response of the medium switch back to the conventional Kerr type response *(i.e.)* $f(\Gamma|E|^2) \to \Gamma|E|^2$. The Eq. (1) corresponds to the instantaneous response of the nonlinearity, but for the case of ultrashort pulses the instantaneous nonlinear response is considered as a limiting case, but in strict sense the response of the medium is indeed delayed in nature. Therefore, to account for the finite response time of the nonlinearity, the above equations have to be suitably modeled. One such model is the so-called Debye relaxational Model [19, 22]. Therefore, the equation corresponding to the propagation of the optical beam in the relaxing saturable nonlinear media is given by

$$i\frac{\partial E}{\partial z} = \pm\frac{\beta_2}{2}\frac{\partial^2 E}{\partial t^2} + \gamma N E, \tag{3a}$$

$$\frac{\partial N}{\partial t} = \frac{1}{\tau}[-N + \tilde{f}_j(\Gamma, |E|^2)]. \tag{3b}$$

where, $\tilde{f}(\Gamma, |E|^2) = f(\Gamma|E|^2)/\Gamma$, and N represents the nonlinear index of the medium [22].

3. Linear Stability Analysis

The stability of the steady-state solution in the presence of small perturbations is studied using linear stability analysis [1]. The steady-state solution, (i.e.) the CW solution is given by

$$E_s = E_0 \exp[-i\tilde{f}_j(\Gamma|E_0|^2)z], \tag{4}$$

$$N_s = \tilde{f}_j(\Gamma, |E_0|^2). \tag{5}$$

Here, $a(z,t)$ and $n(z,t)$ are the perturbation corresponding to the stationary solutions of the electric field and nonlinear index, respectively. The resulting perturbed field is of the form

$$E_p = (E_0 + a(z,t)) \exp[-i\tilde{f}_j(\Gamma|E_0|^2)z], \tag{6}$$

$$N_p = \tilde{f}_j(\Gamma, |E_0|^2) + n(z,t). \tag{7}$$

After some mathematical manipulation, the linearized equation for the perturbation can be written as

$$i\frac{\partial a}{\partial z} = \frac{\beta_2}{2}\frac{\partial^2 a}{\partial t^2} + \gamma\, n E_0, \tag{8}$$

$$i\frac{\partial n}{\partial t} = \frac{1}{\tau}[-n + \Gamma \tilde{f}'(\Gamma, |E_0|^2)\, E_0(a + a^*)]. \tag{9}$$

We assume the following ansatz for the perturbation, where Ω is the frequency detuning, and K will be the wavenumber of the perturbation.

$$a(z,t) = U\exp[-i(Kz - \Omega t)] + V\exp[i(Kz - \Omega t)], \tag{10}$$

where, U and V are the amplitude of the perturbation corresponding to the anti-Stokes and Stokes sidebands, respectively. Solving the above equation leads to two homogenous equation for U and V. Collecting the linear terms in U and V, results in a 2 x 2 stability matrix as follows.

$$\begin{bmatrix} K + \beta_2\frac{\Omega^2}{2} + \tilde{\gamma_j}E_0^2 & \tilde{\gamma_j}E_0^2 \\ \tilde{\gamma_j}E_0^2 & -K + \beta_2\frac{\Omega^2}{2} + \tilde{\gamma_j}E_0^2 \end{bmatrix}, \tag{11}$$

where,

$$\gamma_1 \equiv \frac{\gamma}{(1+\Gamma E_0^2)^2}, \quad \tilde{\gamma_1} \equiv \frac{\gamma_1}{(1+i\Omega\,\tau)} \tag{12a}$$

$$\gamma_2 \equiv \gamma\exp[-\Gamma E_0^2], \quad \tilde{\gamma_2} \equiv \frac{\gamma_2}{(1+i\Omega\,\tau)} \tag{12b}$$

$$\gamma_3 \equiv \frac{\gamma}{(1+\Gamma E_0^2/2)^3}, \quad \tilde{\gamma_3} \equiv \frac{\gamma_3}{(1+i\Omega\,\tau)} \tag{12c}$$

The above matrix leads to non-trivial solutions only when the determinant of the matrix vanishes. The stability matrix determine the wavenumber K of the perturbation given by the dispersion relation

$$K = \pm\frac{1}{2}|\beta_2\Omega|[\Omega^2 + sgn(\beta_2)\Omega_c^2]^{1/2} \tag{13}$$

where, $\beta_2 = \pm 1$ depending upon the sign of β_2, the lower sign (-) corresponds to the anomalous dispersion regime and the upper sign (+) corresponds to the normal dispersion regime. Ω_c is the critical modulation frequency (CMF) given by

$$\Omega_c = \left[\frac{4\tilde{\gamma_j}E_0^2}{|\beta_2|}\right]^{1/2} \tag{14}$$

The MI gain is given by $2Im(K)$ and the corresponding expression for the versatile class of SNL can be written in a general form as

$$G(\Omega) = |\beta_2\Omega|[\Omega_c^2 \pm \Omega^2]^{1/2} \tag{15}$$

The dispersion relation K corresponds the stability of the steady state against the harmonic perturbation. It is a well established fact that in the instantaneous limit ($\tau = 0$) of nonlinear response, the MI occurs only if K takes non-zero imaginary part and therefore the steady

Journal of Physics: Conference Series **605** (2015) 012032 doi:10.1088/1742-6596/605/1/012032

state becomes unstable which leads to the exponential growth of the weak perturbation along the length of the fiber. However, in the case of relaxing nonlinear system, the response is not instantaneous and therefore there is a finite delay in the nonlinear response, which is found to be a crucial factor in the MI dynamics.

This finite delay in nonlinear response is equivalent to assuming complex nonlinearity, which consists both real part and imaginary parts. The real part of the nonlinearity accounts for the parametric MI process through the phase matching condition, on the other hand the imaginary part of the nonlinearity models the Raman like process and leads to instability band due to self phase matching. Thus the instability band in the relaxing nonlinear system is of two kinds, (i) Parametric MI bands and (ii) Raman band, respectively. As a consequence, unlike the instantaneous case, where the unstable frequency is finite ($\Omega < \Omega_c$), here any finite value of nonlinear response time leads to the non-zero imaginary part for all frequencies. The saturation on the other hand suppress the MI due to the depletion in the effective nonlinear index of the medium, which is identified to be dependent on the functional form of the type of SNL system.

4. MI Analysis

4.1. Anomalous Dispersion regime ($\beta_2 < 0$)

Here in this case, the system is in the anomalous dispersion ($\beta_2 < 0$) and the phase matching for the parametric MI band can be achieved through a balance between the nonlinearity and negative dispersion coefficient. For the case of saturable system at the instantaneous limit $\tau \to 0$, the nonlinearity coefficient is $\gamma \to \gamma_j$. For simplicity, we change the notation for the pump power as P instead of E_0^2. From the dispersion relation (13), the MI is realized in the frequency region given by $\Omega^2 < \Omega_c^2$. The critical modulational frequency (CMF) corresponding to the MI is given by

$$\Omega_c = \left[\frac{4\gamma_j P}{|\beta_2|}\right]^{1/2} \tag{16}$$

$$\Omega^2 < \Omega_c^2 \tag{17}$$

The frequency at which the MI gain reaches maximum is known to be the optimum modulation frequency (OMF). The OMF corresponding to the spectral band can be written as

$$\Omega_{OMF} \approx \sqrt{\frac{2\gamma_j P}{|\beta_2|}} \tag{18a}$$

where $j = 1, 2, 3$, takes into account ESN, TSN, CSN respectively. The above two equations find huge value in the technical point of view and provides rich variety of information about the various features of SNL in the MI spectrum. For typical system parameters, the maximum MI gain using the expression Eq. (15) for different SNL can be sorted out as $Gmax_{ESN} > Gmax_{TSN} > Gmax_{CSN}$. This difference in the G_{max} is a consequence of the depletion in the nonlinear index depending upon the functional form of the SNL. This change in the nonlinear index leads to the subsequent change in the nonlinear length ($L_{nl} = \frac{1}{\gamma P_0}$), which is identified to be the reason for the difference in CMF and G_{max} among the individual cases of SNL. Thus among the cases of SNL, ESN possess the maximum value of CMF and G_{max} than the rest.

Now, we consider the case of relaxing nonlinear system, due to the finite value of nonlinear response (τ) the effective nonlinear coefficient becomes complex ($\gamma \to \tilde{\gamma}$), which leads to the non-zero imaginary part for all frequencies. Moreover, unlike the case of instantaneous system,

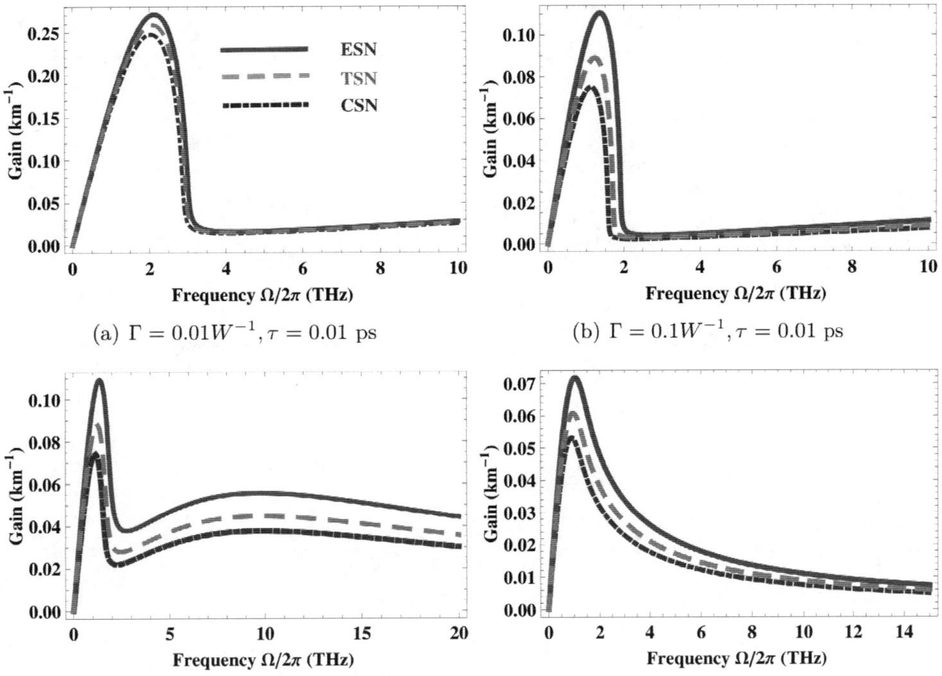

(a) $\Gamma = 0.01W^{-1}, \tau = 0.01$ ps
(b) $\Gamma = 0.1W^{-1}, \tau = 0.01$ ps
(c) $\Gamma = 0.1W^{-1}, \tau = 0.1$ ps
(d) $\Gamma = 0.1W^{-1}, \tau = 1$ ps

Figure 1: The MI spectrum for different combination of Γ and τ in the anomalous dispersion regime. The system parameters are $\beta_2 = -60\,ps^2km^{-1}$, $\gamma = 15W^{-1}km^{-1}$, $P = 10W$.

where there exist unique conventional parametric band, here in the case of relaxing nonlinear system, the unstable frequencies are continuous and run from the center frequency literally down to the infinite frequencies. Interestingly, there exist another band at the higher detuning frequency, which has been recognized as the Raman band. The behavior of SNL is very much in a perceptible manner, where the G_{max} and CMF follows similar order as in the case of instantaneous case. The Raman band due to the delayed nonlinear contribution behaves similar to the conventional band, such that the G_{max} and CMF is maximum for ESN than the rest. To illustrate the influence of relaxation and saturation, the MI spectrum for some representative values of delay (τ) and saturation parameter (Γ) are considered. One can straightforwardly observe from Figs. (1(a)) and (1(b)) that increase in Γ certainly suppresses the MI and also reduces the width of the MI band. This is attributed to the fact that saturation parameter depletes the nonlinear contribution ($\gamma_j < \gamma$), thereby suppress the MI gain and the CMF.

In the limit of slow response time (Fig. 1(d)), the dynamics evolve dramatically in different manner, where there is no signs of distinct instability bands and the MI spectrum consists of only a single coalesced band. The two instability bands (one parametric and one Raman band) that appears for fast response (Figs. 1(a),1(b),1(c)) is no longer distinguishable. The absence of characteristics MI band for higher value of τ is a clear manifestation of the relative dominance of Raman process over the parametric process in the regime of slow response. From Fig. (1(d)), one can also observe that for slow response time, the MI gain strongly decreases and the wide instability band that appears for fast response also decrease in width.

Journal of Physics: Conference Series **605** (2015) 012032 doi:10.1088/1742-6596/605/1/012032

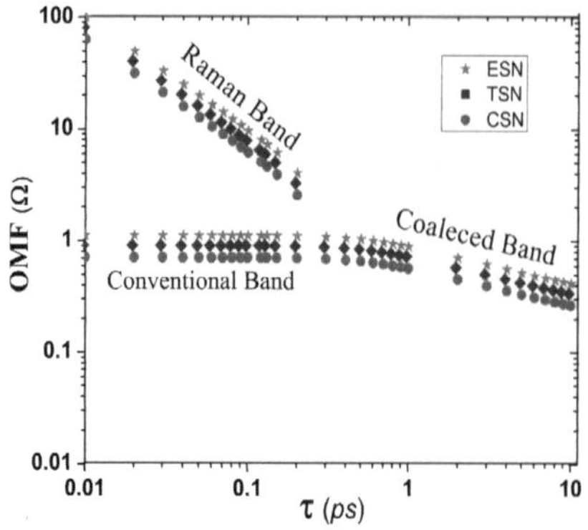

(a) The OMF as a function of delay

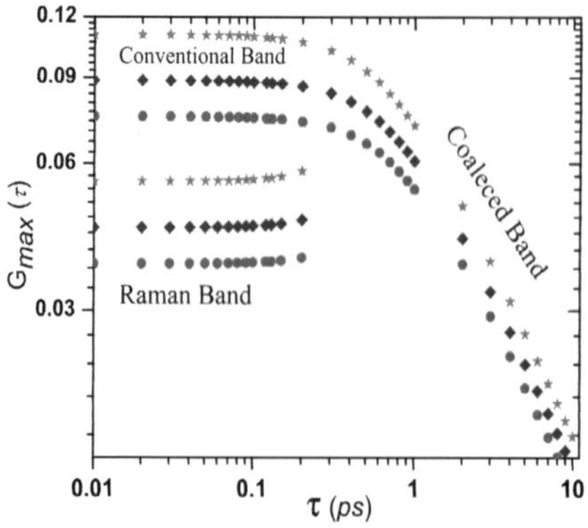

(b) The variation in G_{max} as a function of delay

Figure 2: The variation of OMF and delay for different types of SNL as a function of delay in anomalous dispersion regime.

To put things in perspective, we plot the frequency corresponding to the maximum gain known as optimum modulation frequency (OMF) and the maximum gain (G_{max}) as a function of delay. Fig. (2(a)) shows the OMF as a function of delay time, where one can infer that the nature of the variation of OMF with respected to delay is similar for all class of SNL's. As dealt earlier the OMF of SNL follows the order $OMF_{ESN} > OMF_{TSN} > OMF_{CSN}$. In the regime of fast response (low values of τ) the conventional band remain unaffected and separated from the Raman band. Although the parametric instability bands are found to be independent of τ, the Raman band is extremely sensitive to the delay and decreases by a factor $1/\tau$, as shown in the Fig. (2(a)).

The behavior of MI in the regime of slow response is interesting since the parametric band corresponding to the individual type of SNL vanishes and a single coalesced band can only be observed. From Fig. (2(a)), we infer that with increase in the delay time, the Raman band downshift towards the center frequency and gets coalesced with the instantaneous band at the delay time corresponding to the order of the inverse of the CMF of SNL at the instantaneous limit [22]. Further increase in the delay time leads to the continuous downshift of coalesced band towards the center frequency at a rate of $1/\tau^{1/3}$.

The maximum MI gain associated with the instability band and the Raman band is analyzed in the Fig. (2(b)). As discussed earlier, among the types of SNL, ESN register higher value of G_{max} than the rest. The gain of the Raman band is found to be nearly half of the gain of the conventional band and it is found to be insensitive to the response time. In the regime of the slow response, for modest value of the delay (typically $\tau = 0.1 - 1$ ps), the G_{max} of the conventional band falls rapidly and coalesced with the Raman band as evident from the Fig. (2(b)). Further increase in delay time, G_{max} of the resulting coalesced band decreases by a factor $1/\tau^{2/3}$.

4.2. Normal Dispersion regime ($\beta_2 > 0$)

In the case of normal dispersion regime, the dispersion coefficient β_2 is positive, therefore there is no possibility of phase matching between the nonlinear and dispersion effects and hence, thus ruled out the parametric MI process. However, Saleh *et al.*, demonstrated that the formation

Journal of Physics: Conference Series **605** (2015) 012032 doi:10.1088/1742-6596/605/1/012032

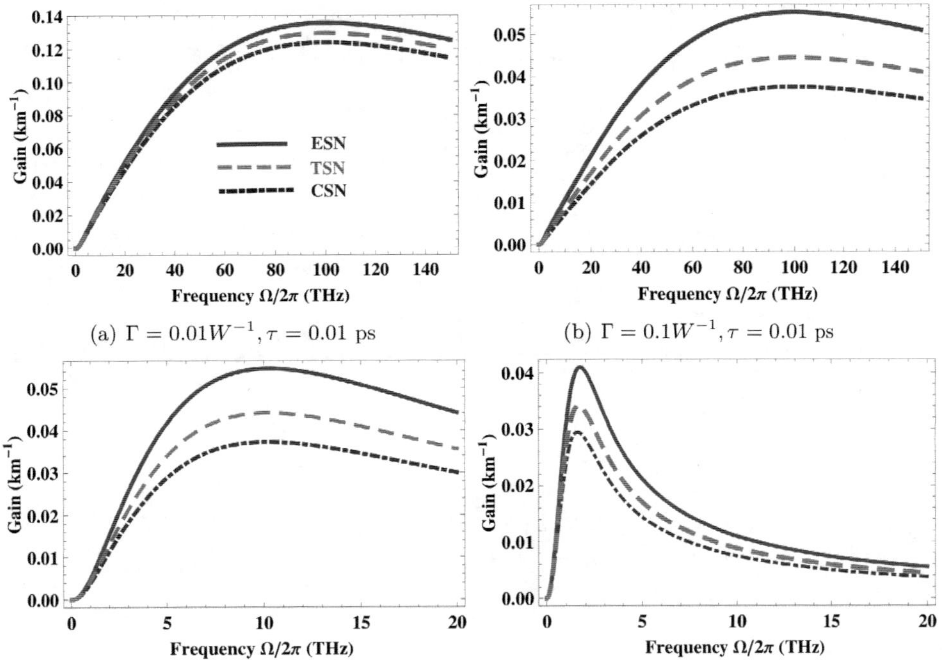

(a) $\Gamma = 0.01W^{-1}, \tau = 0.01$ ps

(b) $\Gamma = 0.1W^{-1}, \tau = 0.01$ ps

(c) $\Gamma = 0.1W^{-1}, \tau = 0.1$ ps

(d) $\Gamma = 0.1W^{-1}, \tau = 1$ ps

Figure 3: The MI spectrum for different combination of Γ and τ in the normal dispersion regime. The system parameters are $\beta_2 = 60\,ps^2km^{-1}$, $\gamma = 15W^{-1}km^{-1}$, $P = 10W$.

of plasma inside gas-filed photonic crystal fibers might yield to the onset of MI in the normal dispersion regime [3]. Here in the case of relaxing system, the incorporation of finite response time leads to instability band (Raman band) even all the possibilities for parametric process is cut out. This is due to the imaginary part of the nonlinearity, which accounts for the MI gain.

Fig. (3) shows the instability band for different combination of delay and saturation parameter. The Fig. (3) illustrates that the G_{max} and the OMF of the Raman band is maximum for ESN than the rest. One can also observe that the G_{max} and the OMF is high for fast response than the slow response. Also the the increase in saturation suppress the instability gain of the Raman band.

For the further insight into the dynamics of MI in the normal dispersion regime of the relaxing nonlinear system we drawn in Fig. (4) the OMF and the G_{max} as a function of delay time. Fig. (4(a)) shows that at fast response the OMF of the Raman band decreases by a factor $1/\tau$ and further increase in the delay takes to the slow response regime. The behavior of G_{max} is quite in a perceptible manner where in fast response case the gain of the Raman band is insensitive to delay, followed by a continuous decrease in the gain at the slow response regime by a factor $1/\tau^{2/3}$ as shown in the Fig. (4(b)).

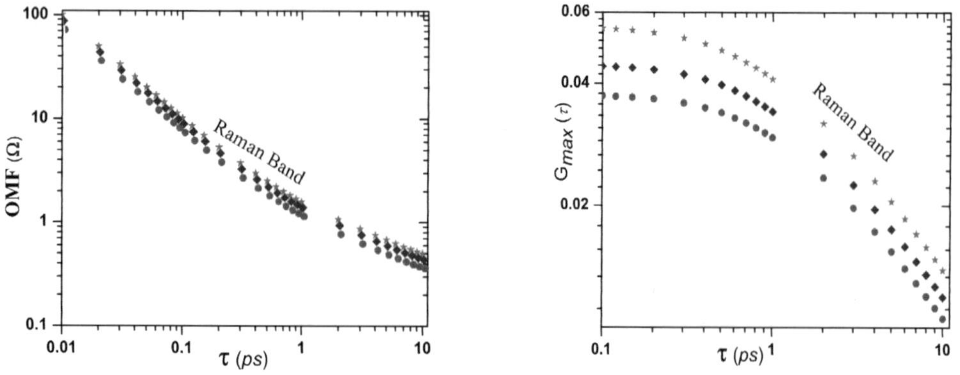

(a) The OMF as a function of delay

(b) The variation in G_{max} as a function of delay

Figure 4: The variation of OMF and delay for different types of SNL as a function of delay in normal dispersion regime.

5. Summary and Conclusion

In summary, we have investigated the MI in the relaxing nonlinear system under the combined action of nonlinear saturation and relaxation. We suitably modeled the standard NLSE, to incorporate the relaxation and saturation of nonlinear response using modified Debye relaxational model. Using the linear stability analysis, the exact dispersion relation corresponding to the stability of the steady solution against harmonic perturbation is found and the MI analysis is performed for different signs of dispersion coefficient. The saturation in general suppresses the MI by depleting the nonlinear contribution and it is found to be dependent on the functional form of the saturable media. In anomalous dispersion regime, there exist two unstable bands (*i.e.*) the conventional MI band due to parametric process and the Raman band due to the imaginary contribution of the nonlinear response. Interestingly, unstable band is still observed even in normal dispersion regime, which is purely due to the delayed nature of the nonlinear response. Among the different functional form of SNL, ESN is found to be dominant and register maximum value of OMF and G_{max}. The general order of MI identities like CMF, OMF, G_{max} can be sorted as $ESN > TSN > CSN$. To conclude, through our theoretical analysis we brings to the light that the characteristic functional form of SNL is a crucial factor in the MI phenomena. We believe that our theoretical result will be useful and can stimulate new experiments in the context of MI in relaxing nonlinear systems.

Acknowledgment

KP thanks DST, CSIR, NBHM, IFCPAR and DST- FCT Government of India, for the financial support through major projects. K. Nithyanandan thanks CSIR, Government of India for providing financial support by awarding Senior Research Fellowship (SRF).

References

[1] Agrawal G 2012 *Nonlinear fiber optics* (San Diego, 5th edition: Academic Press)
[2] T B Benjamin and J E J Fier 1967 *Fluid Mechn.* **27** 417
[3] Saleh M F, Chang W, Travers J C, Russell P S J and Biancalana F 2012 *Phys. Rev. Lett.* **109**(11) 113902
[4] Hasegawa A 1984 *Opt. Lett.* **9** 288–290
[5] Tai K, Hasegawa A and Tomita A 1986 **56** 135–138

[6] Potasek M J 1987 *Opt. Lett.* **12** 921–923
[7] Dudley J M, Genty G and Coen S 2006 *Rev. Mod. Phys.* **78** 1135–1184
[8] K Nithyanandan Vasantha Jayakantha Raja R and Porsezian K 2014 *Laser Physics* **24** 045405
[9] Hammani K, Wetzel B, Kibler B, Fatome J, Finot C, Millot G, Akhmediev N and Dudley J M 2011 *Opt. Lett.* **36** 2140–2142
[10] Kumar A, Labruyere A and Dinda P T 2003 *Optics Communications* **219** 221 – 232 ISSN 0030-4018
[11] Coen S and Haelterman M 2001 *Opt. Lett.* **26** 39–41
[12] Trillo S, Wabnitz S, Stegeman G I and Wright E M 1989 *J. Opt. Soc. Am. B* **6** 889–900
[13] Biancalana F, Skryabin D V and Russell P S J 2003 *Phys. Rev. E* **68**(4) 046603 URL `http://link.aps.org/doi/10.1103/PhysRevE.68.046603`
[14] Gatz S and Herrmann J 1991 *J. Opt. Soc. Am. B* **8** 2296–2302
[15] Chen Y F, Beckwitt K, Wise F W, Aitken B G, Sanghera J S and Aggarwal I D 2006 *J. Opt. Soc. Am. B* **23** 347–352
[16] Hickmann J M, Cavalcanti S B, Borges N M, Gouveia E A and Gouveia-Neto A S 1993 *Opt. Lett.* **18** 182–184
[17] Herrmann J 1991 *J. Opt. Soc. Am. B* **8** 1507–1511
[18] Coutaz J L and Kull M 1991 *J. Opt. Soc. Am. B* **8** 95–98
[19] Liu X, Haus J W and Shahriar S 2008 *Optics Communications* **281** 2907 – 2912
[20] Zhang L, Wen S, Fu X, Deng J, Zhang J and Fan D 2010 *Optics Communications* **283** 2251 – 2257
[21] Cambournac C, Maillotte H, Lantz E, Dudley J M and Chauvet M 2002 *J. Opt. Soc. Am. B* **19** 574–585
[22] Silva G L D, Gleria I, Lyra M L and Sombra A S B 2009 *J. Opt. Soc. Am. B* **26** 183–188
[23] Nithyanandan K, Raja R V J and Porsezian K 2013 *J. Opt. Soc. Am. B* **30** 178–187
[24] Shuang-Chun W, Wen-Hua S, Hua Z, Xi-Quan F, Lie-Jia Q and Dian-Yuan F 2003 *Chinese Physics Letters* **20** 852
[25] Dinda P T and Porsezian K 2010 *J. Opt. Soc. Am. B* **27** 1143–1152
[26] Porsezian K, Nithyanandan K, Raja R V J and Shukla P K 2012 *J. Opt. Soc. Am. B* **29** 2803–2813
[27] K Nithyanandan Vasantha Jayakantha Raja R P K and Uthayakumar T 2013 *Opt. Fib. Tech.* **19** 348
[28] Al-Bader Samir J J H A 1987 *IEEE Journal of Quantum Electronics* **QE-23** 1947–1955
[29] Xian-Qiong Z and An-Ping X 2007 *Chinese Physics* **16** 1683
[30] Wood V E, Evans E D and Kenan R P 1988 *Optics Communications* **69** 156 – 160

Polarization characteristics of nonlinear transmission in rigidly held saturable-dye molecules with random orientations

S Miyanaga and T Sato
College of Design and Manufacturing Technology, Muroran Institute of Technology, 27-1 Mizumoto, Muroran, Hokkaido 050-8585, Japan

E-mail: miyanaga@mmm.muroran-it.ac.jp

Abstract. Polarization-dependent nonlinear transmissions are investigated by a pump-probe method in saturable-dye-doped films in which optically anisotropic saturable dyes are rigidly held with random orientations. The nonlinear transmissions measured by using uranine-doped poly(vinyl alcohol) films are compared with the theoretical predictions that are obtained by considering the effects of pump propagation and molecular orientation on the basis of a rate equation analysis for a four-energy-level model including an excited-state absorption. The measurements were conducted for the two cases of polarization states for which the polarization direction of the probe wave is either parallel or perpendicular to that of the pump wave; the experimental results considerably deviated from the theoretical ones for the probe wave perpendicularly polarized to the pump wave. It is shown that this is explained by modifying the energy level model to include the existence of a nearly-orthogonal component of the transition dipole moment associated with the ground-state absorption in uranine dyes.

1. Introduction

Saturable-dye-doped solid films have been widely used as a phase conjugator chiefly because of their relatively low saturation intensity and easy preparation of large samples in area; their possible applications for real-time image processing have also been demonstrated [1,2]. In dye-doped films, optically anisotropic dye molecules are fixed in solid polymer matrix with random positions and orientations, being isotropic as a whole. Irradiated with linearly polarized intense light, the dye molecules whose transition dipole moments are near directed to the polarization direction are selectively excited and highly saturated by the incident wave. This leads to photo-induced uniaxial anisotropy whose optic axis is defined by the polarization direction. Thus, the nonlinear polarization induced in such films depends not only on the light intensity but also on the polarization states of incident light waves, leading to polarization-dependent saturable absorption characteristics as shown by Kramer et al. [3]. The polarization characteristics of nonlinear transmissions in those media were revealed by Montecchi et al. [4] by comparing the experimental results with the theoretical ones based on a four-energy-level model for the saturable dyes. In our measurements on the nonlinear transmissions by a pump-probe method in uranine-doped poly(vinyl alcohol) (PVA) films, however, the experimental results considerably deviated from the theoretical ones for the case of the probe wave perpendicularly polarized to the pump wave. We investigate a possible mechanism for this observation, in this work, and show that these deviations from the theoretical predictions are explained by the existence of a nearly-orthogonal component of the transition dipole moment associated with the ground-state absorption in uranine dyes.

2. Theory

2.1. Molecular orientation of saturable dyes and Energy-level model

Saturable dyes are basically modeled by a four-energy-level system including an excited-state absorption [4,5], and nonlinear polarization in this system can be derived from a rate equation analysis. In references [4] and [5], the directions of the transition dipole moments associated with a ground- and an excited-state absorption are assumed to be parallel to each other, and each dye molecule is assumed to have only one component of the transition dipole moment associated with the ground-state absorption [6]. In this work, however, we modify the energy-level model to include a nearly orthogonal component of that associated with the ground-state absorption. Thus, we define the directions of the transition dipole moments associated with each absorptive transition in a laboratory coordinate system as shown in figure 1.

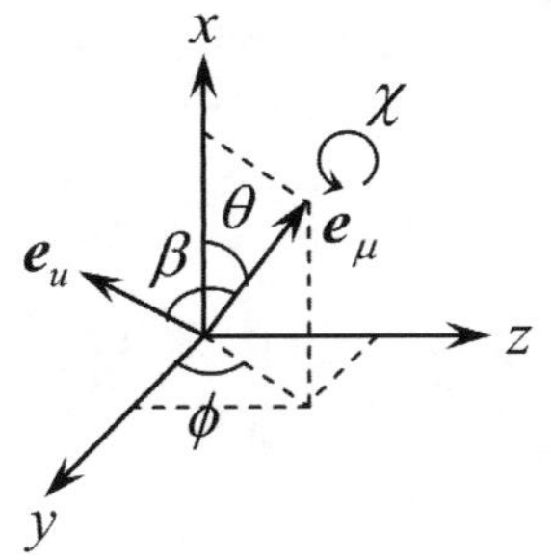

Figure 1. Directions of the transition dipole moments associated with the ground- and the excited-state absorption depicted by a unit vector $\boldsymbol{e}_\mu$, and the nearly orthogonal component of the one associated with the ground-state absorption depicted by a unit vector $\boldsymbol{e}_u$ in a laboratory coordinate system. An angle χ represents the degree of freedom for rotation around the axis in a molecular coordinate system.

In figure 1, $\boldsymbol{e}_\mu$ stands for the unit vector along the transition dipole moments associated with the ground- and the excited-state absorption, and $\boldsymbol{e}_u$ the nearly orthogonal component of that associated with the ground-state one, which is assumed to be at an angle β from $\boldsymbol{e}_\mu$ in a molecular coordinate system. In the present model, the $\boldsymbol{e}_u$ component is assumed not to contribute to the changes in the population densities among the energy levels. The pump wave is assumed to be linearly polarized along the x-direction, and the probe wave to be either parallel or perpendicularly polarized to the pump wave.

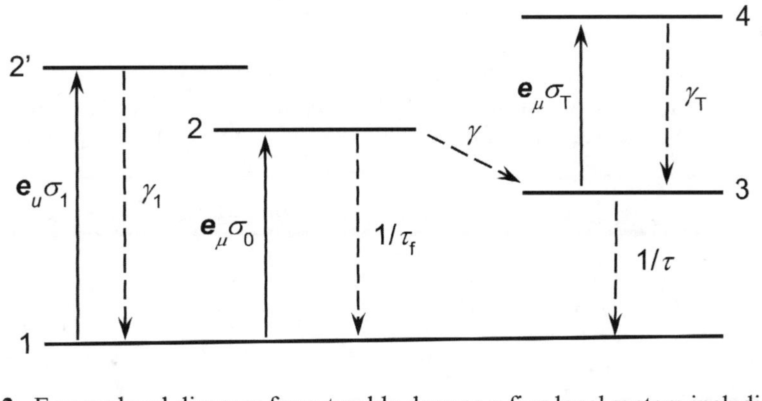

Figure 2. Energy level diagram for saturable dyes as a five level system including the nearly orthogonal component of the ground-state absorption as well as the excited-state absorption. σ_0, σ_1 and σ_T are the absorption cross sections associated with each absorptive transition, τ_f and τ are a fluorescent and a phosphorescent life time, and γ, γ_1 and γ_T are the transition rates associated with an intersystem crossing for γ and internal conversions for γ_1 and γ_T, respectively.

Journal of Physics: Conference Series **605** (2015) 012033 doi:10.1088/1742-6596/605/1/012033

Thus, the saturable dyes are modeled by a five energy-level system including the nearly orthogonal component of the ground-state absorption as well as the excited-state absorption, as shown in figure 2. In figure 2, levels 1, 2 and 2' are singlet states, while levels 3 and 4 triplet states. The molecule excited by the incident light from its ground state, level 1, occupies the lowest excited-triplet state, level 3, with certain probability through intersystem crossing, since the triplet state has relatively long phosphorescent lifetime. Thus, the depletion of the ground-state population density gives rise to the saturable absorption effect. σ_0 and σ_T stand for the absorption cross sections associated with the ground- and the excited-state absorption, respectively, which cause the depletion of the ground-state population density. On the other hand, σ_1 represents the nearly orthogonal component of that associated with the ground-state absorption, which is assumed not to contribute to the changes in the population densities among the energy levels. τ_f and τ are a fluorescent and a phosphorescent life time, respectively. γ denotes the transition rate for an intersystem crossing, while γ_1 and γ_T the ones for internal conversions. As mentioned in figure 1, in this model, the transition dipole moments associated with σ_0 and σ_T are assumed to be parallel each other, having the $\boldsymbol{e}_\mu$ direction, and the one associated with σ_1, having the $\boldsymbol{e}_u$ direction, is nearly orthogonal to them.

2.2. Rate equation analysis for nonlinear transmission

The population density of each level can be derived from the rate equation analysis. In this model, the appreciably populated states are only the ground and the lowest triplet state, levels 1 and 3. The steady-state population densities of these levels are obtained as [5,7]

$$\rho_1 = \frac{1}{1+R_0^2\, I_0/I_{\mathrm{SAT}}}\,,\quad \rho_3 = 1-\rho_1 = \frac{R_0^2\, I_0/I_{\mathrm{SAT}}}{1+R_0^2\, I_0/I_{\mathrm{SAT}}}\,,\quad R_0 = \boldsymbol{e}_\mu \cdot \boldsymbol{e}_0\,, \tag{1}$$

where ρ_1 and ρ_3 are the population densities of the ground and the excited state, respectively, I_0 is the intensity of the pump wave, and $\boldsymbol{e}_0$ stands for the unit vector along the polarization direction of the pump wave. Here, the changes in the population densities are influenced by the $\boldsymbol{e}_\mu$ component of the transition dipole moment alone, and the effects of the molecular orientation are accounted for by the term R_0. In equation (1), with the frequency ν of the incident light and Plank's constant h, a saturation parameter I_{SAT} is defined as

$$I_{\mathrm{SAT}} = \frac{h\nu}{\sigma_0 q \tau}\,,\quad q = \frac{\gamma}{1/\tau_f + \gamma}\,. \tag{2}$$

Here, q is triplet yield, which represents the quantum yield of triplet formation.

The nonlinear absorption properties can be derived from the results of the rate equation analysis. The saturated absorption cross section of each dye molecule for the probe wave polarized to an $\boldsymbol{e}_{\mathrm{Pr}}$ direction is obtained as

$$\boldsymbol{\sigma} = \boldsymbol{e}_\mu R_0^{\mathrm{Pr}}(\sigma_0\rho_1 + \sigma_T\rho_3) + \boldsymbol{e}_u R_1^{\mathrm{Pr}}\sigma_1\rho_1\,,\quad R_0^{\mathrm{Pr}} = \boldsymbol{e}_\mu \cdot \boldsymbol{e}_{\mathrm{Pr}}\,,\quad R_1^{\mathrm{Pr}} = \boldsymbol{e}_u \cdot \boldsymbol{e}_{\mathrm{Pr}}\,. \tag{3}$$

In the case of the pump wave linearly polarized to the x-direction ($\boldsymbol{e}_0=\boldsymbol{e}_x$), the scalar products of the unit vectors are given by

$$R_0 = R_0^{\mathrm{Pr}} = \cos\theta\,,\quad R_1^{\mathrm{Pr}} = \cos\theta\cos\beta - \sin\theta\cos\chi\sin\beta \tag{4}$$

for the parallel polarized probe wave ($\boldsymbol{e}_{\mathrm{Pr}}=\boldsymbol{e}_x$), and

$$R_0^{\mathrm{Pr}} = \sin\theta\sin\phi\,,\quad R_1^{\mathrm{Pr}} = \sin\theta\sin\phi\cos\beta + (\cos\theta\sin\phi\sin\chi + \cos\phi\cos\chi)\sin\beta \tag{5}$$

for the perpendicularly polarized probe wave ($\boldsymbol{e}_{\mathrm{Pr}}=\boldsymbol{e}_y$). The nonlinear absorption coefficient $\boldsymbol{\alpha}$ is obtained by taking orientational average of $\boldsymbol{\sigma}$ as

$$\boldsymbol{\alpha} = N\langle\boldsymbol{\sigma}\rangle = \frac{N}{8\pi^2}\int_0^{2\pi}\int_0^{2\pi}\int_0^{\pi}\boldsymbol{\sigma}\sin\theta\,\mathrm{d}\phi\,\mathrm{d}\chi\,\mathrm{d}\theta\,. \tag{6}$$

Here, N is the number density of the dye molecules, and the bracket denotes orientational averaging.

The nonlinear transmission for each polarization state can be obtained by the following procedures. The pump propagation effect is, first, revealed by numerically solving the nonlinear propagation equation for the pump intensity I_0 as a function of z, penetration depth in the dye-doped film, together with equations (1)-(4) and (6):

$$\frac{\mathrm{d}I_0}{\mathrm{d}z} = -\alpha_\mathrm{S}(I_0)I_0\,. \tag{7}$$

Then, the nonlinear transmissions for the probe wave parallel (T_S) and perpendicularly (T_P) polarized to the pump wave can be calculated from the nonlinear propagation equations for the probe intensity I_Pr with film thickness L:

$$\frac{\mathrm{d}I_\mathrm{Pr}}{\mathrm{d}z} = -\alpha_\mathrm{S}(I_0)I_\mathrm{Pr}\,,\ T_\mathrm{S} = \frac{I_\mathrm{Pr}(L)}{I_\mathrm{Pr}(0)}, \tag{8}$$

by using equation (4) for the parallel polarized probe wave;

$$\frac{\mathrm{d}I_\mathrm{Pr}}{\mathrm{d}z} = -\alpha_\mathrm{P}(I_0)I_\mathrm{Pr}\,,\ T_\mathrm{P} = \frac{I_\mathrm{Pr}(L)}{I_\mathrm{Pr}(0)}, \tag{9}$$

by using equation (5) for the perpendicularly polarized probe wave. In the propagation equations, the nonlinear absorption coefficients for the parallel (α_S) and the perpendicularly (α_P) polarized probe wave are given by

$$\alpha_\mathrm{S}(I_0) = \boldsymbol{\alpha}\cdot\boldsymbol{e}_x = N\langle\boldsymbol{\sigma}\cdot\boldsymbol{e}_x\rangle,\ \alpha_\mathrm{P}(I_0) = \boldsymbol{\alpha}\cdot\boldsymbol{e}_y = N\langle\boldsymbol{\sigma}\cdot\boldsymbol{e}_y\rangle, \tag{10}$$

respectively, as a function of the pump intensity I_0. In this system, as a result of the random orientation of the dye molecules, a small-signal and a large-signal absorption coefficient are given as

$$\alpha_0 = \alpha_\mathrm{S}(0) = N(\sigma_0+\sigma_1)/3\,,\ \alpha_\infty = \alpha_\mathrm{S}(\infty) = N\sigma_\mathrm{T}/3\,, \tag{11}$$

respectively, and saturation intensity I_S is defined by [5]

$$\alpha_\mathrm{S}(I_\mathrm{S}) = (\alpha_0+\alpha_\infty)/2\,. \tag{12}$$

Here, I_S differs numerically from I_SAT of equation (2) by a constant factor.

2.3. Numerical results

Figure 3 shows the numerical results of the nonlinear transmission as a function of the pump intensity I_0 normalized by the saturation intensity I_S for the probe wave parallel (T_S in the figure) and perpendicularly (T_P in the figure) polarized to the pump wave. The solid lines stand for the nonlinear transmissions obtained by considering the effects of the nearly orthogonal component of the absorption cross section σ_1 associated with the ground-state absorption, and the dashed lines those obtained by ignoring the effects of σ_1. In the numerical calculations, the angle β is assumed, for simplicity, to be 90 degrees, and the following parameters are used: the product of the small-signal absorption coefficient α_0 and the film thickness L as $\alpha_0 L$=1.0; the ratios of absorption cross sections as G_0=$\sigma_\mathrm{T}/\sigma_0$=0.46 and G_1=σ_1/σ_0=0.15 for the solid lines, and as G_0=0.40 and G_1=0 (σ_1=0) for the dashed lines.

The effects of the absorption cross section σ_1 are clearly seen in figure 3. While the parallel component T_S is decreased in the region of relatively high pump intensity, the perpendicular one T_P is increased within the wide range of the pump intensity. This can be attributed to the effects of the direction of the transition dipole moment associated with σ_1. The dye molecule closely oriented to the polarization direction of the linearly polarized pump wave highly occupies its excited state with the irradiation of the pump wave; this leads to the simultaneous extinctions of the absorptive transitions associated with σ_0 and σ_1. On the other hand, the dye molecule perpendicularly oriented to the polarization direction of the pump wave remains in its ground state up to considerably high pump intensity. As a result, the nonlinear transmissions, T_S and T_P, are changed by the way observed in figure 3.

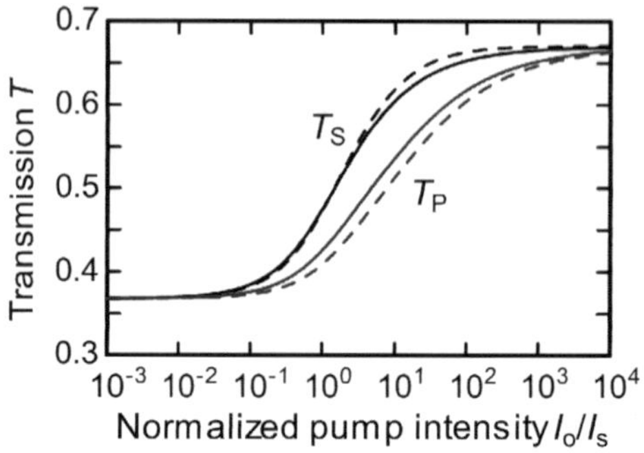

Figure 3. Nonlinear transmission T as a function of the pump intensity I_0 normalized by the saturation intensity I_S for the probe wave parallel (T_S) and perpendicularly (T_P) polarized to the pump wave. The solid and the dashed lines are the corresponding theoretical curves calculated with (solid lines) and without (dashed lines) considering the effects of the nearly orthogonal component of the absorption cross section σ_1 associated with the ground-state absorption.

3. Experimental results and discussion

In the experiment, uranine-doped PVA films with the thickness L of about 80 μm were used as the saturable absorbing media. A strong pump and a weak probe wave from an Ar-ion laser were adjusted to overlap incoherently in the dye-doped film to suppress any wave-mixing processes. The probe intensity was set to be 1/100 as large as the saturation intensity to avoid additional saturation. The polarization direction of the linearly polarized probe wave was at 45 degrees to that of the linearly polarized pump wave. After passing through an interference filter to eliminate fluorescence and phosphorescence emitted from the dye-doped film, the transmitted probe wave was divided into two components that are parallel and perpendicularly polarized to the pump wave by a Wollaston prism. Each polarization component was detected by a photo-diode, from which the nonlinear transmission for the parallel component T_S and for the perpendicular component T_P was derived. The measurements were conducted at the wavelength of 514.5 nm.

Figures 4(a) and 4(b) show the experimental results of the pump-intensity dependence of the nonlinear transmission measured by using the uranine-doped PVA film with $\alpha_0 L$ of 0.50. The symbols are the experimental data for T_S (closed circles) and for T_P (closed triangles), and the solid lines are the corresponding theoretical curves obtained (a) by ignoring the effects of the nearly orthogonal $\boldsymbol{e}_u$ component of the absorption cross section σ_1 associated with the ground-state absorption, and (b) by considering these effects. The parameters of I_S, G_0 and G_1 determined by least-square curve fitting are listed in each figure. It is found from figure 4 that, while the experimental data for the perpendicular components T_P deviate considerably from the theoretical curve without considering the effects of σ_1 in (a), they agree quite well with that obtained by considering these effects in (b). These tendencies were reproducibly observed in the various samples having the different values of $\alpha_0 L$.

Similar deviations of the experimental data from the theoretical curve for the perpendicular component T_P seen in figure 4(a) were also observed in the results shown in reference [4].

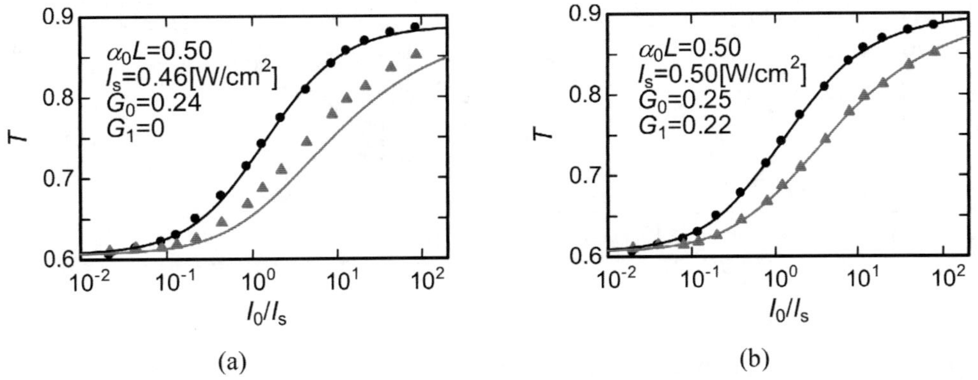

Figure 4. Experimental results of the pump-intensity dependence of the nonlinear transmission measured by using the uranine-doped PVA film with α_0L of 0.50. The closed circles are the experimental data for T_S and the closed triangles for T_P, and the solid lines are the corresponding theoretical curves. The effects of the nearly orthogonal component of the absorption cross section σ_1 associated with the ground-state absorption are ignored in the calculations of (a), while considered in the calculations of (b).

Figures 5 and 6 show the experimental results of the pump-intensity dependence of the nonlinear transmission measured by using the uranine-doped PVA film with α_0L of 1.02 in figures 5(a) and 5(b), and of 2.13 in figures 6(a) and 6(b). As in figure 4, the experimental data for the perpendicular components T_P agree quite well with the theoretical curve obtained by considering the effects of σ_1 in (b). When the direction of the transition dipole moment associated with the excited-state absorption cross section σ_T inclines from that associated with the ground-state one σ_0, the theoretical curves of the nonlinear transmissions, T_S and T_P, are also changed by the angle between the dipole moments associated with σ_0 and σ_T [7]. The numerical calculations, however, revealed that the theoretical curves obtained by considering this effect decreases T_P contrary to the experimental results.

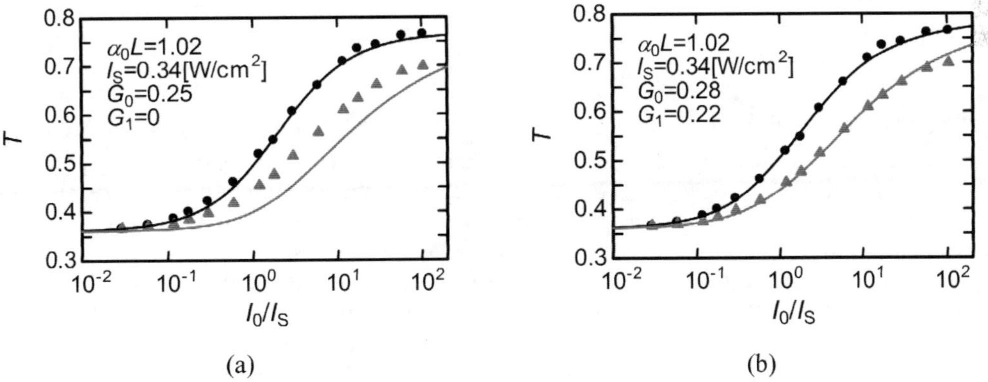

Figure 5. Experimental results of the pump-intensity dependence of the nonlinear transmission measured by using the uranine-doped PVA film with α_0L of 1.02. The meanings of the symbols and the lines are the same as those in figure 4. The effects of the absorption cross section σ_1 are ignored in the calculations of (a), while considered in the calculations of (b).

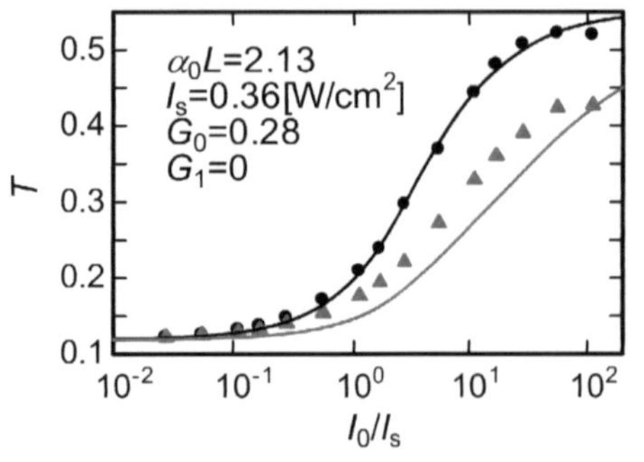

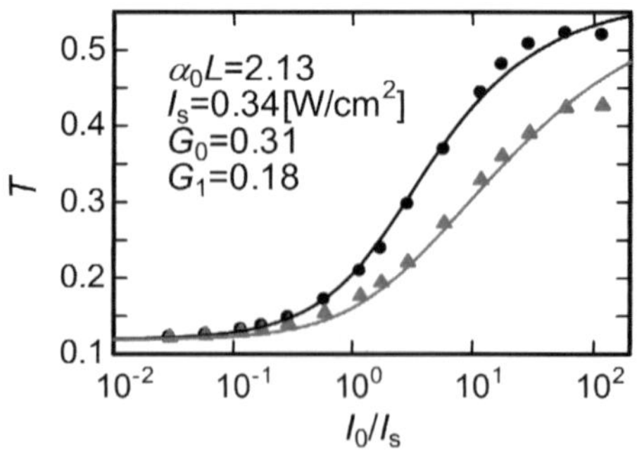

Figure 6. Experimental results of the pump-intensity dependence of the nonlinear transmission measured by using the uranine-doped PVA film with α_0L of 2.13. The meanings of the symbols and the lines are the same as those in figure 4. The effects of the absorption cross section σ_1 are ignored in the calculations of (a), while considered in the calculations of (b).

The saturable absorption parameters of I_S, G_0 and G_1 determined by the least-square curve fitting in (b) of figures 4-6 are listed in table 1. Reproducibility of the determination of these parameters was examined by using the samples with α_0L being approximately equal to 1.0. The mean values with standard deviations among twelve measurements were 0.40 with 0.07 for I_S in W/cm^2, 0.30 with 0.05 for G_0, and 0.21 with 0.03 for G_1, respectively. Although the scattering of data is slightly large, the obtained parameters are consistent with each other among three values of α_0L.

Table 1. Saturable absorption parameters of I_S, G_0 and G_1 determined by the least-square curve fitting in (b) of figures 4-6.

α_0L	I_S [W/cm^2]	G_0	G_1
0.50	0.50	0.25	0.22
1.02	0.34	0.28	0.22
2.13	0.34	0.31	0.18

4. Conclusions

In this work, the polarization-dependent nonlinear transmissions were investigated by the pump-probe method in the saturable-dye-doped films in which optically anisotropic saturable dyes are rigidly held with random orientations. The nonlinear transmissions measured by using the uranine-doped PVA films were compared with the theoretical predictions obtained by considering the effects of the pump propagation and the molecular orientation on the basis of the rate equation analysis. For the saturable dyes, the four-energy-level model with the excited-state absorption previously used was extended, in this work, to include the nearly orthogonal component of the absorption cross section σ_1 associated with the ground-state absorption, which is assumed not to contribute to the changes in the population densities among the energy levels. The measurements were conducted for the two cases of polarization states for which the polarization direction of the probe wave is either parallel or perpendicular to that of the pump wave. The experimental data considerably deviated from the theoretical prediction for the case of the probe wave perpendicularly polarized to the pump wave without considering the effects of the absorption cross section σ_1; it was found, however, that this discrepancy becomes significantly small by considering the effects of σ_1. These tendencies were reproducibly confirmed in the various samples with the different values of $\alpha_0 L$ at the wavelength of 514.5 nm from the Ar-ion laser. In the analysis of the polarization characteristics of wave mixing processes, such as phase conjugation and two-wave mixing, in the materials consisting of saturable-dye molecules rigidly held with random orientations, it is important to include the effects of the nearly orthogonal component shown in this work.

References

[1] Sharma K K, Divakara Rao K and Ravindra Kumar G 1994 Nonlinear optical interactions in dye-doped solids *Opt. Quantum Electron.* **26** 1-23, and references therein (1994)

[2] Takeda T, Mori S and Fujiwara H 1997 Influence of thickness of metyl-orange-doped polyvinyl alcohol film on phase conjugate matched filtering *Opt. Rev.* **4** 490-495

[3] Kramer M A, Tompkin W R and Boyd R W 1986 Nonlinear-optical interactions in fluorescein-doped boric acid glass *Phys. Rev.* A **34** 2026-31

[4] Montecchi M, Settembre M and Romagnoli M 1988 Optically induced birefringence and dichroism in rigidly held dye molecules *J. Opt. Soc. Am.* B **5** 2357-62

[5] Sato T and Miyanaga S 1999 Polarization dependence of phase-conjugate reflectivities in solid films containing randomly oriented saturable dyes *IEEE J. Quantum Electron.* **QE-35** 179-186

[6] Lewis G N, Lipkin D and Magel T T 1941 Reversible photochemical processes in rigid media. A study of the phosphorescent state *J. Am. Chem. Soc.* **63** 3005-18

[7] Sato T and Miyanaga S 1999 Polarization dependent phase-conjugate reflectivity in randomly oriented saturable absorbers: effects of direction of transition dipole moment associated with excited-state absorption *Opt. Rev.* **6** 522-528

Journal of Physics: Conference Series **605** (2015) 012034 doi:10.1088/1742-6596/605/1/012034

Properties of the self-trapped light in an array of subwavelength waveguides

G Mendoza-González and E A Martí-Panameño

Facultad de Ciencias Físico-Matemáticas, Benemérita Universidad Autónoma de Puebla. Avenida San Claudio y 18 sur, Colonia San Manuel, Puebla, CP 72570, Puebla, México.

E-mail: gmendoza1050@gmail.com, erwin.marti@correo.buap.mx

Abstract. Based in the numerical experiment techniques, we study the propagation dynamics of electromagnetic waves along one-dimensional array of parallel cylindrical subwavelength waveguides, possessing Kerr nonlinearity. By means of the proper selection of the array parameters, as well as the incident radiation properties and applying the Finite Difference Time Domain method, we are able to observe the light self-trapping in one single subwavelength waveguide. Furthermore, the position of the output beam can be controlled by the phase difference and the angle of incidence of the input beams. These results could give the possibility to control light by light, with perspectives in applications to implement integrated optics devices at nanoscale.

1. Introduction

Nonlinear optical effects in discrete systems are subjects of both, theoretical and experimental research along several decades. The discretized nature of light propagation gives rise to many phenomena which are not possible in homogeneous bulk media, such as discrete diffraction, diffraction management, temporal and spatial discrete soliton [1–5]. There are many researches focusing in the way to find the dimension properties of a waveguide [6–8], to obtain smaller nonlinear discrete systems and strong confinement of light in waveguides with dimensions smaller than the wavelength of incident light and to propagate in low dimensional periodic structures [9]. Even so, the discrete properties of light in subwavelength arrays are not complete clear, especially in pure dielectric arrays, so, we develop here a numerical study of light propagation and interaction in an one-dimensional array of nonlinear parallel dielectric subwavelength waveguides (SWWGs).

The purpose of our study is to find the physical parameters that allow the self-trapping of light in an array of SWWGs. Among these, we focus our attention in the separation between of the SWWGs and the diameter of the waveguide, the Kerr coefficient value, from one side, and the incident radiation properties, from the other. Once known these parameters, also, we search the possibility to control the position of the output self-trapped beam by changing the angle incidence and phase of the incident beams.

2. Mathematical model

2.1. Nonlinear optics background

Electromagnetic waves consist of time-varying electric and magnetic fields, which are generated one another and giving a resultant wave propagating through the space, vacuum or some material. The electromagnetic radiation is a means of transporting energy that can carry information.

The electromagnetic phenomena are governed by a set of four fundamental coupled partial differential equations known as Maxwell's equations [10, 12], and the constitutive relations equations, which relate the electric field intensity $\mathbf{E}$ to the electric flux density $\mathbf{D}$, and similarly the magnetic field intensity $\mathbf{B}$ to the magnetic flux density $\mathbf{H}$, which are listed below:

$$\mathbf{D} = \varepsilon\mathbf{E} + \mathbf{P} \tag{1}$$

$$\mathbf{B} = \mu\mathbf{H} - \mathbf{M} \tag{2}$$

Where ε is the electric permittivity, μ is the magnetic permeability, $\mathbf{P}$ is the polarization vector in a dielectric and $\mathbf{M}$ is the magnetization vector in a magnetic medium. The permittivity ε and the permeability μ, are macroscopic parameters that describe the relationships among macroscopic field quantities, but they are based on the microscopic behavior of the atoms and molecules in response to the fields. These parameters are constants for linear, homogeneous, time-invariant, and isotropic material media. Otherwise, for nonlinear media these parameters may depend on the magnitudes of $\mathbf{E}$ and $\mathbf{B}$, on spatial coordinates (x, y, z) in inhomogeneous media, on time, on frequency, or on the orientations of $\mathbf{E}$ and $\mathbf{B}$ in anisotropic media. $\mathbf{P}$ and $\mathbf{M}$ vectors account for the presence of matter and express the contributions of the medium to some changes that may suffer an electromagnetic field.

We consider nonlinear media with optical Kerr response, which is a phenomenon where the refractive index of the medium changes by the strong electric field [13]. In the Kerr nonlinear model, the nonlinear contribution to the electric polarization field $\mathbf{P}$ will depend on the electric field $\mathbf{E}$ in the following way:

$$\mathbf{P} = \varepsilon_0(\chi^{(3)}\left|\mathbf{E}\right|^2)\mathbf{E}, \tag{3}$$

where ε_0 is the permittivity in the free space and $\chi^{(3)}$ is the nonlinear electrical susceptibility of the medium. Solving for the displacement field $\mathbf{D}$, equation (1) gives:

$$\mathbf{D} = \varepsilon_0(\varepsilon_r + \chi^{(3)}\left|\mathbf{E}\right|^2)\mathbf{E}, \tag{4}$$

with lineal relative dielectric constant ε_r. In the limit where $\chi^{(3)}\left|\mathbf{E}\right|^2 << \varepsilon_r$, and we know that $n^2 \approx \varepsilon_r$, we can express the refractive index as:

$$n = (\varepsilon_r + \chi^{(3)}\left|\mathbf{E}\right|^2)^{1/2} \cong n_0\left(1 + \frac{1}{2}\frac{\chi^{(3)}\left|\mathbf{E}\right|^2}{\varepsilon_r}\right) = n_0 + n_2 I, \tag{5}$$

where n_0 and n_2 denote linear and nonlinear refractive index, respectively, I is the beam intensity (power per unit area). We have assumed the plane wave relationship between beam intensity and electric field as:

$$I = n_0\sqrt{\frac{\varepsilon_0}{\mu_0}}\left|\mathbf{E}\right|^2 \tag{6}$$

So, the nonlinear refractive index is expressed as:

Journal of Physics: Conference Series **605** (2015) 012034 doi:10.1088/1742-6596/605/1/012034

$$n_2 = \frac{1}{2}\frac{\chi^{(3)}}{n_0^2}\sqrt{\frac{\mu_0}{\varepsilon_0}} \tag{7}$$

Considering a source-free medium, non magnetic ($\mathbf{M} = \mathbf{0}$) and isotropic nonlinear media, we can reduce the Maxwell's equations to following two equations:

$$\frac{\partial \mathbf{H}}{\partial t} = -\frac{1}{\mu}\nabla \times \mathbf{E}, \tag{8}$$

$$\frac{\partial \mathbf{D}}{\partial t} = \nabla \times \mathbf{H} \tag{9}$$

Considering the constitutive relation (4), the equations (8-9) are the differential equations that represent the six electromagnetic field components that need to be solved in low dimensionality problems.

2.2. *Methodology*

The field propagation can be found out by substituting (4) in (9) and solving simultaneously the equations (8-9), with proper boundary conditions applying the Finite Difference Time Domain (FDTD) method [10, 11, 14–16]. The FDTD method is used because we are interested in the interactions of electromagnetic waves with complex structures that cannot be easily described analytically.

The FDTD method belongs in the general class of grid-based differential time domain numerical modeling methods. The time-dependent Maxwell's equations (in partial differential form) are discretized using central-difference approximations for the space and time partial derivatives, this approach is the discretization scheme proposed by Yee [16], which can be visualized using a cubic unit cell, where the electric and magnetic fields are defined at specific points in the cell at alternative time steps. The resulting finite difference equations are solved by first solving the electric field vector components in a volume of space at a given instant in time, and then solving the magnetic field vector components in the same spatial volume at the next instant in time. This process is repeated until the desired transient or steady-state electromagnetic field behavior is fully evolved. The electric and magnetic fields are thus determined at every point in space over the physical region in which the simulation is performed.

For a more accurate representation of the device, we can use the smallest mesh size but at a substantial cost of more simulation time. Then, the mesh size should be such that electromagnetic fields do not change substantially from one point to the next in the mesh. This means that to have significant results the size of the grid will be a fraction of the wavelength, for that, we use the follow relation:

$$\Delta t \leq \frac{1}{c_{max}(1/(\Delta x)^2 + 1/(\Delta y)^2 + 1/(\Delta z)^2)^{1/2}} \tag{10}$$

where Δt is time step size, c_{max} is the speed light in the medium and Δx, Δy and Δz are the spatial mesh size.

The relation showed in the (10) is known as Courant's condition [17], which is both necessary and sufficient for numerical stability in an FDTD code that considers linear effects and nonlinear Kerr materials.

3. Simulations and results

To simulate the propagation of the optical field through an array of SWWGs, a 3D rectangular region is used, with dimension of $30 \times 4 \times 40$ μm on x, y and z direction, respectively. The cell size is between 15 and 40 nm, time step size is $t = 0.07086 fs$, fulfilling the Courant's condition. We use boundary conditions of 40 perfectly matched layers (PML) in the simulation region.

We consider a one-dimensional SWWGs array in the direction of the x axis, the direction of propagation in the z axis, formed with different number m of SWWG, however, for $m = 23$ we observe that the light propagation maintains the identical properties observed for larger arrays $m > 50$, all the SWWGs are modeled with SiO_2 with linear refractive index $n_0 = 1.455$ and Kerr nonlinear coefficient $\chi^{(3)} = 2.0 \times 10^{-18}\ m^2V^{-2}$, each has circular cross section of radius $r = 300nm$ and the separation among waveguides is $d = 250nm$ coupled to its neighbor by means of evanescent field.

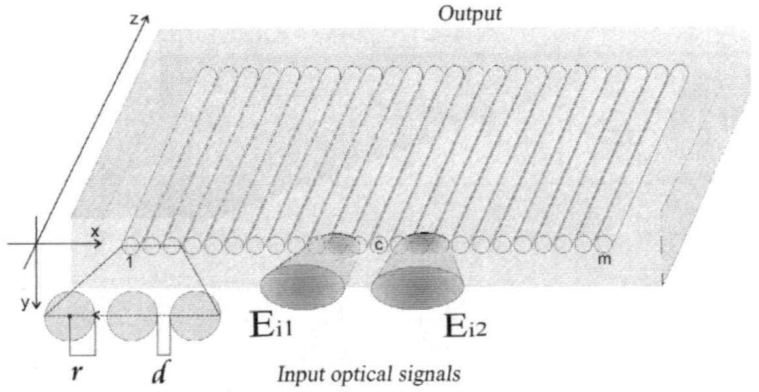

Figure 1. Diagram of the array and incident beams.

The incident radiation are two linearly polarized monochromatic beams in the x axis with Gaussian profile, constant peak amplitude of $4.5 \times 10^8 V/m$ and wavelength $\lambda = 800nm$, focusing on three SWWGs at $z = 0$, each one, $\mathbf{E_{i1}}$ and $\mathbf{E_{i2}}$ are shown in the figure 1.

With the physical parameters above described we are able to achieve self-trapped light in a central SWWG. For steering the output beam position, we introduce difference phase ($\Delta\phi$) and incident angle (θ) of the input beams. To get a deep insight into the dynamics of such a process, we study three different cases.

3.1. Normal incidence

We consider that $\mathbf{E_{i1}}$ and $\mathbf{E_{i2}}$ are normally incident over cross section of the array with $\Delta\phi = 0$.

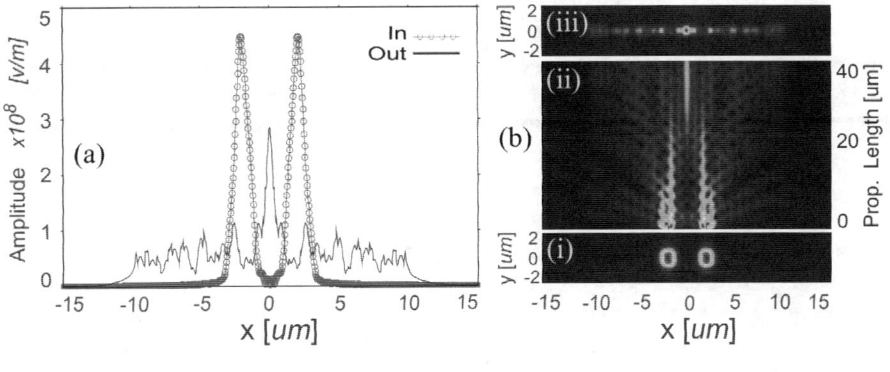

Figure 2. Normal incidence of $\mathbf{E_{i1}}$ and $\mathbf{E_{i2}}$.(a) Input and output beam profile and (b) energy distribution: input(i) propagation (ii) and output (iii).

Journal of Physics: Conference Series **605** (2015) 012034 doi:10.1088/1742-6596/605/1/012034

In the figure 2(a), we show the input beam profile (line with circles) and the output beam profile (continuous line) with peak amplitude of $2.8 \times 10^8 V/m$, and it is trapped in the central waveguide. In 2(b), show the energy distribution; (i) input beams, (ii) the propagation of $\mathbf{E_{i1}}$ and $\mathbf{E_{i2}}$ beams for interact in the middle of the array and produce one resulting beam in the central SWWG, (iii) output beam, where we can see that all SWWGs have energy, but the more intense is in the central.

3.2. Separately influence of phase difference and angle of incidence
Now, we considered normal incident waves with $\Delta\phi = \pi/2$, in figure 3(a), we have the input and output beam profile (line with circles and continuous line, respectively), when incident wave $\mathbf{E_{i1}}$ has phase 0 and $\mathbf{E_{i2}}$ has phase $\pi/2$, we can see that the output beam is in a waveguide shifted to right of the central, with amplitude of $2.35 \times 10^8 V/m$. The direction of the output displacement depends on what of the input fields experiences the phase change.

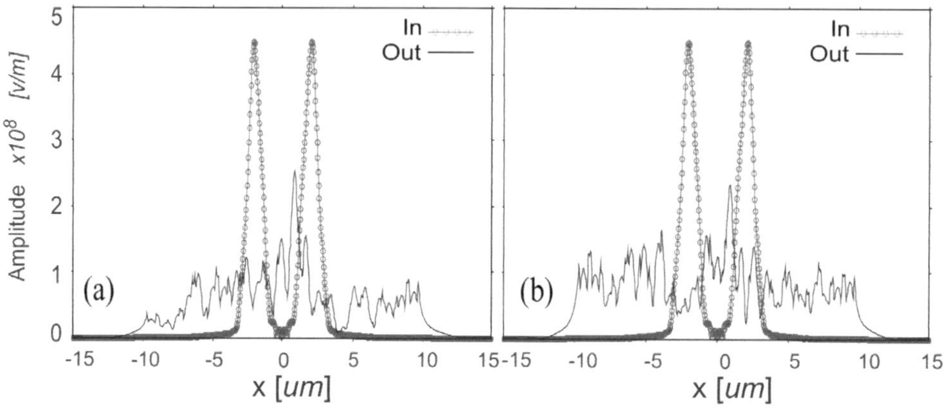

Figure 3. (a) Input and output beam profile under the influence of phase difference $\Delta\phi = \pi/2$, (b) Input and output beam profile, influence of incident angle ($\theta = 20^o$). Red line– input beam profile (line with circles). Blue line– output beam profile (continuos line)

However, a similar result is obtained, if we consider the wave $\mathbf{E_{i1}}$ has an incident angle $\theta = 20^o$ and $\mathbf{E_{i2}}$ is normal incident both with with phase 0. In the figure 3(b) we have the input and output beam profile. We can see the output beam is in a single waveguide to right of the central one with amplitude of $2.25 \times 10^8 V/m$. This amplitude is smaller than the showed in 3(a), so, with the modify of phase or incidence angle in the input electromagnetic field, we can change the amplitud and the output beam position.

3.3. Simultaneously influence of both phase and incident angle
Now, we introduce phase in one input beam and incident angle in the other.

We obtain a similar result like shown in figure 3, where the output beam is shifted to right the central SWWG. In this case, we have the wave $\mathbf{E_{i1}}$ with incident angle $\theta = 10^o$ and phase 0, $\mathbf{E_{i2}}$ is normal incidence with phase $\pi/2$. In figure 4 show the energy distribution; (a) input beams, (b) output beam, (c) input and output beam profile. With this parameters we have the output beam with peak amplitude of $2.87 \times 10^8 V/m$ and is to right the central SWWG, we can see a better self-trapped of the light in this SWWG.

This self-trapping effect is presented because the evolution of the field is affected by the self-focusing that the Kerr media presents when an intense field is launched. This phenomenon

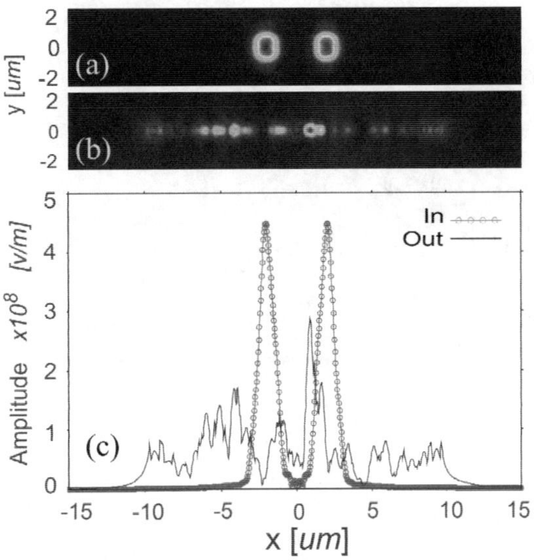

Figure 4. (a) Input and (b) output energy distribution, (c) input and output beam profile with simultaneously influence of incident angle and phase: $\theta = 10^o$ and $\Delta\phi = \pi/2$. In (c), red line–input beam profile (line with circles). Blue line– output beam profile (continuos line).

produces a change of phase shift in the wavefront, due to the change in the refractive index of the SWWGs.

4. Conclusions

We have studied the dynamics behavior of the discrete self-trapping beam as a result of the interaction between two input optical fields in a subwavelength waveguide arrays with Kerr nonlinearities. We have shown the dependence of the output position in function of the phase difference and the angle of incidence. We observed the change the position of the output beam, these results could give the possibility to control light by light with perspectives in applications to integrated optics at nanoscale.

Acknowledgments

GMG thanks CONACYT-México for the Ph.D. scholarship. This work was partially supported by VIEP-BUAP 2014 grant.

References

[1] Kivshar Y S and Agrawal G 2003 *Optical Solitons: From Fibers to Photonic Crystals*, (Boston, Ma: Academic Press).
[2] Christodoulides D N and Joseph R I 1988 *Opt. Lett.***13** 794-6
[3] Eisenberg H S, Silberberg Y Morandotti R and Aitchison J S 1998 *Phy. Rev. Lett.* **81** 3383-6
[4] Eisenberg H S, Silberberg Y, Morandotti R and Aitchison J S 2000 *Phys. Rev. Lett.* **85** 1863-6
[5] Morandotti R, Eisenberg H S, Silberberg Y , Sorel M and Aitchison J S 2001 *Phy. Rev. Lett.* **86** 3296-9
[6] Foster M A, Moll K D and Gaeta A L 2004 *Opt. Express* **12** 2881-7
[7] Tong L and Sumetsky M 2010 *Subwavelength and nanometer diameter optical fibers* (New York: Springer)
[8] Foster M A, Turner A C, Lipson M and Gaeta A L 2008 *Opt. Express* **16** 1300-20
[9] Conforti M, Guasoni M and De Angelis C 2008 *Opt. Lett.* **33** 2662-4
[10] Sullivan D M 2000 *Electromagnetic simulation using the FDTD method* (New York: IEEE Press)

[11] Taflove A and Hagness S C 2000 *Computational electrodynamics: the finite difference time domain method* (Norwood: Artech House)
[12] Agrawal G P 2004 *Nonlinear Fiber Optics* (New York: Academic Press)
[13] Okamoto K 2006 *Fundamentals of optical waveguides* (San Diego, Ca: Academic Press)
[14] Inan U S and Marshall R A 2011 *Numerical electromagnetics: the FDTD method* (New York: Cambridge University Press)
[15] Zhaoming Z and Thomas G B 2002 *Opt. Express* **10** 853-64
[16] Yee K 1966 *IEEE Trans. Ant. Prop.* **14** 302-7
[17] Reinke C M *et al* 2006 *J. Lightwave Technol.* **24** 624-34

Preparing arbitrary pure states of spatial qudits with a single phase-only spatial light modulator

J J M Varga [1]*, M A Solís-Prosser[2], L. Rebón[3], A Arias[4], L Neves[5], C Iemmi[1], S Ledesma[1]

[1]Departamento de Física, FCEyN, Universidad de Buenos Aires, Buenos Aires 1428, Argentina
[2]Center for Optics and Photonics, Universidad de Concepción, Casilla 4016, Concepción, Chile
[3]Instituto de Física de La Plata, Universidad Nacional de La Plata, C.C. 67, 1900 La Plata, Argentina
[4]Center for Research, Instituto Tecnológico Metropolitano, A.A 54954, Medellín, Colombia
[5]Departamento de Física, Universidade Federal de Minas Gerais, Belo Horizonte, MG 30123-970, Brazil

E-mail: miguel@df.uba.ar

Abstract. We present a new method for preparing multidimensional spatial qudits by means of a single phase-only spatial light modulator (SLM). This method improves previous ones that use two SLMs, one working in amplitude regime and the other in phase regime. To that end, we addressed diffraction gratings on the slits that define the state and then we performed a spatial filtering in the Fourier plane. The amplitude of the coefficients of the quantum state are determined by the modulation deep of the diffraction gratings, and the relative phase is the mean phase value of the diffraction gratings. This encoding result to be more compact, less expensive and use the photons more efficiently.

Quantum systems are the information carriers in quantum information processing and computing protocols. While qubits (two-level systems) are the usual and most basic systems to carry out such tasks, qudits [D-level systems ($D>2$)] have been attracting growing interest due to their greater potential for those applications [1]. In particular, single photons are the natural choice for communications since they are easily transportable and have several degrees of freedom (DOFs) to encode information. In the case of using the transverse momentum-position encoding [2, 3], the simplest approach is achieved by discretizing the one-dimensional transverse modes of the photons when they are made to pass through an aperture with D slits which sets the dimension of these so-called *spatial qudits*.

Spatial qudits are relatively simple to generate and offer the possibility of working in high dimensions without cumbersome optical setups. Recently, this encoding has drawn interest for miscellaneous applications such as quantum information protocols [4], quantum games [5], and quantum key distribution [6]. Therefore, the ability to prepare arbitrary pure states of such systems represents an important step toward realizations of quantum optics experiments based on it. Static amplitude and phase masks could be used for this purpose, but in practice it would be extremely difficult and time-consuming as each state intended to be prepared would require to set up the corresponding masks in the apparatus. Programmable spatial light modulators (SLMs) can dramatically simplify this process, allowing real-time manipulations of the state coefficients without physically aligning any optical components. In this regard, it has been

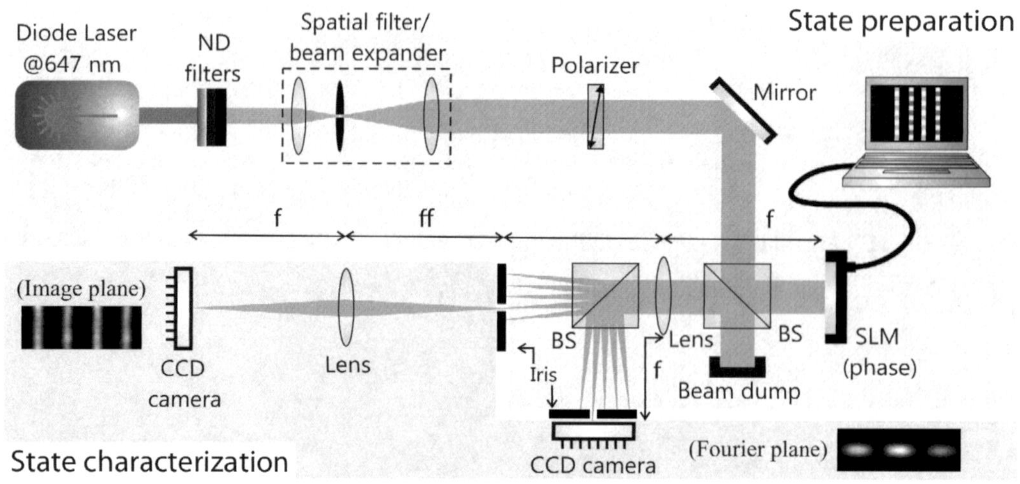

Figure 1. Experimental setup

shown [7] that by imaging the output beam of an amplitude-only SLM onto a phase-only SLM allows one to get complete and independent control of the amplitude and phase of the complex coefficients that define the qudit state. However, the use of two SLMs, besides being more expensive, entails two drawbacks: (i) the overall diffraction efficiency is very low, and (ii) in order to avoid even more losses, the image of the first SLM must match the second one pixel by pixel, which is difficult in practice.

Here, we present a demonstration of a method for preparing arbitrary pure states of spatial qudits with a single phase-only SLM [8], which has a much higher diffraction efficiency and does not require imaging systems. Among the alternative methods to represent a complex function in a single SLM [9, 10] we choose, for simplicity, a technique where the amplitude information is encoded in a phase-only filter [11, 12]. On one hand, the required amplitude is achieved by programming a phase grating, with an appropriate modulation into the slit regions. On the other hand, the required phase value is obtained by adding a constant phase value to the phase grating. Then, the first diffraction order is selected at the Fourier plane and the filtered information is antitransformed on the final plane (Fig. 1). The result is a light distribution corresponding to an image of the slits containing the complete complex modulation. Regarding the type of selected grating, either binary or blazed, the overall quality of the prepared states is very similar. However, blazed gratings provide better diffraction efficiency than binary ones. This enhanced efficiency can be important to increase the signal-to-noise ratio when working with single photon sources.

We have analyzed the performance of the proposed method by preparing a large number of arbitrary pure qubit states uniformly distributed on the surface of the Bloch sphere. To quantify the quality of the preparation we used the fidelity between the state intended to be prepared and the density matrix of the state actually prepared which is reconstructed by tomography (Fig. 2). The tomographic process is performed by measurements at different transverse positions in the near and in the far field [13]. In both cases, either binary or blazed grating, we obtained fidelities above 96% and average fidelities of 99.6%. We have found for $D = 3$ and $D = 7$ that the fidelities were above 94%. All these relevant results are summerized in the Table 1.

We compared the luminous efficiency of our setup with the one that employed two SLMs [7]. In the case of using a blazed grating, our scheme is $\frac{1}{\eta}$ times more efficient, where η represents the light attenuation due to polarizing elements needed to obtain pure amplitude modulation

Table 1. Average fidelities of preparation for D-level qudits.

D	2	3	4	5	7
# of states/grating	561	24	70	25	94
$\bar{F}_{\text{bin}}$	0.996	0.995	0.985	0.968	0.970
$\bar{F}_{\text{bla}}$	0.996	0.996	0.991	0.971	0.977

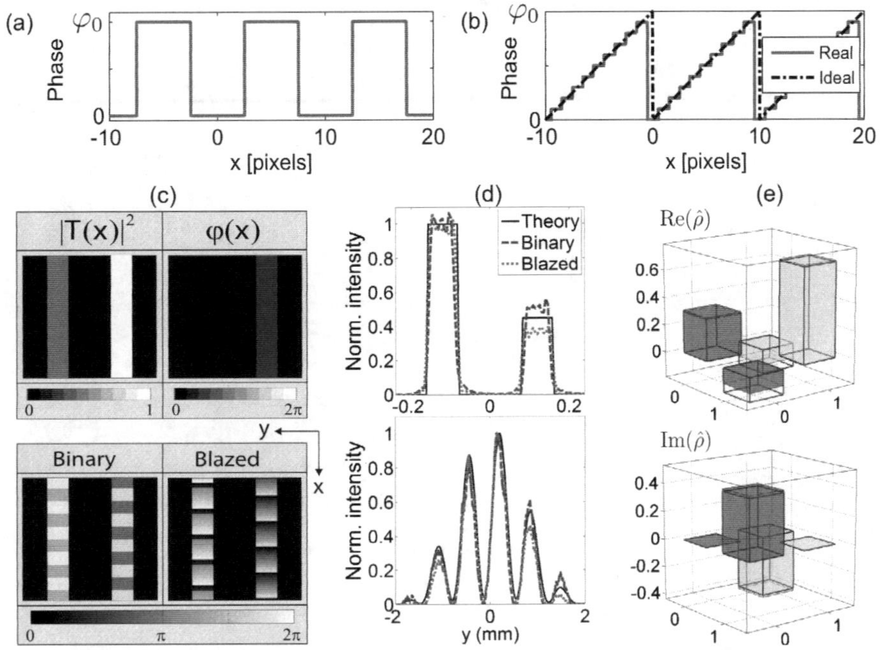

Figure 2. Ten-pixels period gratings: (a) binary and (b) blazed. (c) (Upper panels) amplitude and phase components of the aperture $T(\mathbf{x})$ for preparing the state $0.56|0\rangle+0.83e^{i0.63\pi}|1\rangle$. (Lower panels) corresponding phase masks for preparing this state at the first diffraction order. (d) Normalized intensity distributions and theoretical prediction for measurements of this state at near (upper panel) and far field (lower panel). (e) Real and imaginary parts of the experimentally reconstructed density matrix using the blazed grating. The ideal density matrix is shown as a blue wire grid and the obtained fidelity is $F = 0.996$.

with a twisted nematic SLM [15, 16]. The factor η depends on the parameters involved in each particular case (wavelength, twist angle, birefringence of the SLM, etc.), but a typical value is about 0.1. Therefore, the method presented here appears as a valuable tool for experiments based on this encoding. In addition, it can be useful for assisting the measurement of high-dimensional photonic qudits, also including mixed states, enabling not only the storing of information but also the read out of it. It enables the implementation of the strategy developed in [14] by proyecting the image of the state to be measured at SLM and addresing at this device the phase grating function corresponding to the measurement state. A detection at the center of the filtered diffraction order at the Fourier plane will acomplish the process allowing one to obtain the statistics of arbitrary observables.

Finally, it is important to remark that although we have obtained a good quality of

Table 2. Selected arbitrary states

state 1	$\|\psi_1\rangle = 0.70\|0\rangle + 0.70e^{i1.06\pi}\|1\rangle$
state 2	$\|\psi_2\rangle = 0.70\|0\rangle + 0.70e^{i0.23\pi}\|1\rangle$
state 3	$\|\psi_3\rangle = -0.69\|0\rangle + 0.73e^{i1.35\pi}\|1\rangle$
state 4	$\|\psi_4\rangle = -0.92\|0\rangle + 0.40e^{i1.90\pi}\|1\rangle$
state 5	$\|\psi_5\rangle = 0.48\|0\rangle + 0.88e^{i0.87\pi}\|1\rangle$
state 6	$\|\psi_6\rangle = -0.48\|0\rangle + 0.88e^{i1.55\pi}\|1\rangle$

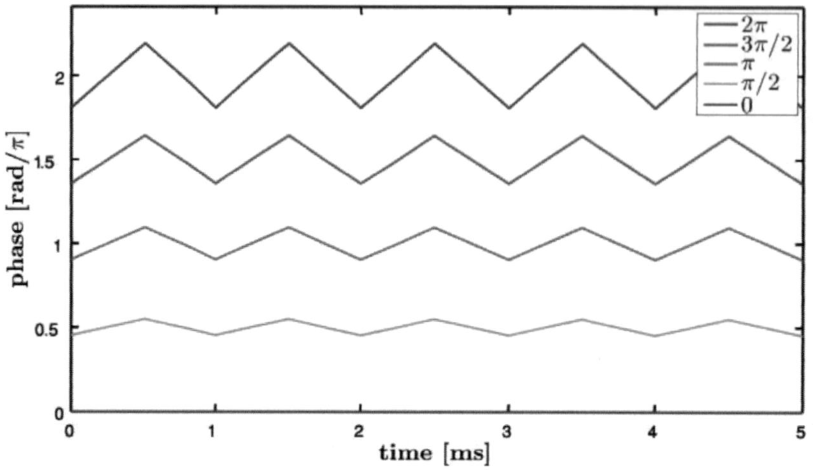

Figure 3. Temporal phase fluctuation model for the SLM.

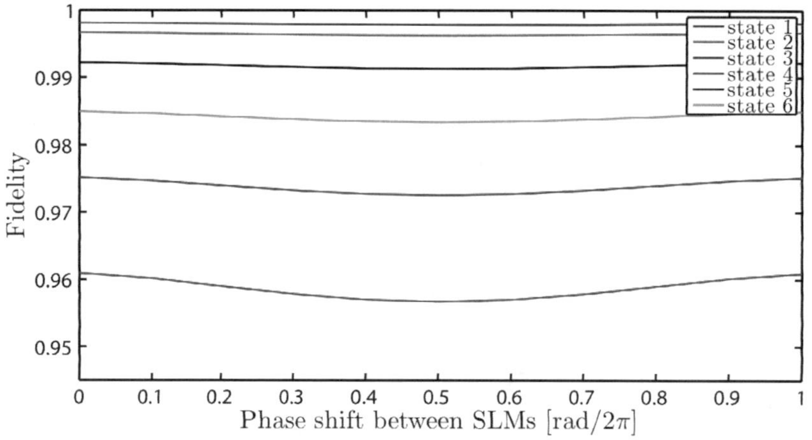

Figure 4. Fidelity vs. phase shift between SLMs, obtained by means of numerical simulation for six arbitrary states ($D = 2$) using the phase adition method.

preparation, the method presents an inherent phase fluctuation associated to the used SLMs, in our case liquid crystal on silicon displays (LCoS). We have selected LCoSs due to their high spatial resolution and very high light efficiency. Nevertheless, these displays may lead to flicker in the optical beam because of their addressing scheme (pulse width modulation) which introduces temporal phase fluctuations [17, 18] that affect the quality of the encoded state. In order to study this effect we have analyzed the performance of the proposed method by simulating the preparation of some selected arbitrary states in one SLM and their tomographic reconstruction in a second SLM. For this purpose we have used a simplified model for the typical temporal fluctuations of the SLMs that consists on a triangular shaped fluctuation as is depicted in Fig. 3. Given that the optical configuration contain two similar SLMs that could be desynchronized, we have considered different phase shift between both modulators. Numerical simulations of the fidelity vs. the phase shift are shown in Fig. 4 for six arbitrary states. These states are shown in Table 2 As it can be noted fidelities are almost independent of the phase shift between the phase fluctuations in each SLM. This would indicate that it is not necessary to deal with the temporal synchronization of the modulators.

Conclusions

We have shown that a single phase-only SLM enables the preparation of arbitrary pure states of spatial qudits. Mean values of fidelities of preparation above 94% were obtained and a higher efficiency (at least 10 times) with respect to other methods is expected. It is important to note that this method can be useful not only for preparation but also for assisting the measurement of such systems. It enables the implementation of a strategy based on imaging the state represented on the first SLM onto the second SLM. In this case the first SLM is used to represent the state to be analized and the second one to represent the reconstruction basis. Therefore, the method presented here may become a valuable tool for experiments using spatial qudits.

Acknowledgment

We thank P. Kolenderski and A. Calatayud for helpful discussions. This work was supported by CONICYT PFB08-24 and Milenio ICM P10-030-F (Chile) and by UBACyT 20020100100689, CONICET PIP 112-200801-03047, and ANPCYT PICT 2010-02179 (Argentina).

References

[1] S. M. Barnett, *Quantum Information* (Oxford University Press, Oxford, 2009).

[2] L. Neves, G. Lima, J. G. Aguirre-Gómez, C. H. Monken, C. Saavedra, and S. Pádua, "Generation of entangled states of qudits using twin photons", Phys. Rev. Lett. **94**, 100501 (2005).

[3] M. N. O'Sullivan-Hale, I. Ali-Khan, R. W. Boyd, and J. C. Howell, "Pixel Entanglement: Experimental Realization of Optically Entangled $d = 3$ and $d = 6$ Qudits", Phys. Rev. Lett. **94**, 220501 (2005).

[4] M. A. Solís-Prosser and L. Neves, "Remote state preparation of spatial qubits", Phys. Rev. A **84**, 012330 (2011).

[5] P. Kolenderski, U. Sinha, L. Youning, T. Zhao, M. Volpini, A. Cabello, R. Laflamme, and T. Jennewein, "Aharon-Vaidman quantum game with a Young-type photonic qutrit", Phys. Rev. A **86**, 012321 (2012).

[6] S. Etcheverry, G. Cañas, E. S. Gómez, W. A. T. Nogueira, C. Saavedra, G. B. Xavier, and G. Lima, "Quantum key distribution session with 16-dimensional photonic states", Sci. Rep. **3**, 2316 (2013).

[7] G. Lima, L. Neves, R. Guzmán, E. S. Gómez, W. A. T. Nogueira, A. Delgado, A. Vargas, and C. Saavedra, "Experimental quantum tomography of photonic qudits via mutually unbiased basis", Opt. Express **19**, 3542 (2011).

[8] M. A. Solís-Prosser, A. Arias, J. J. M. Varga, L. Rebń, S. Ledesma, C. Iemmi, and L. Neves, "Preparing arbitrary pure states of spatial qudits with a single phase-only spatial light modulator", Opt. Lett. **38**, 4762 (2013)

[9] R. W. Gerchberg and W. O. Saxton, "A practical algorithm for the determination of the phase from image and diffraction plane pictures", Optik **35**, 237 (1972).

[10] E. G. van Putten, I. M. Vellekoop, and A. P. Mosk, "Spatial amplitude and phase modulation using commercial twisted nematic LCDs", Appl. Opt. **47**, 2076 (2008).

[11] J. A. Davis, D. M. Cottrell, J. Campos, M. J. Yzuel, and I. Moreno, "Encoding amplitude information onto phase-only filters", Appl. Opt. **38**, 5004 (1999).

[12] V. Bagnoud and J. D. Zuegel, "Independent phase and amplitude control of a laser beam by use of a single-phase-only spatial light modulator", Opt. Lett. **29**, 295 (2004).
[13] G. Lima, F. A. Torres-Ruiz, L. Neves, A. Delgado, C. Saavedra, and S. Pdua, "Measurement of spatial qubits", J. Phys. B **41**, 185501 (2008).
[14] G. Lima, L. Neves, R. Guzmán, E. S. Gómez, W. A. T. Nogueira, A. Delgado, A. Vargas, and C. Saavedra, "Experimental quantum tomography of photonic qudits via mutually unbiased basis", Opt. Express **19**, 3542 (2011).
[15] A. Márquez, C. Iemmi, I. Moreno, J. Davis, J. Campos, and M. Yzuel, "Quantitative predictions of the modulation behavior of twisted nematic liquid crystal displays based on a simple physical model", Opt. Eng. **40**, 2558 (2001)
[16] I. Moreno, C. Iemmi, A. Márquez, J. Campos, and M. Yzuel, "Modulation light efficiency of diffractive lenses displayed onto a restricted phase-mostly modulation display", Appl. Opt. **43**, 6278 (2004)
[17] A. Lizana, I. Moreno, A. Márquez, C. Iemmi, E. Fernández, J. Campos, and M. J. Yzuel, "Time fluctuations of the phase modulation in a liquid crystal on silicon display: characterization and effects in diffractive optics," Opt. Express **16**, 16711 (2008)
[18] A. Lizana, A. Márquez, L. Lobato, Y. Rodange, I. Moreno, C. Iemmi, J. Campos "The minimum Euclidean distance principle applied to improve the modulation diffraction efficiency in digitally controlled spatial light modulators" Opt. Exp. **18**, 10581, (2010)

Journal of Physics: Conference Series **605** (2015) 012036 doi:10.1088/1742-6596/605/1/012036

Verification of the Violation of WWŻB Inequality Using Werner States

Álvaro J Almeida[1,2], Luís P Martins[3], Paulo S André[2,4] and Armando N Pinto[2,5]

[1]Department of Physics, University of Aveiro, Campus Universitário de Santiago, 3810-193 Aveiro, Portugal

[2]Instituto de Telecomunicações, University of Aveiro, 3810-193 Aveiro, Portugal

[3]Department of Physics and Astronomy, University of Sheffield, S3 7RH, United Kingdom

[4]Department of Electrical and Computer Engineering, Instituto Superior Técnico, 1049-001 Lisboa, Portugal

[5]Department of Electronics, Telecommunications and Informatics, University of Aveiro, 3810-193 Aveiro, Portugal

E-mail: aalmeida@av.it.pt

Abstract. The generation and manipulation of entangled particles presents itself as one of the most important results in quantum mechanics. With the work from Bell, it was possible to prove the nonlocal nature of quantum mechanics, which is nowadays widely accepted. Apart from being possible to prove entanglement from Bell's inequality, it is difficult to compute the system as it increases the number of particles. Such a system containing N qubits can be described by Werner-Wolf-Żukowski-Brukner (WWŻB) inequality. In this work, we show how to obtain the maximum of violation of WWŻB inequality using a Werner state, simplifying the problem considerably. We get two different results; one for a system containing an odd number of particles and other for a system containing an even number.

1. Introduction

In quantum key distribution (QKD) systems the security is based on the impossibility to copy a photon with an arbitrary quantum state [1]. However, due to losses that occur in the quantum channel (optical fiber), the distance to which key distribution can be achieved is limited, since it is not possible to use optical amplification to extend it [2]. Therefore, QKD can be performed up to a few hundreds of kilometers, even if we are able to significantly improve single-photon detectors [3, 4]. By using entangled-photon pairs, in theory it is possible to double the distance to which key distribution can be achieved, being extremely difficult to surpass 1000 km [5, 2]. Through the use of quantum repeaters it is expected to be possible to surpass this distance, but the problem is that they are are still in a preliminary stage of research [6]. As an alternative to quantum repeaters stands entanglement swapping [7], which promises to be helpful in extending the distance for quantum communications beyond 1000 km [8, 9, 10].

The entanglement of particles spatially separated is one of the most noticeable results of quantum mechanics. As a consequence, it generates nonclassical correlations between the particles of the entangled system. These are the correlations that Einstein, Podolsky and Rosen claimed to give rise to a paradox that shows the incompleteness of quantum mechanics, as a

Journal of Physics: Conference Series **605** (2015) 012036 doi:10.1088/1742-6596/605/1/012036

local theory [11]. The nonlocal nature of quantum mechanics was only widely accepted after the demonstration of the experiment suggested by Bell [12, 13, 14]. Bell presented his experiment with an inequality which obeys to local theories but is incompatible with quantum mechanics, in a seminal result inside quantum information science [12]. But even if the original theory of Bell considered only an entangled state with two photons, if one wants to use an N-qubit quantum system where each part is allowed to choose independently between two dichotomic observables (i.e. each one has only two possible results), Bell's theorem can be generalized to the so-called Werner-Wolf-Żukowski-Brukner (WWŻB) inequality [15, 16].

In this work, we verify the maximum for violation of WWŻB inequality using a Werner state [17]. This type of state can be interesting for quantum communications since it traduces a mixed state which includes noise and that approximates more accurately to what we obtain in the laboratory. We show in detail how to find the maximum for violation of WWŻB inequality from the correlation function and evaluate it for odd and even systems of particles. Finally, we present our conclusions for this work.

2. Theoretical Description

2.1. WWŻB Inequality and Werner States

In the conditions of the standard Bell experiment [13, 14], i.e. if we consider a scheme in which each local observer can choose between two dichotomic observables, the restrictions given by local theories can be expressed through the WWŻB inequality as,

$$\left| \sum_{k_1,\ldots,k_N=-1,1} S(k_1,\ldots,k_N) \times \sum_{s_1,\ldots,s_N=1,2} k_1^{s_1-1}\ldots k_N^{s_N-1} E(s_1,\ldots,s_N) \right| \leq 2^N . \tag{1}$$

In Eq. (1), $S(k_1,\ldots,k_N) = \pm 1$ and represents an arbitrary function of the sum of the indices $k_1,\ldots,k_N$. The term s_j traduces the two possible measurement configurations for the observer j and $E(s_1,\ldots,s_N)$ is the correlation function [16]. Equation (1) compiles a set of 2^{2^N} Bell inequalities for the correlation function and summarizes all possible local realistic constraints on this function for an N-qubit system.

Let us now define a Werner state, which will be used in the verification of the violation of WWŻB inequality. This type of mixed state can be represented by the density matrix and can take the form:

$$\rho_{\mathrm{W}} = v|\Psi_{\mathrm{GHZ}}\rangle\langle\Psi_{\mathrm{GHZ}}| + (1-v)\rho_{\mathrm{noise}} , \tag{2}$$

where $|\Psi_{\mathrm{GHZ}}\rangle$ is a Greenberger-Horne-Zeilinger state that allows maximum entanglement and v can be interpreted as the interferometric visibility observed in a multi-particle experiment [17]. In an experiment of interferometry this scaling parameter is directly related to the visibility of the interference pattern. Visibilities lower than 1 generally traduce the existence of some noise contribution to the state. The term $\rho_{\mathrm{noise}} = \mathbb{I}/2^N$ represents the completely uncorrelated noise contribution, i.e. describes a totaly mixed state where $\mathbb{I}$ is an unitary matrix of size $2^N \times 2^N$. The aforementioned critical v gives the threshold beyond which no local hidden variables model can resemble the quantum predictions [18].

2.2. Maximum of the Correlation Function

The correlation function for two particles can vary between -1 and 1 and reflects the classical correlation level that they present. However, it does not tells us anything about quantum correlations. Nevertheless, it is from the correlation function that we will be able to calculate the maximum of violation of WWŻB inequality, and it is this quantity that will traduce the desired quantum correlations.

In the most general formulation, the correlation function can be seen as a function that depends on the spin state and the measurement directions. The correlation function can be written as,

$$E = \mathrm{Tr}[\rho_{\mathrm{W}}\Omega] , \tag{3}$$

where ρ_{W} from Eq. (2) can be written also as,

$$\rho_{\mathrm{W}} = \begin{pmatrix} a & 0 & \cdots & 0 & b \\ 0 & c & \cdots & 0 & 0 \\ \vdots & \vdots & \ddots & \vdots & \vdots \\ 0 & 0 & \cdots & c & 0 \\ b & 0 & \cdots & 0 & a \end{pmatrix} , \tag{4}$$

with $a = \frac{v}{2} + \frac{1-v}{2^N}$, $b = \frac{v}{2}$ and $c = \frac{1-v}{2^N}$ [19]. The parameter Ω is the Kronecker product of Pauli vectors. For the plane $\hat{x} - \hat{z}$, the Pauli vector of the ith observer is $\sin(\theta_i)\sigma_x + \cos(\theta_i)\sigma_z$, where σ_x and σ_z are Pauli matrices. Therefore, Ω takes the form of a product of cosines and sines and can be expressed as,

$$\Omega = \begin{pmatrix} \xi_1 \prod_{i=1}^N \cos\theta_i & \cdots & \cdots & \cdots & \prod_{i=1}^N \sin\theta_i \\ \vdots & \ddots & \vdots & \iddots & \vdots \\ \vdots & \cdots & \xi_j \prod_{i=1}^N \cos\theta_i & \cdots & \vdots \\ \vdots & \iddots & \vdots & \ddots & \vdots \\ \prod_{i=1}^N \sin\theta_i & \cdots & \cdots & \cdots & \xi_{2^N} \prod_{i=1}^N \cos\theta_i \end{pmatrix} , \tag{5}$$

where ξ_j is used to explicit the signal of the jth diagonal element and θ_i is the angle that defines the measurement direction of the observer i. In Eq. (5), ξ is given by,

$$\xi = \bigotimes_{i=1}^{N} \begin{pmatrix} + \\ - \end{pmatrix} . \tag{6}$$

Note that the matrix in Eq. (5) shows only the elements necessary for the results to be presented in the next section, i.e., only the corner and diagonal elements are used. Also, the diagonal elements are always the same, $\prod_{i=1}^N \cos\theta_i$, where only the signal represented by ξ changes.

Next, we can apply Eq. (6) to two different cases, i.e., when the number of particles is odd and when it is even. When the number of particles is odd we will have one result, which is the same for all odd cases; when the number of particles is even we will get another result, which is also the same for all even cases. This is due to the fact that Eq. (6) has always as many '+' as '–', but in the odd case it starts with '+' and ends with '–' and in the even case it starts and ends with '+'. As an example, we can use $N = 3$ to represent the odd case, and $N = 4$ to represent the even case.

Starting from the odd case, when $N = 3$, we get from Eq. (6),

$$\xi = (+ - - + - + + -) , \tag{7}$$

which means that the diagonal of Eq. (5) is null. Therefore, solving Eq. (3), we obtain,

$$E = 2b \prod_{i=1}^{N} \sin\theta_i . \tag{8}$$

Journal of Physics: Conference Series **605** (2015) 012036 doi:10.1088/1742-6596/605/1/012036

We can now calculate the maximum of E from Eq. (8), which we call $\mathcal{L}$, as,

$$\mathcal{L} = \sqrt{(2b)^2} = v\,. \tag{9}$$

Moving now to the even case, when $N = 4$, Eq. (6) is written as,

$$\xi = (+--+-++--++-+--+)\,. \tag{10}$$

Applying this result to Eq. (3), we now get,

$$E = 2a\prod_{i=1}^{N}\cos\theta_i + 2b\prod_{i=1}^{N}\sin\theta_i - 2c\prod_{i=1}^{N}\cos\theta_i\,, \tag{11}$$

whose solution for the maximum is found as,

$$\mathcal{L} = \sqrt{(2(a-c))^2 + (2b)^2} = \sqrt{2}v\,. \tag{12}$$

From Eqs. (9) and (12) we can describe the maximum for violation of WWŻB inequality as,

$$\mathcal{L} = \begin{cases} v, & \text{if } N \text{ is odd} \\ \sqrt{2}v, & \text{if } N \text{ is even} \end{cases}. \tag{13}$$

When $\mathcal{L} > 1$ it is observed the violation of WWŻB inequality in Eq. (1).

3. Numerical Results

In this section we present the results for the verification of the violation of WWŻB inequality, both when using an odd and an even number of particles.

In Fig. 1 are shown the results given by Eq. (13), where it is possible to distinguish between an even and an odd system of particles. First it can be seen that the odd system can be described

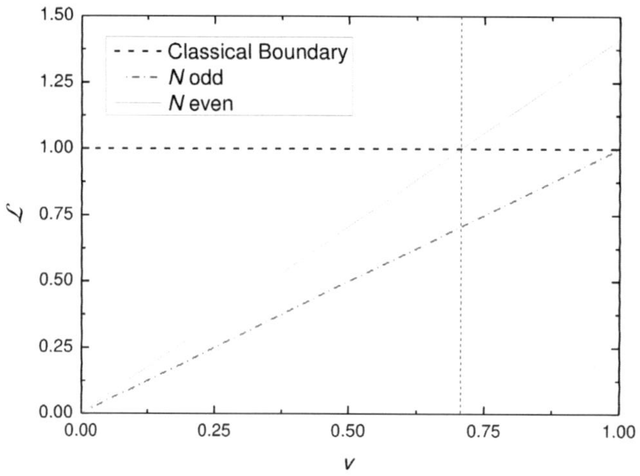

Figure 1. Representation of Eq. (13) as a function of the parameter v, when N is odd (dash-dotted red line) and N is even (solid blue line). The vertical dashed line at $v = 1/\sqrt{2}$ represents the point from where a system with an even number of particles violates Eq. (13).

by local theories, since it does not violates the WWŻB inequality in Eq. (1) for any value of v. From a system with an even number of particles, one can see that it violates WWŻB inequality from $v = 1/\sqrt{2}$.

The result shown in Fig. 1, although relatively simple, tells us that ρ_{noise} does not has any influence in the maximum of violation. This comes from the fact that the result of Eq. (13) is equal to the result for the GHZ state multiplied with the weight of the parameter v [19].

In Fig. 2 is displayed the result from Eq. (1) for two particular cases, i.e., when $N = 3$ and when $N = 4$. The value of v was set to 0.8, which is above the threshold of violation when N is even. When $N = 3$, from Eq. (1) are obtained $2^{2^3} = 256$ inequalities and when $N = 4$ we obtain $2^{2^4} = 65536$, as shown in Fig. 2. From Eq. (13), in the even case, it is possible to determine that when $v = 0.8$, $\mathcal{L} = 0.8\sqrt{2} \approx 1.13$, which is the maximum that we observe in Fig. 2(b). We verified also that from the 65536 inequalities which can be obtained from Eq. (1) when $N = 4$, 1794 violate the classical boundary. Additionally, we determined that from the 1794 inequalities, 120 allow the maximum of violation. These results are important due to the following reason: if one wants to know if a given state cannot be written by a local theory, it is

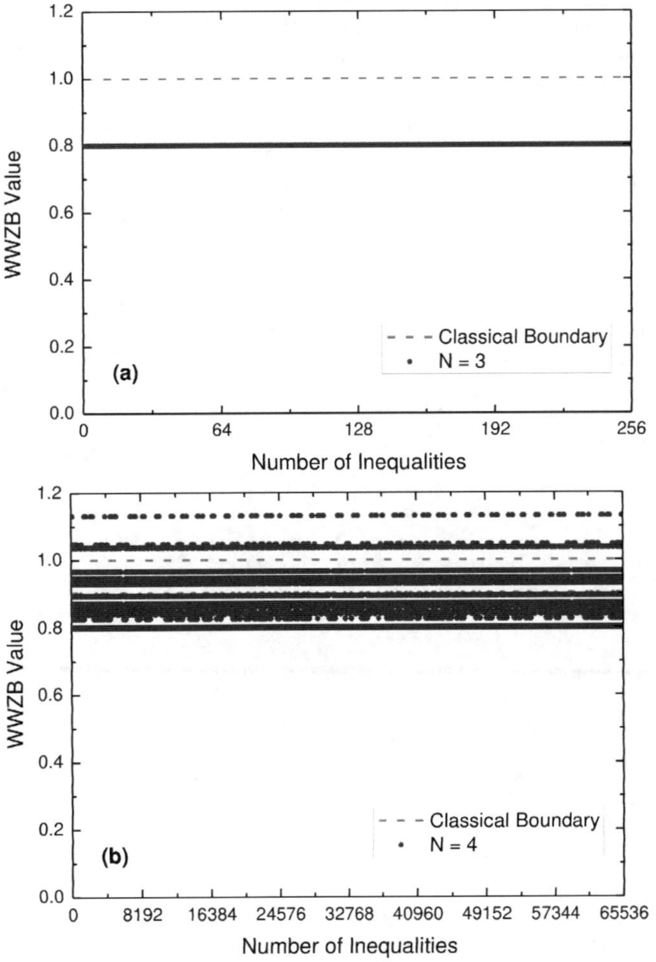

Figure 2. Values of WWŻB inequalities when $N = 3$ (up) and $N = 4$ (down).

necessary to find at least one inequality of Eq. (1) that violates the classical boundary. However, from Eq. (1) we observe that the number of inequalities increases exponentially with N, and as larger it is, more difficult is to find a solution for the system. This way, from Eq. (13), which is a single equation applicable to all cases, it is possible to obtain the same information in a much more easier way.

4. Conclusions

We have shown that entangled quantum states have a major role in QKD systems, mainly in the possibility that they allow to increase the distance of transmission by several orders.

Since Werner states are entangled states which include a term of noise, they traduce more accurately what is obtained in experiments. Thus, the understanding of their behavior is fundamental to improve the systems of quantum technologies. Also, we verified that these states can present a simple behavior, being possible to discriminate two cases, one for N even and other for N odd. In what concerns the parameter v, we have shown that it is the one that determines the threshold for violation of WWŻB inequality when N is even.

The generation of multipartite entangled quantum states is a key scientific endeavour only by itself, but also an enabling technology for quantum communication systems. Up to now it was possible to demonstrate the generation of eight-photon entanglement, which is a breakthrough achievement for science [20, 21].

Acknowledgments

A.A. thanks the Fundação para a Ciência e a Tecnologia (FCT) for support through Ph.D. Grant No. SFRH/BD/79482/2011. The authors also acknowledge the FCT and the Instituto de Telecomunicações under the PEst-OE/EEI/LA0008/2013 program, project "CV-Quantum".

5. References

[1] Gisin N, Ribordy G, Tittel W and Zbinden H 2002 *Rev. Mod. Phys.* **74** 145–195
[2] Takesue H 2011 *NTT Tech. Review* **9**(9) 50–54
[3] Collins D, Gisin N and de Riedmatten H 2005 *J. Mod. Opt.* **52** 735–753
[4] Wang S, Chen W, Guo J F, Yin Z Q, Li H W, Zhou Z, Guo G C and Han Z F 2012 *Opt. Lett.* **37** 1008
[5] Takesue H and Miquel B 2009 *Opt. Express* **17** 10748
[6] Sangouard N, Simon C, de Riedmatten H and Gisin N 2011 *Rev. Mod. Phys.* **83** 33–34
[7] Żukowski M, Zeilinger A, Horne M A and Ekert A K 1993 *Phys. Rev. Lett.* **71** 4287–4290
[8] Abruzzo S, Bratzik S, Bernardes N K, Kampermann H, van Loock P and Bruß D 2013 *Phys. Rev. A* **87** 052315
[9] Bratzik S, Abruzzo S, Kampermann H and Bruß D 2013 *Phys. Rev. A* **87** 062335
[10] Khalique A and Sanders B C 2014 *Phys. Rev. A* **90**(3) 032304
[11] Einstein A, Podolsky B and Rosen N 1935 *Phys. Rev.* **47**(10) 777–780
[12] Bell J S 1964 *Physics* **1** 195–200
[13] Freedman S J and Clauser J F 1972 *Phys. Rev. Lett.* **28** 938–941
[14] Aspect A, Grangier P and Roger G 1982 *Phys. Rev. Lett.* **49** 91–94
[15] Werner R F and Wolf M M 2001 *Phys. Rev. A* **64**(3) 032112
[16] Żukowski M and Brukner Č 2002 *Phys. Rev. Lett.* **88** 210401
[17] Werner R F 1989 *Phys. Rev. A* **40**(8) 4277–4281
[18] Weinfurter H and Żukowski M 2001 *Phys. Rev. A* **64**(1) 010102
[19] Martins L P, Almeida Á J, Silva N A, André P S and Pinto A N 2014 *Eur. Phys. J. D* **68** 228
[20] Huang Y F, Liu B H, Peng L, Li Y H, Li L, Li C F and Guo G C 2011 *Nat. Commun.* **2** 546
[21] Yao X C, Wang T X, Xu P, Lu H, Pan G S, Bao X H, Peng C Z, Lu C Y, Chen Y A and Pan J W 2012 *Nature Photon.* **6** 225–228

Journal of Physics: Conference Series **605** (2015) 012037 doi:10.1088/1742-6596/605/1/012037

Optical clock transition in a rare-earth-ion-doped crystal: coherence lifetime extension for quantum storage applications

Robert-Christopher Tongning[1], Thierry Chanelière[1], Jean-Louis Le Gouët[1] and María Florencia Pascual-Winter[2*]

[1] Laboratoire Aimé Cotton, CNRS, Université Paris-Sud and ENS Cachan, 91405 Orsay, France

[2] Centro Atómico Bariloche e Instituto Balseiro, CNEA, (8400) Bariloche, Río Negro, Argentina

E-mail: florencia.pascualwinter@cab.cnea.gov.ar

Abstract. Atomic clock transitions are desirable for quantum information storage and processing thanks to the protection from decoherence they provide. In the context of rare-earth-ion-doped crystals for quantum information storage, clock Zeeman or hyperfine transitions have been identified and exploited for long-lived storage in spin degrees of freedom. We present a theoretical and experimental analysis on the existence of an optical clock transition in Tm^{3+}:YAG, in view of storage in optical coherences. The combination of a Zeeman-like term and a quadratic electronic Zeeman term in the Hamiltonian, lead to the existence of a magnetic field amplitude (12 mT) for which the derivative of the optical transition energy with respect to the field amplitude vanishes, regardless of the magnetic field orientation. We have verified this prediction through hole-burning spectroscopy experiments. In addition to that, a study of the behavior of the Hamiltonian as a function of the magnetic field orientation yields the direction for which both derivatives with respect to the magnetic field angular coordinates also vanish. The condition for an optical clock transition with three vanishing partial derivatives is met.

1. Introduction

In atom-based quantum memory systems, quantum information is stored in atomic superposition states, that is, in atomic coherences. Thus, the storage time is limited by the atomic coherence lifetime T_2. Long T_2 systems are desirable for such applications. Rare-earth-ion-doped crystals (REIDC) display satisfying T_2 values at low temperature (< 3 K). Typical T_2's are of the order of some tens of microseconds for optical coherences [1], that is, a superposition state between two levels separated by energies in the optical electromagnetic spectrum. Therefore, REIDC have been intensively studied as candidates for quantum memories. Aside their long optical T_2, they constitute compact systems in which the storage centers are immobile within a solid matrix, simplifying their experimental manipulation. If longer T_2's are desirable, REIDC offer the possibility of transferring the quantum information from optical coherences to Zeeman or hyperfine ones, of longer lifetime of the order of some hundreds of microseconds. This can be performed by applying a control pulse tuned to one arm of a Λ-system (two sub-levels of the ground level optically linked to an excited one) [2]. Of course, back-conversion is compulsory before launching the retrieval of the stored information.

Additional strategies to further extend T_2 have been demonstrated in REIDC. One of them, known as dynamical decoupling (DD), is based on the application of successive RF π pulses all through the time the information is stored in the Zeeman (or hyperfine) coherences [3, 4]. If the pulses are applied at a rate faster than the correlation time of the magnetic environment fluctuations (magnetic moments of the lattice ions) responsible for the decoherence, the effect of the environment on the rare-earth ion coherence is canceled out. This way, Zeeman coherence extension of more than two orders of magnitude has been demonstrated in Pr^{3+}:Y_2SiO_5 [4], Tm^{3+}:YAG [5] and other REIDCs. Another strategy for T_2 extension is to work on an atomic clock transition [6, 7], that is, a transition whose frequency is insensitive (to first order) to fluctuations in the magnetic environment because of vanishing partial derivatives with respect to one or more magnetic field coordinates. In other words, if we note Δ_g as the Zeeman splitting of the ground level, and B, θ_B, φ_B the spherical coordinates of the magnetic field $\boldsymbol{B}$, an atomic clock transition is met when there exists some $\boldsymbol{B}_0$ such that $\partial\Delta_g/\partial B\ |_{\boldsymbol{B}_0}= 0$ and/or $\partial\Delta_g/\partial\theta_B\ |_{\boldsymbol{B}_0}= 0$ and/or $\partial\Delta_g/\partial\varphi_B\ |_{\boldsymbol{B}_0}= 0$. That way, Δ_g is insensitive, to first order, to fluctuations of some or all of the magnetic field coordinates introduced by the environing magnetic moments. Therefore, decoherence is minimized. A clock transition with three vanishing partial derivatives will be henceforth called a *full* clock transition. Such a clock transition is also known, in the context of REIDC, as the ZEFOZ (*zero first order Zeeman effect*) technique [4, 8]. The combination of DD and ZEFOZ has yielded impressive results in Pr^{3+}:Y_2SiO_5, reaching the value of 40 s for the Zeeman T_2 [9].

The results of both techniques for T_2 extension just described have been obtained on Zeeman or hyperfine coherences. However, no demonstration has been presented in the case of optical coherences. DD is not well-suited for the optical case since a photon echo would be emitted after the first π pulse, leading to immediate retrieval of the stored information (this is irrelevant in the Zeeman or hyperfine case since the radiated-energy-to-$\hbar\omega$ ratio scales as ω^3). On the contrary, it would be worth it to implement an optical clock transition, proposed in the context of REIDC initially in Ref. [10]. If long enough optical T_2's were obtained that way, there would be no need to resort to storage in the Zeeman coherences, and the loss in efficiency that comes along would be avoided. Then, quantum storage protocols such as the atomic frequency comb (AFC) [11, 12] or the revival of silenced echo (ROSE) [13] could be implemented with higher efficiencies.

In the present communication, we investigate the viability of implementing a full optical clock transition in Tm^{3+}:YAG. Aside from its advantages in terms of diode-laser-compatible wavelength or large inhomogeneous broadening for broadband storage discussed elsewhere [14], such REIDC presents a rather long optical level lifetime T_1, typically 800 μs [15]. As $2T_1$ is the theoretical limit for T_2, Tm^{3+}:YAG is an interesting candidate for implementing an optical clock transition. In the following section we will develop a theoretical analysis for identifying an optical clock transition with three vanishing partial derivatives. In section 3, we will present results from hole-burning spectroscopy that validate our theoretical findings.

2. Theoretical analysis for identifying a full optical clock transition

Trivalent thulium ions substitute for trivalent yttrium ions in sites of D_2 symmetry of the YAG lattice. The facts that Tm^{3+} is a non-Kramers ion (even number of electrons), has nuclear spin 1/2 and lies in a low symmetry site lead to an effective Hamiltonian of the form [16]

$$H_{eff} = -\hbar \boldsymbol{B}\cdot \bar{\bar{\gamma}}^{(J)}\cdot \boldsymbol{I} - \boldsymbol{B}\cdot \bar{\bar{\Lambda}}^{(J)}\cdot \boldsymbol{B}, \qquad (1)$$

where $\boldsymbol{B}$ is the magnetic field, $\boldsymbol{I}$ is the nuclear spin operator, and $\bar{\bar{\gamma}}$ and $\bar{\bar{\Lambda}}^{(J)}$ are diagonal tensors whose elements are given by

$$\Lambda_{\alpha\beta}^{(J)} = \mu_B^2 g_J^2 \delta_{\alpha\beta} \sum_{n=1}^{2J+1} \frac{|\langle 0|J_\alpha|n\rangle|^2}{E_n - E_0}, \tag{2}$$

$$\gamma_{\alpha\beta}^{(J)} = \left(\gamma_n + \frac{2A_J \Lambda_{\alpha\alpha}^{(J)}}{\hbar \mu_B g_J}\right) \delta_{\alpha\beta}. \tag{3}$$

In the above expressions γ_n is the nuclear gyromagnetic ratio, μ_B is the Bohr magneton, g_J is the Landé factor, A_J is the hyperfine interaction parameter, $\boldsymbol{J}$ is the total angular momentum operator, J is its eigenvalue, $|n\rangle$ are the eigenstates in the J multiplet, E_n are their corresponding energies, and $\delta_{\alpha\beta}$ is the Kronecker delta. The tensors in (2) and (3) are expressed in the coordinate system defined by the three orthogonal C_2 axes of the site symmetry group, D_2. The parameters γ_{xx}, γ_{yy} and γ_{zz} in the effective Hamiltonian have already been measured for the ground [$^3H_6(0)$] and excited [$^3H_4(0)$] states. In what follows, we will use the values reported in Ref. [17]. From them, the values for Λ_{xx}, Λ_{yy}, Λ_{zz} can be deduced.

The eigenstates of H_{eff} are two sublevels for the ground state, that we will note $|1\rangle$ and $|2\rangle$, and two sub-levels for the excited state, $|3\rangle$ and $|4\rangle$. The first term in H_{eff} is a Zeeman-like one, that would lead to a linear increase of the ground and excited level splittings with growing magnetic field amplitude. We choose $|1\rangle$ and $|3\rangle$ ($|2\rangle$ and $|4\rangle$) to be the sub-levels whose energies decrease (increase) with increasing magnetic field. The second term does not depend on $\boldsymbol{I}$, therefore, it rigidly shifts both sub-levels of the ground or excited states. The amount of that shift depends quadratically on the magnetic field amplitude, and its curvature differs for the ground and excited levels. The latter fact ensures that, regardless of the orientation of the magnetic field, there always exists a magnetic field amplitude B_0 for which $\partial\Delta_{13}/\partial B|_{B_0} = 0$, where Δ_{13} is the energy difference between levels $|1\rangle$ and $|3\rangle$. Thus, one of the conditions for a full clock transition is met. This can be seen in Fig. 1(a), where we plot the calculated Δ_{13} and Δ_{24} offset to their values at zero magnetic field as a function of the field amplitude (the field orientation relative to the crystal axes $\{[100], [010], [001]\}$ has been chosen as $\theta_B = 90°$, $\varphi_B = 45°$ for reasons described later on). A vanishing derivative is found for Δ_{13} at $B_0 = 13$ mT. As expected, no vanishing derivative is observed for Δ_{24}.

At this point of the discussion we must clarify that Y^{3+} ions, and thus Tm^{3+} ions, in a YAG lattice occupy six sites that are equivalent in terms of their local environment but distinct in terms of the orientation of this environment, i.e. the axes of the D_2 point group are differently oriented (see Ref. [18] and references therein). When an external field is applied, the equivalence between sites is broken. Therefore, curves such as the one plotted in Fig. 1(a) differ from site to site. In Fig. 1(a) we have computed Δ_{13} and Δ_{24} for the site usually labeled "1" in the literature [18], that is, the site whose C_2 axes are $\{[0\bar{1}\bar{1}], [01\bar{1}], [100]\}$. Similar plots can be obtained for the remaining sites.

Still, the magnetic field orientation for a full clock transition has to be identified. In other words, we need to find values for the magnetic field angular coordinates for which the remaining two partial derivatives of Δ_{13} vanish. For that, in Fig. 2 we plot $\Delta_{13}(B = 13\text{ mT})$ as a function of θ_B and φ_B, which are spherical coordinates with respect to the crystal axes $\{[100], [010], [001]\}$. Here again, we consider site 1. We observe that both derivatives vanish for $\theta_B = 90°$, $\varphi_B = 45°$ (this orientation is indicated in Fig. 2 as the intersection between the two lines). Hence, a full clock transition exists for site 1 and is given in spherical coordinates by $\boldsymbol{B}_0 = (13\text{ mT}, 90°, 45°)$.

For that particular magnetic field orientation, it happens that sites 3, 4, 5 and 6 remain equivalent (it can be easily seen that the projection of $\boldsymbol{B}$ onto each of the three orthogonal C_2 axes of one site is the same for all sites 3 to 6.) This means that the Zeeman-like sub-level splittings are the same for sites 3-6, both in the ground and excited levels, which results in equal Δ_{13} for all sites 3-6, as well as equal Δ_{24}. Of course, these values differ from those of site 1. As

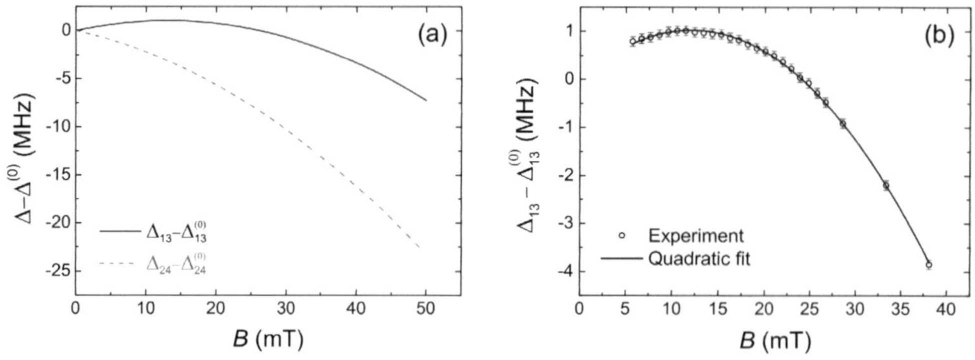

Figure 1. (a) Calculated energy difference between ground and excited sub-levels $|1\rangle$ and $|3\rangle$ (full line), and $|2\rangle$ and $|4\rangle$ (dashed line) of site 1. An offset corresponding to the values at zero magnetic field has been subtracted. (b) Hole-burning spectral position of the hole burnt in site 1's level $|1\rangle$ as a function of the magnetic field amplitude. The line is a quadratic fit.

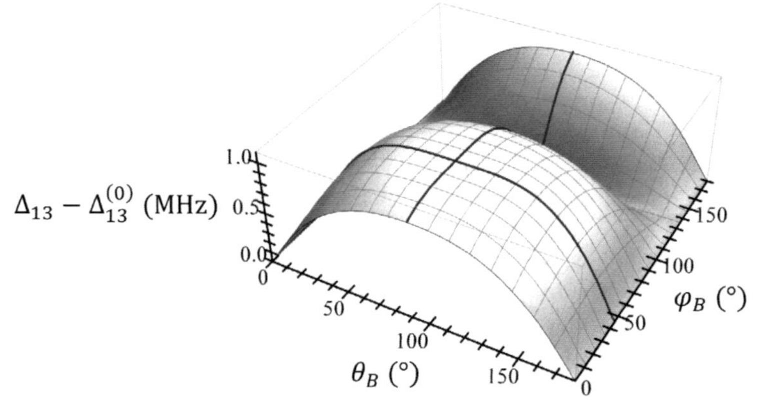

Figure 2. Calculated energy difference between ground and excited sub-levels $|1\rangle$ and $|3\rangle$ as a function of the magnetic field spherical angular coordinates θ_B and φ_B defined with respect to the Cartesian frame $\{[100],[010],[001]\}$. The magnetic field amplitude has been fixed to 13 mT. An offset corresponding to the transition frequency at zero magnetic field has been subtracted.

regards site 2, it does not participate at all in the experiments because the electric field vector ([110]) is perpendicular to the ${}^3H_6(0) \rightarrow {}^3H_4(0)$ transition dipole moment ($[1\bar{1}0]$).

3. Hole-burning spectroscopy results

We use the spectral hole burning (SHB) technique to measure the optical transition frequency variation with the external magnetic field amplitude B. After burning a hole at $B = B_i = 19$ mT, we probe the hole at $B = B_f$. We have enough time to vary the field amplitude from B_i to B_f as the holes are burnt into long lifetime (~ 1 min) ground state sub-levels. Repeating the operation with different B_f values, and the same initial B_i, we are able to follow the dependence of the hole spectral position with B. The B_f values range from 7 to 27 mT, by steps of 0.95 mT. As the hole is probed though the optical transitions $|1\rangle \rightarrow |3\rangle$ and $|2\rangle \rightarrow |4\rangle$, this gives us access to the dependence of Δ_{13} and Δ_{24} with B. The experiment is performed with $\boldsymbol{B}$ orientation fixed at $\theta_B = (90 \pm 5)°$, $\varphi_B = (45 \pm 5)°$.

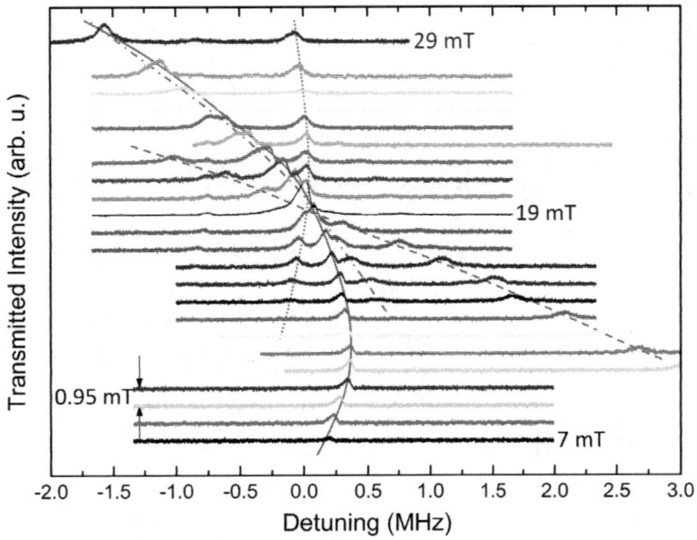

Figure 3. Hole burning spectra. In each curve the hole is burnt at $B = B_i = 19$ mT and the spectrum is read at $B = B_f$. B_f is set to values ranging from 7 mT to 29 mT by steps of 0.95 mT (the last step is twice as large). The horizontal axis indicates the spectral detuning from the position the hole takes when $B_f = B_i = 19$ mT. Four different holes are identified with guides to the eye. They correspond to: site 1, hole burnt in $|1\rangle$, read through transition $|1\rangle \rightarrow |3\rangle$ (——); site 1, hole burnt in $|2\rangle$, read through transition $|2\rangle \rightarrow |4\rangle$ (– – –); sites 3-6, hole burnt in $|1\rangle$, read through transition $|1\rangle \rightarrow |3\rangle$ (······); sites 3-6, hole burnt in $|2\rangle$, read through transition $|2\rangle \rightarrow |4\rangle$ (— · —).

The results are shown in Fig. 3. When $B_f = B_i = 19$ mT, a single hole is observed (a single peak). As B_f is shifted away from B_i we see that the initial hole splits in four holes that display different dependences with B_f. These dependences are indicated with the four guides to the eye. After some analysis of the Hamiltonian in (1), we are able to assign the four holes to:

- Site 1, hole burnt in $|1\rangle$, read through transition $|1\rangle \rightarrow |3\rangle$ (——).
- Site 1, hole burnt in $|2\rangle$, read through transition $|2\rangle \rightarrow |4\rangle$ (– – –).
- Sites 3-6, hole burnt in $|1\rangle$, read through transition $|1\rangle \rightarrow |3\rangle$ (······).
- Sites 3-6, hole burnt in $|2\rangle$, read through transition $|2\rangle \rightarrow |4\rangle$ (— · —).

(Crossed transitions $|1\rangle \rightarrow |4\rangle$ and $|2\rangle \rightarrow |3\rangle$ are negligibly weak). As the optical transition inhomogeneous broadening ($\sim$ 20 GHz) is much larger than the sub-level splitting ($\sim$ 1 MHz), for some Tm^{3+} ions, the optical pump energy is resonant to Δ_{13}, while for others, it is resonant to Δ_{24}. In the former case, the hole is burnt in sub-level $|1\rangle$; in the latter, the hole is burnt in $|2\rangle$. As this happens for site 1 and for sites 3-6, four different holes are burnt. The full line in Fig. 3 shows the dependence of site 1's Δ_{13}, the transition expected to be a full clock transition. In fact, a vanishing derivative is observed around $B_f = 12$ mT (seventh curve from the bottom).

The spectral positions of the holes burnt in $|1\rangle$ for site 1 are plotted in Fig. 1(b) as a function of B_f. We observe that the behavior is perfectly quadratic, as expected, and displays a vanishing derivative at $B_f = 12$ mT. The linear (quadratic) term coefficient of the fit in Fig. 1(b) differs from the predicted one in 6% (15%). Our experimental setup does not allow fast variation of the magnetic field orientation. Therefore, an analog study as a function of the magnetic field angular coordinates was not possible. However, the good agreement of the experimental data to the predicted behavior suggests that a full optical clock transition should be possible in site 1 of Tm^{3+}:YAG for extending the optical T_2.

Our search for a clock transition has been limited to site 1. Although not shown here, it can be demonstrated that sites 3-6 display very small angular derivatives at the field orientation found for site 1, and that the derivative with respect to the amplitude vanishes for a somewhat higher field (~ 20 mT). The latter statement can be verified by observing the dotted line in Fig. 3. Therefore, even if not on a proper clock transition, T_2 will also be extended for sites 3-6.

For no site, clock transition conditions are met in the case of transition $|2\rangle \rightarrow |4\rangle$. In a quantum information storage scheme, one would wish to avoid storage in that transition, since decoherence is not prevented. A simple preparation stage can be implemented prior to the storage, where sub-level $|2\rangle$ would be emptied through optical pumping.

4. Summary

We have performed a theoretical and experimental study of the viability of gaining profit from a full clock transition to extend the optical T_2 in a REIDC system compatible with quantum information storage. From the analysis of the Hamiltonian we have identified the magnetic field coordinates for a full optical clock transition. Experimentally, we have verified that the theoretical prediction for the amplitude coordinate is in good agreement with hole-burning spectroscopy results.

5. Acknowledgements

We thank John G. Bartholomew for fruitful discussions on the advantages of Tm^{3+}:YAG regarding the possibility of an optical clock transition. We also acknowledge Rose Ahlefeldt for discussions on the experimental results.

References

[1] Macfarlane R M 1993 *Opt. Lett.* **18** 1958–1960
[2] Lovrić M, Suter D, Ferrier A and Goldner P 2013 *Phys. Rev. Lett.* **111**(2) 020503
[3] Yang W, Wang Z Y and Liu R B 2011 *Frontiers of Physics* **6**(1) 2–14 ISSN 2095-0462 10.1007/s11467-010-0113-8
[4] Fraval E, Sellars M J and Longdell J J 2004 *Phys. Rev. Lett.* **92**(7) 077601
[5] Pascual-Winter M F, Tongning R C, Chanelière T and Le Gouët J L 2012 *Phys. Rev. B* **86**(18) 184301
[6] Bollinger J J, Prestage J D, Itano W M and Wineland D J 1985 *Phys. Rev. Lett.* **54**(10) 1000–1003
[7] Wolfowicz G, Tyryshkin A M, George R E, Riemann H, Abrosimov N V, Becker P, Pohl H J, Thewalt M L W, Lyon S A and Morton J J L 2013 *Nat. Nano.* **8**(8) 561–564
[8] Lovrić M, Glasenapp P, Suter D, Tumino B, Ferrier A, Goldner P, Sabooni M, Rippe L and Kröll S 2011 *Phys. Rev. B* **84**(10) 104417
[9] Heinze G, Hubrich C and Halfmann T 2013 *Phys. Rev. Lett.* **111**(3) 033601
[10] McAuslan D L, Bartholomew J G, Sellars M J and Longdell J J 2012 *Phys. Rev. A* **85**(3) 032339
[11] de Riedmatten H, Afzelius M, Staudt M U, Simon C and Gisin N 2008 *Nature* **456**(7223) 773–777
[12] Afzelius M, Simon C, de Riedmatten H and Gisin N 2009 *Phys. Rev. A* **79**(5) 052329
[13] Damon V, Bonarota M, Louchet-Chauvet A, Chanelière T and Le Gouët J L 2011 *New Journal of Physics* **13** 093031
[14] Bonarota M, Le Gouët J L and Chanelière T 2011 *New Journal of Physics* **13** 013013
[15] Armagan G, Buoncristiani A and Bartolo B D 1992 *Opt. Mater.* **1** 11 – 20 ISSN 0925-3467
[16] Macfarlane R M and Shelby R M 1987 *Spectroscopy of Solids Containing Rare Earth Ions* vol 21 ed Kaplyanskii A and Macfarlane R M (Amsterdam: North-Holland)

[17] de Seze F, Louchet A, Crozatier V, Lorgeré I, Bretenaker F, Le Gouët J L, Guillot-Noël O and Goldner P 2006 *Phys. Rev. B* **73**(8) 085112
[18] Sun Y, Wang G M, Cone R L, Equall R W and Leask M J M 2000 *Phys. Rev. B* **62**(23) 15443–15451

Preparing cold atomic samples in different density regimes for Rydberg studies

M M Valado[1,2], M D Hoogerland[4], C Simonelli[2], E Arimondo[1,2,3], D Ciampini[1,2,3], O Morsch[1,2]

[1] INO-CNR, Via G. Moruzzi 1, 56124 Pisa, Italy
[2] Dipartimento di Fisica E. Fermi, Università di Pisa, Largo Pontecorvo 3, 56127 Pisa, Italy
[3] CNISM UdR Pisa, Dipartimento di Fisica E. Fermi, Università di Pisa, Largo Pontecorvo 3, 56127 Pisa, Italy
[4] Department of Physics, University of Auckland, Private Bag 92019, Auckland, New Zealand

E-mail: valado@df.unipi.it

Abstract. Here we present two different experimental techniques to generate atomic clouds in different density regimes that could be used as a tool to study Rydberg systems in different scenarios. As an example, we discuss an experiment in which the dynamics of Rydberg excitations were studied both in the few atom regime and in the high density regime.

1. Introduction

Rydberg excitations in ultra-cold gases have been studied extensively in the past years [1–6]. Strongly interacting Rydberg systems offer the possibility to study different problems like many-body collective effects [7, 8], or glassy features that Rydberg systems exhibit in the incoherent regime [9]. A systematic study of these effects can be done, e.g., by varying the density of the atomic sample. Here we describe two different experimental techniques which allow us to work in different density regimes, from the few atom regime, with typically $N \leq 10$ atoms, to the high density regime.

2. Experimental techniques

Our experiments are performed on magneto-optical traps (MOT) of ultra-cold 87-Rb atoms. The MOT parameters (size of the MOT beams, loading flux from a primary two-dimensional MOT, magnetic field gradient) are chosen such as to obtain clouds of sizes ranging from: $\sigma_{x,y,z} \approx 30\,\mu$m to $150\,\mu$m, containing up to $N \approx 10^5$ atoms.

In order to excite atoms to the Rydberg state we use a coherent two-step excitation scheme with a 421 nm laser beam detuned by a range from $\Delta = +500$ MHz to $\Delta = +720$ MHz in order to avoid population of the intermediate state. A 1013 nm laser allows us to reach Rydberg states with principal quantum numbers ranging from $n = 55$ to $n = 120$. Depending on Δ, the effective two-photon Rabi frequencies range from $\Omega = 240$ kHz to $\Omega = 1.2$ MHz. The waists of the beams are focused to $40\,\mu$m and $110\,\mu$m respectively. After the excitation pulse of duration between 0.2 and $100\,\mu$s, an electric field is applied for $2\,\mu$s in order to field ionize the Rydberg atoms and to accelerate the resulting ions towards a channeltron. The counting distributions are obtained by performing 200 repetitions of the experiments. The overall detection efficiency is $\eta \approx 50\%$. During the excitation and detection sequence the MOT beams are switched off. We are also able to directly photoionize atoms [10]. To do so, the MOT beams remain switched on thus exciting atoms to the $5P_{3/2}$-level. The excited atoms are then photoionized with a $1-2\,\mu$s pulse of 421 nm laser light. The ions produced are detected using the same electric field pulses as in the Rydberg detection. We checked that the number of ions created is proportional to the number of atoms in the cloud [11]. Hence can be used to estimate N in conditions where a direct measurement using our CCD camera is not possible.

In order to vary the density of the atomic cloud by several orders of magnitude we use two different techniques.

The first technique consists in depumping some of the atoms of the sample into a hyperfine level of the ground state not coupled to the Rydberg transition. This is achieved by illuminating the atomic cloud with a short pulse of laser light of intensity $I = 3.6 \cdot I_{sat}$ (where I_{sat} is the saturation intensity of the transition $|5S_{1/2}\rangle \rightarrow |5P_{3/2}\rangle$ transition), resonant with the transition $|5S_{1/2}; F = 2\rangle \rightarrow |5P_{3/2}; F' = 2\rangle$, after the MOT and the repumping beams are switched off. The $|5P_{3/2}; F' = 2\rangle$ state has a lifetime $\tau = 30$ ns and decays both to the $|5S_{1/2}; F = 2\rangle$ and to the $|5S_{1/2}; F = 1\rangle$ states, with a branching ratio of 1/5 [12]. Although atoms in the state $|5S_{1/2}; F = 1\rangle$ are still present in the atomic cloud, they are no longer coupled to the Rydberg transition due to the large energy difference between the sub-levels of the ground state ($\approx 6, 8$ GHz). This leads to a decrease in the effective density n_{eff}, i.e, the density of atoms in the $|5S_{1/2}; F = 2\rangle$ state. In order to measure n_{eff} using this technique, after the depumping pulse goes off, we directly ionize the atoms of the cloud using the 421 nm laser and measure the number of ions created.

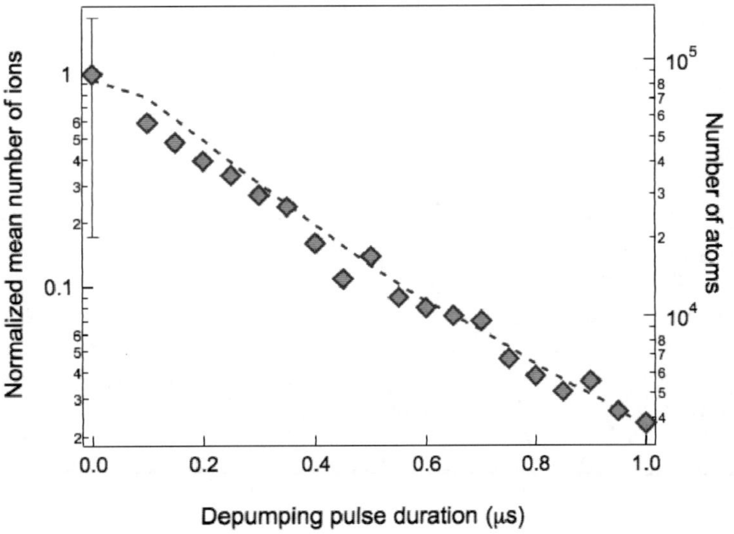

Figure 1. (Color online) Variation of the number of ions as a function of the depumping pulse duration. On the left axis, the normalized number of ions obtained as a function of the duration of the depumping pulse (blue diamonds) is plotted. The dashed line is a guide to the eye. On the right axis the number of atoms for each point is indicated.

Using this technique we managed to decrease the effective density of the sample, while its size remained constant. In Fig.1, we show data up to $1\,\mu$s of depumping pulse duration.Within this time, the effective density of the cloud decreases by 2 orders of magnitude. Increasing the duration of the depumping pulse up to 2 μs, we were able to decrease the effective density of the sample by $3-4$ orders of magnitude, generating clouds with ≈ 10 atoms. For depumping pulse duration $> 2\,\mu$s the efficiency of this technique decreases, making it difficult to reach a regime of (on average) a single atom in the cloud.

The few atom regime could be reached using the second technique, in which we reduced the loading time (τ_{loading}) of the MOT.

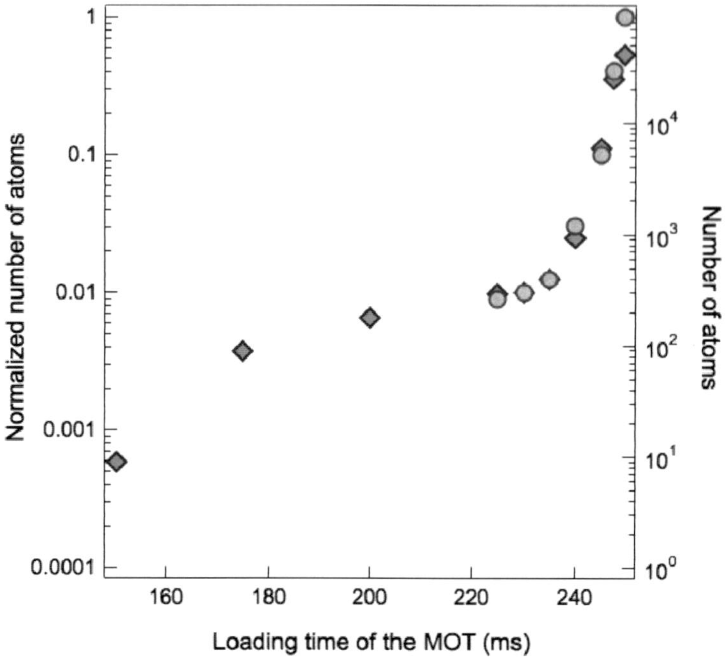

Figure 2. (Color online) Variation of the number of atoms in the MOT as a function of the loading time. On the left axis, the normalized number of atoms in the MOT is plotted as a function of τ_{loading} in ms. On the right axis it is shown the absolute value for each point. Grey circles indicate the number of atoms directly measured by the CCD camera (statistical error of $\sim 10\%$, systematic error estimated to be around a factor of 2). Blue diamonds correspond to the number of atoms calculated based on the number of ions measured when the sample is directly ionized. For both cases, the error bars are smaller than the symbols.

In Fig.2, grey circles indicate measurements of the number of atoms of the MOT taken by using a CCD camera. This device collects a portion of the light scattered by the atoms and from this signal we calculate the number of atoms in the MOT as well as its size, with a statistical error of $\sim 10\%$ (the systematic error is estimated to be around a factor of 2 due to uncertainties in the calculation of the atom number from the scattered light collected by the camera). Below $N \approx 300$ atoms, background noise made it impossible to obtain an accurate estimation of the atom number in this way. From this point (i.e., when $\tau_{\text{loading}} < 220$ ms), N was deduced by photoionizing the sample and measuring the number of ions generated. Knowing N from CCD measurement and the number of ions created in a photoionization experiment for a certain τ_{loading}, we could extrapolate the value of N in the regime where the CCD signal measurement is not possible (blue diamonds in Fig.2). Regarding the cloud size, for τ_{loading} up to 220 μs we did not observe significant variations, which we assume to be valid also for lower values of

τ_{loading}. Hence, using this technique we were able to reach a regime with $N \leq 10$ atoms.

Both techniques offer the possibility to generate clouds from several thousands to a few atoms without changing the size of the cloud. The variation of the loading time of the MOT allows us to reach the few atom regime, whereas the depumping technique makes possible to vary the effective density of the sample at any time of the experiment.

3. Results

We now use the techniques described in the previous section to study the dynamics of Rydberg excitations in a MOT from the few atom regime up to the high density regime. The variation of the MOT τ_{loading} was used to generate single atom conditions. At $\tau_{\text{loading}} = 250\,\text{ms}$, the cloud size was $\sigma_{x,y,z} \approx 29\,\mu\text{m}$ and $N = 7.1 \cdot 10^4$, almost single atom conditions.

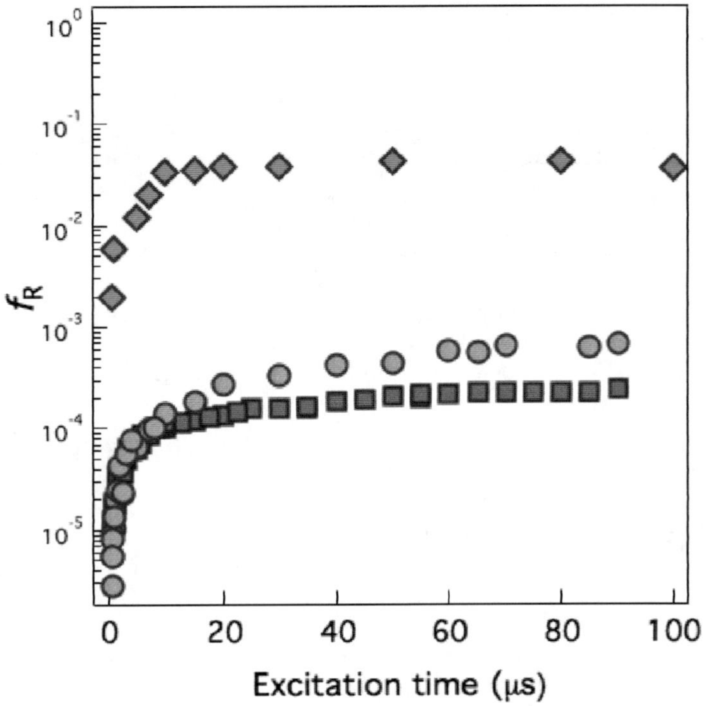

Figure 3. (Color online) Measured Rydberg fraction (f_R) as a function of the pulse duration. The densities of the different samples are: $n_{\text{eff}} \approx 6.0 \cdot 10^{10}\,\text{cm}^{-3}$ (red squares), $n_{\text{eff}} \approx 1.8 \cdot 10^{10}\,\text{cm}^{-3}$ (grey circles), $n_{\text{eff}} \approx 4.2 \cdot 10^{7}\,\text{cm}^{-3}$ (blue diamonds).

The higher effective densities were prepared using the depumping technique starting from a MOT of $n \approx 6.0 \cdot 10^{10}\,\text{cm}^{-3}$ ($\sigma_{x,y,z} \approx 160\,\mu\text{m}$, $N = 2.4 \cdot 10^5$ atoms). Because this cloud size is larger than the beams waists, only the number of atoms within the volume of interaction with the Rydberg transition lasers was considered in the calculation of the Rydberg fraction, f_R). The detuning of the intermediate state and Rabi frequencies were: $\Delta = +500\,\text{MHz}$ and $\Omega = 1.2\,\text{MHz}$, for the experiment in the few atom regime; and $\Delta = +720\,\text{MHz}$ and $\Omega = 240\,\text{kHz}$, for the high density regime experiment. In both cases, atoms were excited to the $70S$ Rydberg state, for which the blockade radius is $r_b \approx 10\,\mu\text{m}$.

In Fig.3 we show number of Rydberg excitations divided by the total number of atoms of the sample, f_R, as function of time. Red squares correspond to effective densities of:

$n_{\text{eff}} \approx 6.0 \cdot 10^{10}\,\text{cm}^{-3}$, grey circles: $n_{\text{eff}} \approx 1.8 \cdot 10^{10}\,\text{cm}^{-3}$ and blue diamonds: $n_{\text{eff}} \approx 4.2 \cdot 10^{7}\,\text{cm}^{-3}$, corresponding to interparticle distances of: $\sim 2.5\,\mu\text{m}$, $\sim 5.7\,\mu\text{m}$ and $\sim 28.8\,\mu\text{m}$, respectively .

In the few atom regime, atoms oscillate between the ground and the excited state with Rabi frequency Ω. After some time, the system reaches the stationary state with a fraction of atoms in the Rydberg state $f_R = 1/2$. Taking into account our detection efficiency, $\eta \approx 50\%$, this observed fraction is expected to be $f_R = 1/4$. The measured value $f_R \approx 0.05$ (see blue diamonds in Fig.3) is significantly lower. A possible explanation can be a systematic error in the estimated number of atoms induced by systematic error in the measurement of the the CCD camera. The resolution in the atom counting could be improved either by improving the measurement technique itself, or in the subsequent image processing. In the first case, different imaging techniques, ranging from the use of beams of high intensity to maximize the photon scattering rate [13] to the reduction of the energy released in light-assisted collisions between atoms [14] could be implemented; whereas an advanced postprocessing of the images through the combination of a fringe removal algorithm and maximum likelihood estimation[15] provides also an enhanced sensitivity on the readout of the atom number.

Another possible source of error could be that the expected f_R considers that all the atoms are driven with the same Rabi frequency and thus, reach the stationary state at the same time, which is clearly not the case for the gaussian intensity distribution of the excitation lasers of our experiment.

In the higher density regimes, f_R is much lower than in the few atom case because the number of atoms able to be excited to the Rydberg state is much smaller than the number of atoms in the ground state. This happens mainly due to the effect of dipole blockade [16], i.e., a suppression of excitation of an atom by an already excited one within the blockade sphere [17–19]. Alternatively, this effect can be interpreted as a manifestation of kinetic constraints that slow down the dynamics [9].

4. Conclusion

In conclusion we have presented two different experimental techniques to prepare atomic samples of different density regimes. Both of them offer the possibility to generate clouds from several thousands to few atoms. This could be applied to explore the behavior of Rydberg samples in an incoherent driving scenario, where kinetic constraints are expected to become manifest. Additionally, the depumping technique allows us to vary the density of the sample at any time of the experiment which permits us to, e.g., generate Rydberg atoms in a high density regime and, after some time, to remove the ground state atoms from the cloud.

Acknowledgments

We acknowledge financial support from the EU Marie Curie ITN COHERENCE Network.

References

[1] Gallagher T F 1994 *Rydberg Atoms* (Cambridge University Press)

[2] Gallagher T F and Pillet P 2008 *Advances In Atomic, Molecular, and Optical Physics* **56** 161 – 218

[3] Jaksch D, Cirac J I, Zoller P, Rolston S L, Côté R and Lukin M D 2000 *Phys. Rev. Lett.* **85** 2208

[4] Lukin M D, Fleischhauer M, Côté R, Duan L M, Jaksch D, Cirac J I and Zoller P 2001 *Phys. Rev. Lett.* **87** 037901

[5] Urban E, Johnson T A, Henage T, Isenhower L, Yavuz D D, Walker T G and Saffman M 2009 *Nat. Phys.* **5** 110

[6] Gaëtan A, Miroshnychenko Y, Wilk T, Chotia A, Viteau M, Comparat D, Pillet P, Browaeys A and Grangier P 2009 *Nat. Phys.* **5** 115

[7] Dudin Y O, Li L, Bariani F and Kuzmich A 2012 *Nat. Phys.* **8** 790

[8] Gärttner M, Whitlock S, Schönleber D and Evers J 2014 *eprint arXiv:1408.2453*

[9] Lesanovsky I and Garrahan J P 2013 *Phys. Rev. Lett.* **111**(21) 215305

[10] Viteau M, Radogostowicz J, Chotia A, Bason M G, Malossi N, Fuso F, Ciampini D, Morsch O, Ryabtsev I and Arimondo E 2010 *J. Phys. B: At. Mol. Opt. Phys.* **43** 155301

[11] Valado M M, Malossi N, Scotto S, Ciampini D, Arimondo E and Morsch O 2013 *Phys. Rev. A* **88**(4) 045401

[12] Steck D A URL `http://steck.us/alkalidata/rubidium87numbers.pdf`

[13] Muessel W, Strobel H, Joos M, Nicklas E, Stroescu I, Tomkovič J, Hume D and Oberthaler M 2013 *Applied Physics B* **113** 69

[14] McGovern M, Hilliard A J, Grünzweig T and Andersen M F 2011 *Opt. Lett.* **36** 1041

[15] Ockeloen C F, Tauschinsky A F, Spreeuw R J C and Whitlock S 2010 *Phys. Rev. A* **82** 061606

[16] Comparat D and Pillet P 2010 *J. Opt. Soc. Am. B* **27** A208

[17] Vogt T, Viteau M, Chotia A, Zhao J, Comparat D and Pillet P 2007 *Phys. Rev. Lett.* **99** 073002

[18] Tong D, Farooqi S M, Stanojevic J, Krishnan S, Zhang Y P, Côté R, Eyler E E and Gould P L 2004 *Phys. Rev. Lett.* **93** 063001

[19] Singer K, Reetz-Lamour M, Amthor T, Marcassa L G and Weidemüller M 2004 *Phys. Rev. Lett.* **93** 163001

Webcams as a tool for teaching in Optometry training

A Gargallo and J Arines
Departamento de Física Aplicada (Área de Optometría)
Universidade de Santiago de Compostela
Facultade de Óptica e Optometría (Campus Vida)
15782 Santiago de Compostela (SPAIN)

E-mail: ana.gargallo@usc.es - justo.arines@usc.es

Abstract. Clinical Optometry lab training is devoted to develop the student´s skills needed in eye healthcare professional practice. Nevertheless, students always find difficulties in the management of some optometric instruments and in the understanding of the evaluation techniques. Moreover, teachers also have problems in explaining the eye evaluation tests or making demonstrations of instruments handling. In order to facilitate the learning process, webcams adapted to the optometric devices represent a helpful and useful tool. In this work we present the use of webcams in some of the most common clinical test in Optometry as ocular refraction, colour vision test, eye health evaluation with slip-lamp, retinoscopy, ophthalmoscopy and contact lens fitting. Our experience shows that with this simple approach we can do things easier: show the instrument handling to all the students at the same time; take pictures or videos of different eye health conditions or exploratory routines for posterior visualization with all the students; recreate visual experience of the patient during optometric exam; simulate colour vision pathologies; increase the interactions between students allowing them to help and correct each other; and also record the final routine exam in order to make possible its revision with the students.

1. Introduction

Clinical Optometry lab training is one of the pillars in the Degree in Optics and Optometry program. It is devoted to develop the student´s skills needed in eye healthcare professional practice. They should manage all the optometric instruments with ability and control the evaluation techniques. Nevertheless, students always find difficulties in this task, in particular for some eye evaluation tests. On the other hand, teachers also have problems in explaining the management of the instruments or making demonstrations of its handling.

In order to facilitate the learning process, webcams adapted to the optometric devices represent a helpful and useful tool. This system allows teachers to make demonstrations of instrument handling, to recreate visual experience of the patient during optometric exam, to record the student practice examination and to simulate colour vision pathologies, among others. Besides, our experience shows that this method increases students' self-confidence and promotes interaction between them. In other words, they are able to see what a colleague is doing in the eye evaluation and they realize the errors committed and discuss about them.

We present in this work the use of webcams in some of the most common clinical test in Optometry. There are included ocular refraction, colour vision test, eye health evaluation with slip-lamp, retinoscopy, ophthalmoscopy and contact lens fitting [1-4].

2. Uses of webcams in Optometry lab

Following we explain the uses of webcams in different optometric test. We used an intermediate quality webcam model for all the tests, the *Microsoft Lifecam HD-5000* connected to a PC running Linux, which reduces the cost of the setup.

2.1. Subjective refraction and refractive error simulator

The evaluation of refractive error is part of the usual optometry practice. Subjective refraction is the technique of comparing one lens against another, using changes in vision as the criterion, to arrive at the dioptric lens combination that results in maximum visual acuity. Optometrist uses trial frames or a phoroter combined with optotypes for this task [1].

In this experience we modified the webcam by removing its lens and adapting a 2.54 cm focal length lens with a variable aperture diaphragm. The modified webcam resembles an "artificial eye". The ametropies of the eye were introducing by modified its axial length or placing an ophthalmics lens before the diaphragm. Then the "artificial eye" was placed behind the aperture of the phoroter that the patient sees through. An observer interpreted the image provided by the webcam in order to performe the "artificial eye" refraction, using common procedure of subjective refraction with current Snellen optotype.

This experience was used to explain how the patient sees the optotypes depending on the ametropy and sees the optotypes during the different steps of the refraction protocol. As we said, with the aid of different ophthalmic lenses placed before the diaphragm we induced different refraction errors to the "artificial eye", and then a student executed the subjective refraction protocol, allowing the teacher to evaluate the ability of the student, by watching the images provided by the "articial eye" (see in figure 1 three images provided by the "artificial eye" in three different steps of the subjective refraction protocol).

Figure 1 Optotype cards seen thought the "artificial eye" during refraction. In (a) we can see how the "artificial eye" sees the optotype card in the refraction test with letters optotype and spherical lenses. In (b) the astigmatism is being tested with the clock test. The vision at the end of the refraction with the better correction is shown in (c)

2.2. Colour vision test

The pseudoisochromatic charts are the most popular colour vision test [1, 2]. They consist in a book with several charts; each one contains a number or figure, or even an irregular line. The figure is composed of points that differ from the background in tone and bright, this disposition tends to hide

Journal of Physics: Conference Series **605** (2015) 012039 doi:10.1088/1742-6596/605/1/012039

the figure to those patients with a determined colour vision deficiency [2]. In order to show the visual experience of a patient with different colour vision deficiencies, we placed coloured filters before the webcam and evaluated a pseudoisochromatic chart test. That allow us to explain our alumni how patients with a colour vision disorder perceive the colours and why they fail the test. We simulated three chromatic deficiencies with red (protan), green (deutan) and blue (tritan) filters (figure 2).

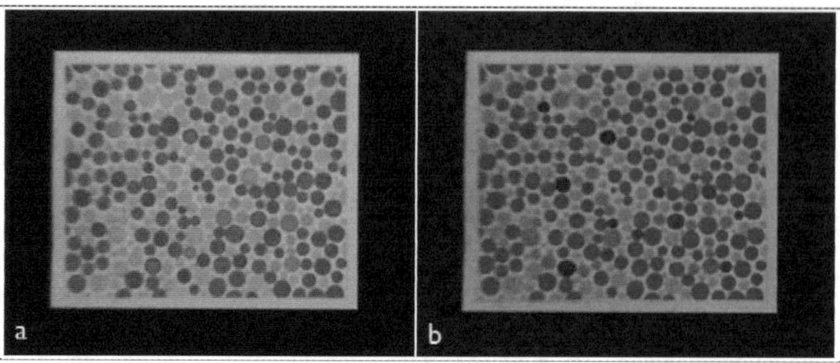

Figure 2 Ishihara´s test seen with the webcam without a filter (a) and with a red filter simulating a red-green deficiency (b).

2.3. *Slit-Lamp management*

Slit-lamp, also known as biomicroscope, is one of the most important tools in optometry practice. It is a low-power microscope consisting of an observation system, which is a binocular microscope; an illumination system, which is a bright light source with a mechanism that allows to control the width an orientation of the slit; and the mechanical support for its coordination [1-4]. The training in the use of the slit lamp has always been difficult for Optics and Optometry students. Instruments with associated cameras help a lot in this task. However, these devices are more expensive than those that do not have an integrated camera connected to a display unit.

With the aim to improve students´ skills in the management of slip lamp, we fabricated a device to adapt to one of the oculars of the slit-lamp, a webcam connected to a PC running Linux [5]. We used it to make demonstrations of an eye health evaluation routine and to control the student in its management (figure 3). Our experience shows that this simple method has several advantages. It allows us to take pictures with a good quality of different conditions of the eye health; we can record videos of eye evaluation and make demonstrations. It increases the interactions between students because they could see what their colleagues are doing and take conscious of the mistakes, help and correct each other. It is a useful tool for teachers for monitoring the exam of slit-lamp handling too. We think that the method supports the training in optometry practice and increase the student´s confidence without a huge outlay.

Figure 3 Slit lamp webcam adaptor.

Journal of Physics: Conference Series **605** (2015) 012039 doi:10.1088/1742-6596/605/1/012039

2.4. Direct Ophthalmoscopy

Ophthalmoscope is an instrument designed to view anatomical details of the human retina, it is composed by a light source, a moveable disc that held an assortment of lenses and a set of mirrors [1]. In this experience we adapted the webcam to a direct ophthalmoscope with the aim of making video recordings of an eye fundus exploration. That recording was used in class to show how retina is seen through this instrument and how to perform the eye fundus evaluation. In figure 4 we present an example of the pictures obtained.

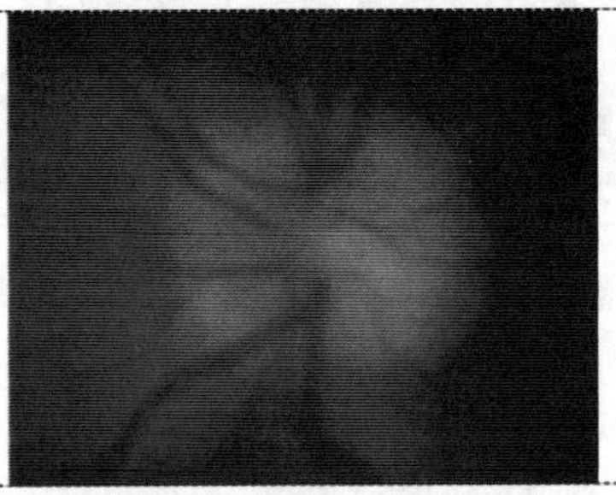

Figure 4 Picture of a retina, seen through a direct ophtalmoscope.

2.5. Retinoscopy

The most commonly used objective refraction instrument in clinical practice is the retinocospe. Before subjective refraction, optometrists usually performed retinoscopy to determine patient´s refractive status. A retinoscope is a small, handheld device that emits visible white light toward the pupil of the eye being analyzed and allows the operator to view the red reflex of light reflected back through the pupil from the ocular fundus [1]. Depending on the kind of movement of the backreflected light (or shadow) an experienced optometrist can evaluate the kind of refractive error present in the examined eye. In order to explain this test, we adapted the webcam to a retinoscope and recorded the shadows in the pupil. Our students could understand how the shadows change their movement during the performance of this test.

2.6 Contact lens fitting

Contact lens practice requires the uses of a slit-lamp in order to evaluate the fitting. Quantification of contact lens movement and the technique of push up cannot be understood without a correct visualization of it. In this context slit lamps with associated cameras are very helpful, they allow teachers to observe and control if the students contact lens fitting evaluation on real time. The webcams adapted to slit-lamps described before also were used to take pictures of lenses in the eye and record lens movement or fluorogram [3, 6]. Moreover, it was used to record the insertion of the lens in the eye or the removal proceeding. In figure 5 we present an example of a picture of a fluorogram used to evaluate rigid contact lens fitting.

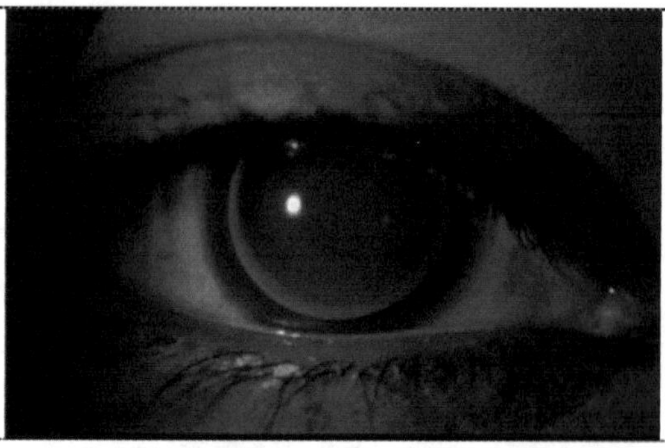

Figure 5 Webcam picture of the fluorogram of a rigid gas permeable contact lens in an eye seen through the slit-lamp.

3. Discussion

Our experience has shown that webcams are a helpful tool in Optometry lab training. They have several uses and more potential application in teaching task.

In subjective refraction the examiner determines the patient´s refractive error based on the subjective perception of different visual targets. That situation is a problem because the ability to discriminate between dioptric presentations varies among individuals. With the aid of the webcam we can show the stimulus presented to the patient and how it changes with the addition of corrective lenses. It was a helpful tool to simulated refractive errors and present our students how is the vision under a myopia, hyperopia and astigmatism. It was useful in particular in the explanation of Jackson´s cross cylinder which is used to adjust the axis and power of astigmatism [1,2].

There is a bad understanding of colour vision deficiencies, they are called colour blindness and people think that those patients don´t see the colours, but they see them differently. Most of the colours are seen as being more washed out or paler as compared with the way that those with normal colour vision see them. With the aid of webcams and filters we explained this to our alumni. We performed in the pseudoisochromatic charts test but could be used in others colour vision test as Farnsworht´s [2]. It can also help to show how some patients with these deficiencies can be beneficiated from a coloured filter.

Ophthalmoscopy is a simple technique to perform but it is difficult to explain, the webcam recording helps a lot in this task. We can show how the posterior pole is evaluated and what we can see. The ophthalmoscope yields a great amount of magnification and this characteristic makes the field of vision narrower, so the examiner should move the head of the ophthalmoscope with precision to evaluate the patient´s retina. A potential use of the webcam adapted to this instrument could enhance the ability of students in the management and control of the ophthalmoscope.

Retinoscopic findings may be very important in the final prescription of optical correction in some patients as children, low-vision patients, uncooperative patient, or mentally disorder patients. Then, it is essential a good comprehension of this technique by optometrist. At the beginning of the training, the identification of the movement of the shadows is very difficult, so that, webcam recording can help students to realize how they are, they and they move in presence of different ametropies and during the neutralization of the ametropy of the rack lenses used for that purpose.

With the association of webcams in the slit-lamps of our contact lenses laboratory we improve the teaching in the eye health evaluation and contact lens fitting evaluation with this instrument. That is low-cost method in comparison with these devices with associated cameras and showed several advantages. Our alumni revealed a good assessment of our method by means of self-confidence increase and reduction in the training time [5].

4. Conclusions

In general terms, webcams have several advantages in the Optometry teaching. They are inexpensive and could be adapted to a considerable number of optometric instruments for eye health or ocular refraction evaluation. Also they are able to simulate some vision anomalies as ametropies or colour deficiencies and make the student conscious of the vision quality in these patients.

Acknowledgments

This work supported by the University of Santiago de Compostela, grant Convocatoria de axudas para o desenvolvemento de accións de mellora da docencia 2011, Isidro Parga Pondal Programme 2009 and Plan I2C Programme Apoyo a la etapa Predoctoral (Xunta de Galicia).

References

[1] Benjamin W J 2006 *Borish´s Clinical Refraction* (Sant Louis, Missouri: Butterworth Heinemann Elsevier, 2nd Ed,)
[2] Grosvenor T 2007 *Primary Care Optometry* (Boston: Butterworth Heinemann Elsevier, 5th Ed.)
[3] Carlson N B and Kurtz D 2003 *Clinical Procedures for Ocular Examination* (Third Edition McGraw Hill Professional)
[4] González-Cavada Benavides J 2000 *Atlas de Lámpara de Hendidura (Biomicroscopia ocular)* (Madrid: Editorial Complutense)
[5] Gargallo A and Arines J 2014 *Opt. Pura Apl.* **47** (2) 163-167 DOI: 10.7149/OPA.47.2.163
[6] Sulley A, Young G, Lorenzo K O and Hunt C 2013 *Ophthalmic Physiol. Opt.* **33** 94-113 DOI: 10.1111/opo.12028

Journal of Physics: Conference Series **605** (2015) 012040 doi:10.1088/1742-6596/605/1/012040

Spreading Optics in the primary school

Ana Gargallo[1], Ana I. Gómez-Varela, Héctor Gónzalez-Nuñez, Tamara Delgado, Citlalli Almaguer, Ferran Cambronero, Ángel García-Sánchez, David Pallarés, María Aymerich, Ángel L. Aragón and Maria T. Flores-Arias

Departamento de Física Aplicada (Área de Óptica), Universidade de Santiago de Compostela, Facultade de Óptica e Optometría (Campus Vida) 15782 Santiago de Compostela (SPAIN)

[1] E-mail: usc.osa.sc@gmail.com

Abstract. The USC-OSA is a student chapter located at the University of Santiago de Compostela (Spain) whose objective is to bring optics and photonics knowledge closer to general public. In order to arouse kids´ interest in Optics we developed an activity called Funny Light. This activity consisted on a visit of some USC-OSA members to a several local primary schools where we organized several optics experiments. In this work we present the optics demonstrations and the reaction of the 6 years-old students. The activities with greater acceptance include an explanation of light properties as polarization, refraction or reflection, and the workshop where they learnt how to build their own kaleidoscope and made a chromatic disk. Besides, they also participated in a demonstration and explanation of color properties and some optical illusions. We think that this activity has several benefits including spreading Optics through children meanwhile they have fun and experiment science in real life, as well as helping teachers to explain some complex properties and Physics phenomena of light. Given the broad acceptance of this activity, we are intending to make it a routine event of our student chapter repeating it every year.

1. Introduction

International scientific societies focused in Optics and Photonics promote the association of students in universities worldwide in order to involve them in the task of spreading the Optics knowledge into the community. These students groups are called Student Chapters or Student Clubs and are formed by students of different careers related to Optics. They receive funding and networking from an international organization with the goal of increasing the interest in Optics in pre-college students and provide at the same time professional development of its members.

The USC-OSA student chapter was founded in 2013 at the University of Santiago de Compostela and is supported by the Optical Society of America (OSA); this is an organization whose mission is to promote the generation, application and archiving of knowledge in optics and photonics and to disseminate this knowledge worldwide since 1916. In 2002 the OSA established a non-profit organization for keeping with that mission, the OSA Foundation, which supports programs that advance youth science education, enriches Optics and Photonics education in developing nations, provides professional development resources to college-level students and recognizes technical and business excellence. The OSA Foundation funds several grant programs, competitions and events that benefit the Student Chapters and their members [1].

Journal of Physics: Conference Series **605** (2015) 012040 doi:10.1088/1742-6596/605/1/012040

One of the goals of the USC-OSA student chapter is to arouse kids´ interest in Optics. In order to attain this objective we developed an activity called Funny Light, which was carried out at some primary schools in Santiago. This activity consisted on a visit of some USC-OSA members to local primary schools where we organized several optics experiments with children. In this work we present the simple optics demonstrations and the reaction of the 6 years-old students. The activities were grouped in two blocks; the first included demonstrations made by chapter´s students and the second block consisted in some workshops where children could make up optics handicrafts. Those with greater acceptance include an explanation of light properties as polarization, refraction or reflection, and the workshop where they learnt how to build their own kaleidoscope to understand reflections of light and made a chromatic disk to feel how white light is compound of colors. Besides, they also participated in a demonstration and explanation of color properties and some optical illusions.

2. Optics experiments for children included in Funny Light activity

In this section we present the activities carried out in three primary schools located in Santiago de Compostela with children of 1st and 2nd course.

We separated pupils in two groups of 25-30 children. One of the groups realized the first activity called *Let´s play with light*, which consisted on Optics demonstrations of refraction, reflection, polarization, etc, meanwhile the other group built the kaleidoscope and the chromatic disk.

In the global activity, around 200 kids between ages 6 and 7 of different schools were involved. All the activities were developed in a 4 hours morning school journey in separate days for each school.

2.1 Let's play with light.

In this activity, we usually started with a presentation and then the children were separated in small groups of 6 or 7 people and distributed in 4 stands where an USC-OSA member made a demonstration. Each group spent 15 minutes in each stand and then they changed their location with their colleagues in order to participate in all of the experiments. *Let´s play with light* consisted on the following activities to show how the light works:

2.1.1. Visual perception. We started with a presentation explaining with funny cartoons that all we see is light. With the aid of the Optics Suitcase, provided by OSA, the children recognized the colors in white light with a little lantern and a diffractive grating (figure 1). Some curious aspects about human vision as binocularity, contrast and sharp perception or color vision were demonstrated with optical illusions cards. Optical illusions or visual illusions can induce images that are different from the objects that create them. For example, some of the illusions can show the ability to see in three dimensions even though the image in the retina is only two-dimensional. Others affect the contrast of a homogeneous object, appearing darker against a black field compared to a white field.

We made a selection of different cards to explain contrast and sharp perception, binocularity, motion perception or color vision.

Figure 1. Children attending to an explanation about vision and optical phenomena.

2.1.2. Light painting. Light painting is a photographic technique in which exposures are made by moving a hand-held light source or by moving the camera. With a reflex camera Canon EOS 7D and led lanterns we took pictures using this technique. In groups of 5, each child draw a letter to form a five-letter word while the other children had to guess what they wrote (figure 2).

Figure 2. Light painting.

2.1.3. Color experiment. We demonstrated the color composition of white light by creating a rainbow in the classroom with the aid of a prism and a white light source and explaining children how the raindrops behave as a prism.

In order to make children learn about primary and secondary colors we used an old overhead projector, three movable mirrors and a plate with three filters (green, red and blue). We placed a stand with the three movable mirrors close to the projector and a plate with the three filters (green, red and blue) over the light source. This set up allowed us to project color circles on a screen or a white wall and mix them in pairs by moving the mirrors; at the end, we mixed up the three color circles to obtain white light. Moreover, the children could make colored shadows with their hands (Figure 3). We also used a template with a part of a house in each color and the complete figure was formed when the three colors were mixed.

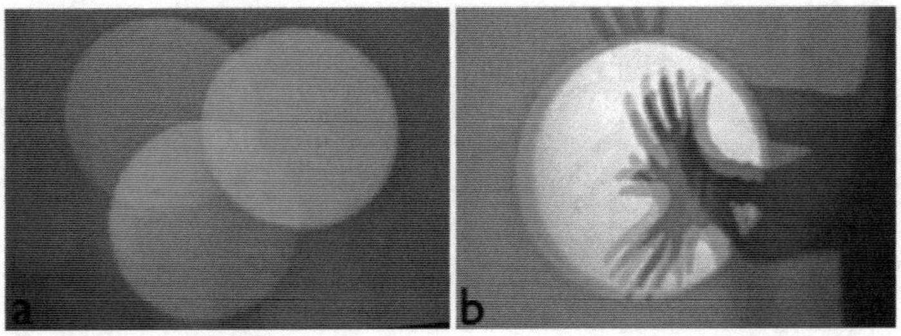

Figure 3. Color experiment. In (a) we can see the projection of the three primary colors in a white wall and the secondary color as a result of the mixing in pairs. In (b) there is a picture of colored shadows.

2.1.4. Polarization. This property of the light is very difficult to understand for children and also for adults. We designed an experiment to show how polarization is a useful property in life. We used an overhead projector with a polarized filter in the light source, some polarized filters and different objects made of transparent plastic material as a rule, toothbrush or cookies envelope (Figure 4). The Optics Suitcase includes individual packs of polarizer filters and plastic materials and each kid could experiment by itself this characteristic of light.

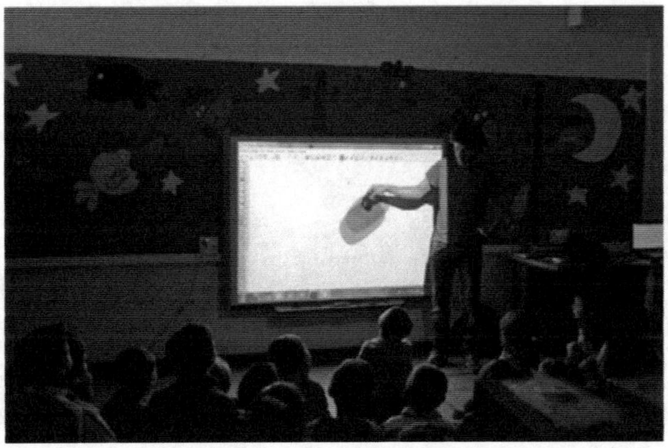

Figure 4. An USC-OSA member showing to children the change in color of a projector with a polarized filter.

2.1.5. Refraction and reflection demonstrations. We explained refraction with two glasses and a pencil. One of the glasses was filled with water and the other with oil, thus children could see how the pencil seems to break and twist more in oil than in water because of their different refraction indexes. Another experiment was carried out with a glass over a coin; the coin looks to disappear when the glass is filled with water because of the change in the angle of the light.

With a special box made by two opposite concave mirrors and an aperture in upper mirror we generated a real image of a well-known physical object (Mirage® PARABOLIC MIRROR HOLOGRAM MAKER). The object is placed in the center of the bottom concave mirror and a real image instantly projects up through the aperture, appearing to the viewer as a truly solid object. Any small objects can be used e.g. a coin, marble, sweets, etc. This is an ideal way to demonstrate and teach the concept of real images [2].

2.1.6. Optical fiber simulation with water. Based on Colladon, Babinet and Tyndall [3] experiments, we built a waveguide device with a bottle of water, a laser and a bucket. We made a hole in a lateral of the bottle close to the bottom; the bottle was filled with water and illuminated through from the other side of the bottle with a red laser beam. Water stream was collected in the bucket. Kids could see light being guided through the water and reaching the bottom of the bucket, instead of propagating in a straight line to the wall. Water plays as a waveguide in this simple experiment and explains the total reflection of the light.

2.2. Optics handicrafts

This part of the event involved fifteen 6-years old children from the 1st course of primary school. It took place in a manual arts classroom.

2.2.1. Kaleidoscope construction. Brewster invented the kaleidoscope in 1816 [4] and this device still leaves an impression on young children. We organized a workshop in the school where every child built his own kaleidoscope. We used three mirrors arranged in an equilateral triangle, taking into account the age of the children, we substituted crystal mirrors for plastic plates covered with reflective paper. The plates were situated inside a toilet paper tube; one transparent plastic circle was located in one side of the cylinder; over it, the children put some glass beads of different colors and finally covered all with a piece of one translucent plastic bag. The other side of the tube was covered up with a ring made of cardboard. The outside was decorated with color paper and figures (Figure 5). We had to help some of them with cutting and sticking tasks because of the young age or child ability. USC-OSA members could explain reflection property of the light with this enjoyable workshop.

Figure 5. The kaleidoscope workshop. The finished kaleidoscopes are shown in (a); in (b) children are looking through them.

2.2.2. Chromatic disk. In 19th century the use of chromatic disk mixture was the only mean for defining color stimuli and matching. Nowadays it is only a curiosity but could help in children comprehension of colors [5]. We prepared a template in order that the kids could paint it following a pattern, then they cut the circle and stuck it in a pencil to allow them to spin the disk around and observe the color mixture (Figure 6).

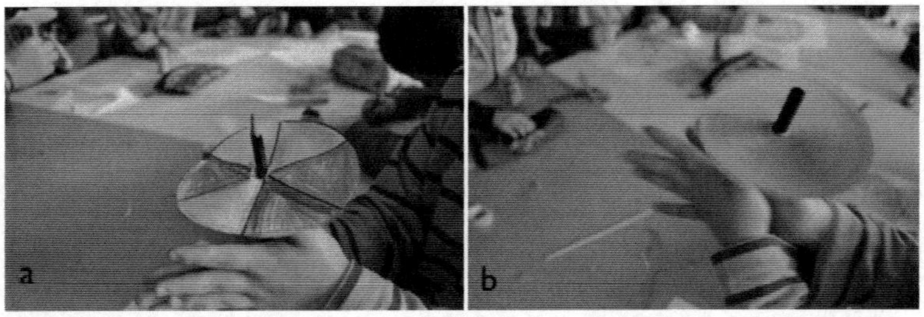

Figure 6. In this figure a child is trying out the color mixture with a chromatic disk in a not moving (a) and moving state (b).

3. Discussion

It is important the knowledge of basic experiments and techniques for spreading Optics and Photonics in schools and high schools, but it is also important the people who develop these activities into the community and bring them closer to teachers and students. We think that an organized group works better than an individual person and takes less effort for its members, so we created the USC-OSA student chapter.

Physics concepts of light properties are often difficult for kids, also for adults who are not familiarized with that field. With simple experiments we could catch the attention of the youngest schoolchildren. We are conscious that they did not understand in deep some of the light properties presented, but they discovered some aspects of Optics that aroused their interest. We want to broaden the activity into older children and pre-university students in order to make visible the uses of Optics and Photonics in the daily life.

Concerning the first part of the event, the children enjoyed a lot with all the experiments and show an active participation. In particular, the Polarization demonstration was very welcome, in contrast with our expectations because it is a complex concept of Optics. The Optics handicrafts activity was developed only in one course because it required us to work in small groups with the kids in order to help in the task of cutting, assembling pieces, sticking, etc.

4. Conclusions

We think that this activity has several benefits including spreading Optics to children meanwhile they have fun and experiment science in real life, as well as helping teachers to explain some complex properties and Physics phenomenon of light. Given the broad acceptance of this activity, we are intending to make it a routine event of our student chapter repeating it every year.

Acknowledgments

We want to acknowledge The Optical Society of America for the financial support and the collaboration of the University of Santiago de Compostela for the material contribution.

References

[1] http://www.osa.org/en-us/foundation/
[2] Sieradzan A 1990 Teaching geometrical optics with the "optic mirage", *Phys. Teach. 28, 534*
[3] Testa G, Huang Y, Zeni L; Sarro P M and Bernini R 2012 *Proc. SPIE 8264, Integrated Optics: Devices, Materials, and Technologies* XVI, 826415 (February 9, 2012)
[4] Brewster D 1858 *The kaleidoscope* 2nd Ed. (London, 1858) http://books.google.es/books?id=vjIDAAAAQAAJ
[5] Kuehni G 2010 *Color Res. Appl.* **35**: 110–121

Explore the concept of "light" and its interaction with matter: an inquiry-based science education project in primary school

P Varela[1] **and M F Costa**[2]
[1] Research Centre on Child Studies, University of Minho, 4710-057 Braga, Portugal.
[2] Centre of Physics, University of Minho, 4710-057 Braga, Portugal.

E-mail: pibvarela@ie.uminho.pt

Abstract. The exploration process leading to the understanding of physical phenomena, such as light and its interaction with matter, raises great interest and curiosity in children. However, in most primary schools, children rarely have the opportunity to conduct science activities in which they can engage in an enquiry process even if by the action of the teacher. In this context, we have organised several in-service teacher training courses and carried out several pedagogic interventions in Portuguese primary schools, with the aim of promoting inquiry-based science education. This article describes one of those projects, developed with a class of the third grade, which explored the curricular topic "Light Experiments". Various activities were planned and implemented, during a total of ten hours spread over five lessons. The specific objectives of this paper are: to illustrate and analyse the teaching and learning process promoted in the classroom during the exploration of one of these lessons, and to assess children's learning three weeks after the lessons. The results suggest that children made significant learning which persisted. We conclude discussing some processes that stimulated children' learning, including the importance of teacher questioning in scaffolding children's learning and some didactic implications for teacher training.

1. Introduction

Very early on, children manifest a natural curiosity and interest in knowing and making sense of the world that surrounds them. The science teaching should take advantage and enhance these natural qualities of children, as they constitute the necessary support for active and meaningful learning in the classroom [1, 2, 3]. The goal is to "educate" the children's natural curiosity in order to develop more systematic, deeper and autonomous thinking patterns [4]; stimulate them to pose questions and look for possible answers for what they do and see; enable them to devise ways to test their ideas and thought strategies; to share and discuss their own theories and explanations with others [5, 2]. Unfortunately, the traditional science teaching works in a way that generally discourages the natural process of inquiry. Therefore, the inquiry activities are a privileged path to convert classrooms into places of leisure, satisfaction and personal fulfilment, as they allow the creation of a learning environment where children learn and do things they really enjoy [3, 6].

Inquiry-based teaching has been suggested by the science curriculum guidelines of many countries and recommended by several international reports and studies [7, 8, 9]. For example, the US *National Science Education Standards* were developed by the *National Research Council* to "promote a scientifically literate citizens". These Standards frequently encourage the use of inquiry in science classrooms, defining it as "... a multifaceted activity that involves making observations; posing questions; examining books and other sources of information to see what is already known; planning investigations; reviewing what is already known in light of experimental evidence; using tools to gather, analyse and interpret data; proposing answers, explanations and predictions; and communicating the results" [7, p. 23]. Inquiry can also be defined as the "intentional process of diagnosing problems, critiquing experiments and distinguishing alternatives, planning investigations, researching conjectures, searching for information, constructing models, debating with peers and forming coherent arguments" [10, p. 518].

The *National Research Council* [7] identifies five essential features for classroom inquiry: a) learners are engaged by scientifically oriented questions; b) learners give priority to evidence; c) learners formulate explanations from evidence, d) learners evaluate their explanations in light of alternative explanations; and e) learners communicate and justify their explanations. These features may vary in the amount of detailed guidance provided by the teachers. Thus, there may be different inquiry approaches, depending on the degree to which teachers structure what students do. These approaches are sometimes referred to as "guided" versus "open" inquiry. Guided inquiry teaching can be more focused on the development of particular science concepts. Open inquiry, on the other hand, will afford the best opportunities for cognitive development and scientific reasoning.

In science education, inquiry teaching reflects the concerns of Dewey, who, as early as the beginning of last century, considered that there was too much emphasis on facts and not enough emphasis on science in the development of thought and attitude of the mind. For Dewey, the student should be actively involved in the learning process and the teacher should take on a role of facilitator and guide. According to Drayton and Falk [11, p. 25], "The inquiry-based approach to science education [...] introduces students to science contents, including the process of investigation, in a context of reasoning, which gives science its dynamic nature and provides the logical framework that enables the understanding of scientific innovation and the evaluation of scientific claims. Inquiry is not process versus content; it is rather a way of learning content". Inquiry teaching is an approach that enables the learning of concepts and the development of process skills [2]. Thus, the alleged opposition between content and science process skills is a false dichotomy, as: "science process skills, on the one hand, and knowledge and comprehension, on the other, intensify each other in an interdependency that generates higher levels of process skills and higher levels of knowledge and comprehension" [3, pp. 58-59].

An inquiry-based learning environment promotes opportunities for children "to learn science, to learn how to do science and to learn about science" [7, p. xv]. Inquiry-based science education in the early years of schooling is vital to help the children: understand the world around them; learn to obtain and organise information; develop ways to discover; test ideas and use evidence; and develop positive attitudes towards science [2, 12]. On the other hand, it can also help children develop very different thinking skills early on [13], e.g., scientific thought, critical thinking, autonomous problem solving, communication and metacognitive skills, which are likely to be transferred and applied to other contexts and learning situations [14, 15]. Inquiry-based science education not only contributes to a better understanding of scientific concepts and skills but, because science inquiry in the classroom is carried out in a social context, it also contributes to the children's social and intellectual development [14]. Research shows that, when involved in inquiry activities, children are more actively engaged in their learning; they use and develop skills from other curricular areas, including language and mathematics [14, 15, 16] and they develop a positive attitude towards science [2]. In addition, inquiry helps children to create "habits of mind" [14], which are transferred to other experiences and learning contexts. This science teaching practice has proven effective in increasing the students' interest and achievement levels at both primary and secondary levels [17], while, at the same time, stimulating the teachers' motivation [18].

However, in the majority of European countries, the reality of classrooms is that these approaches to science teaching and learning are still only implemented by relatively few primary school teachers [19, 20]. Several causes have been identified, including the realisation that teachers have insufficient scientific knowledge on the contents they need to approach with the students and limited conceptions on inquiry-based science education and its pedagogical approach in the classroom [2, 21, 22].

In Portugal, the situation is no different. Although the science curriculum of primary education suggests a teaching practice in which students should be "active observers, with the ability to discover, investigate, experiment and learn" [23, p. 102], such a teaching practice is still only occasional, with only a residual expression in the teachers' pedagogical practices [3].

In this context, we have been promoting many teachers training courses and developing small pedagogical intervention projects at some primary school classes, with the aim of promoting an inquiry-based science teaching practice. This article describes one of these projects, centred in an approach to the curricular topic "Light Experiments". For that purpose, several lessons were planned and implemented in the classroom. Thus, the specific objectives of this paper are: a) to describe and analyse the teaching and learning process promoted in the classroom during the exploration of one of these lessons, and b) to assess the learning acquired by the children.

2. Methodology

The pedagogical intervention project adopted an action research methodology and was carried out with a class of the 3rd year in a Portuguese primary school, situated in the suburbs of the city of Braga. The class was composed by 16 students, 10 boys and 6 girls, aged between 7 and 8 years. For two months, 5 lessons were taught on the curricular topic "Light Experiments", amounting to a total of 10 hours of intervention in the classroom, as presented in the following table:

Table 1. Lessons subject and duration.

Lesson subject	Duration
1. What is light? Why can't we see objects in the dark?	2 hours
2. How does light travel?	2 hours
3. Which materials let light travel through them?	2 hours
4. Reflection of light.	2 hours
5. Refraction of light.	2 hours
Total	10 hours

For each topic addressed, a teaching and learning plan was prepared, containing the following elements: i) learning goals; ii) materials needed for the groups to implement the planned activities; iii) guidelines for the teaching and learning process, and iv) an individual record sheet for each student. Each lesson, which corresponds to one action research cycle, begins with a teaching and learning plan, which is implemented flexibly, according to the teaching and learning processes generated and promoted in the class reality. The lessons were taught by the first author of this paper, who, in collaboration with the class teacher, played the role of both researcher and teacher.

The data generated in this intervention were collected using two complementary methods: the field notes made by the researcher and the audio recordings of the lessons. This raw data were subsequently compiled in the form of detailed narratives including the most relevant events that took place in the classroom – the class diaries. These constituted the principal method of recording data and, simultaneously, a strategy for reflection and for the modelling of the teaching and learning process [24, 25]. With the purpose of assessing the learnings achieved by the children, was built and applied a survey with true or false items, three weeks after the pedagogical intervention.

3. Results

The following presents the analysis of the class diary concerning "reflection of light" and the results of the applied questionnaire that intended to assess students learning, after three weeks of the pedagogical intervention in sciences.

3.1. Content analysis of the class diary

Students are arranged in small collaborative groups. The lesson begins with the following questions:

- What happens when the light of a flashlight hits a mirror? And a cardstock?

Each group has on their table a mirror and a cardstock target, propped up with plasticine.

  - *The groups make predictions.*
    - In relation to the mirror, the prevailing prediction is that the light will be reflected: "the light hits the mirror and comes back" (Bruna); "it hits it and comes back" (Luís); "it reflects back" (Simão). Others make predictions using the knowledge acquired in previous classes: "the mirror is opaque and the light comes back" (Diogo); "opaque materials reflect light" (Lara).
    - In relation to the cardstock, predictions are divergent. Some argue that it does not reflect the light: "On the cardstock, it does not come back" (Eva); "the light will stay there" (Guilherme); "it hits and stays on the cardstock" (Gonçalo); while others maintain that the cardstock also reflects the light: "If the cardstock is opaque, the light also has to come back" (Joel).
- What must we do in order to see what will happen?
  - *The students suggest ways to test their predictions. Excerpt from the class diary:*

  [The majority of the students simply suggest "try it". However, some suggest a way to test their predictions: "We can point the flashlight at it and see if the light comes back. First, we point it at the mirror and if the light appears here on the table, it means it comes back" (Leonardo)].

  - *They test the predictions and make observations.*

  Each group of students is given a flashlight. The flashlight is covered with opaque paper, which has a slit in the middle. The students focus the light that passes through the slit onto the mirror and then onto the cardstock. Their observations are consistent with the idea that light "comes straight back" when pointed at the mirror; but in relation to the cardstock, they have doubts.
- What are the differences between what is happening with the mirror and with the cardstock?
  - *They reflect on their observations. Excerpt from the class diary:*

  [There is an apparently unanimous idea that both materials reflect light, but in a different way. "It is different. With the mirror, we see a clear line of light on the table, but with the cardstock, that doesn't happen" (Joel); "with the mirror, the light comes straight back and, with the cardstock, it comes back only a little bit" (Daniela); "the light hits the mirror and then comes straight back to the same place, in a straight line. With the cardstock, it does not seem to be hitting only this place (the table)" (Diogo), "With the mirror, the light comes straight back, you can see it here on the table" (Ângelo); "it goes straight, we learned this in the last lesson. Then it also comes back in a straight line" (Bárbara)].
- Will you be able to receive the light reflected from the mirror on a cardstock target?
  - *They test their ideas and communicate observations.*

  On their tables, groups assemble a device similar to that depicted on Figure 2, in order to make the cardstock target receive the light reflected from the mirror.

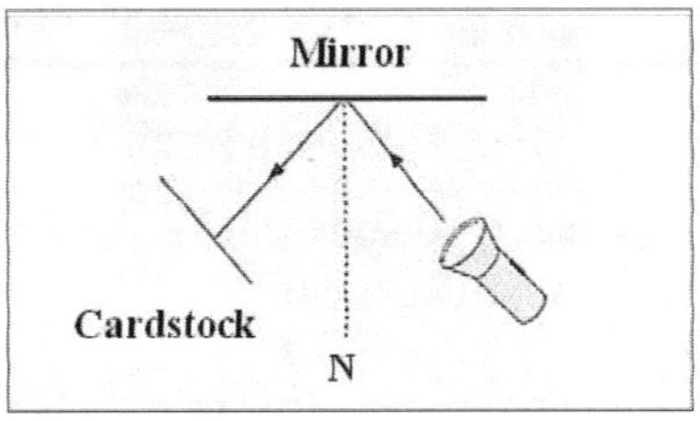

Figure 1. The students testing their ideas in groups.

After the test, the following comments were made: "the mirror has to be in front of the cardstock and the flashlight has to be in front of the mirror" (Francisca); "We pointed the flashlight at the mirror and the light reflected onto the cardstock" (Eva); "the cardstock has to be in front of the mirror, because the mirror reflects back" (Simão).

- Will you be able to receive, on a second cardstock, the light reflected by the mirror?
  - *They infer that the cardstock does not reflect light like a mirror. Excerpt from the class diary:*

  [After several attempts, the groups are unanimous in stating that they are not able to receive the light reflected by the mirror on a second cardstock: "It does not work, but with another mirror it would. There would have to be two mirrors" (Eva). "Why do you think it is not possible?" – I ask. "Because the cardstock does not reflect" (Eva); "First, the light hits the mirror and reflects onto the cardstock but, afterwards, the light does not reflect from one cardstock to the other" (Daniela). "The mirror only reflects onto one cardstock and the cardstock is not the same as the mirror" (Leonardo). At this time, I explained that, because the cardstock is not smooth like the mirror, it reflects light in all directions. Therefore, we are not able to make the light appear on the other cardstock – diffuse reflection – whereas the mirror reflects light in a well-defined direction – specular reflection].
- Do you think we can reflect light from one mirror to another?
  - *They predict the path of the light between the two mirrors. Excerpt from the class diary:*

  [The class is unanimous in answering "yes". A second mirror is distributed to the groups. When questioned, the students correctly predict the path of light between the two mirrors: "The light hits the mirror and then comes back, and we are going to put the mirror… here!" (Joel's group); "the light comes from the flashlight this way, hits this mirror (the first one) and then it goes onto this one (the second mirror)" (Daniela)].
  - *They test their predictions and interpret the observations made. Excerpt from the class diary:*

  ["The light goes this way, this way and this way" – referring to the trajectory of the light between the two mirrors. "Then what figure does the light form?" – I ask. "It is like a reversed "V" (Diogo); "It always makes a "V" and the mirror is always at the tip of the "V". "If we put another mirror, we get another "V" (Simão)].
  - *They record the light path on their individual record sheets.*

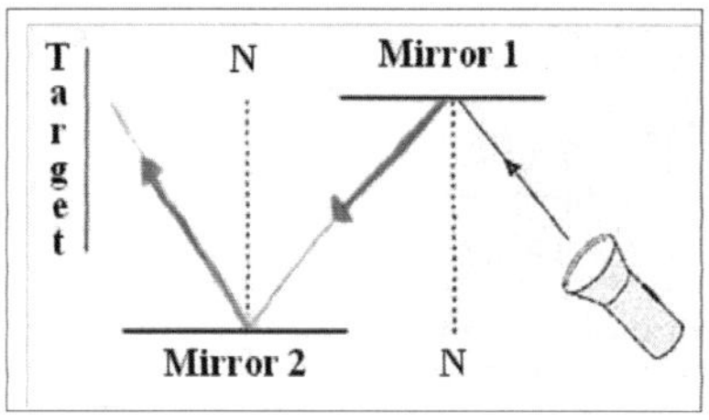

Figure 2. Example of a record made by the student.

  - *They infer and communicate the path of light between the two mirrors. Excerpt from the class diary:*

  ["How did you draw the path of the light?" – I ask. "We drew it from the flashlight to mirror 1. And then from 1 to 2 and then onto the target" (Leonardo). "And how are the lines?" – I ask. "The lines are straight" (Bruna). "They are straight and make sort of a "V" (Ângelo). The majority of the students drew the light path correctly. The comments within the groups revealed a good understanding: "The light comes and goes like this, perpendicularly" (Joel); "it reflects from the mirror and makes a perpendicular line onto the other" (Lara); "the line that goes one way is similar to the one that comes back" (Bruna); "from one mirror, it reflects onto the other and then it goes from that one onto the cardstock" (Diogo)].

- *They identify regularities, recognising that light is reflected according to a certain rule.*

Through successive questioning, the children are stimulated to think about the rule that governs the reflection of light. From the drawings they made in their records, they infer that the dotted line drawn perpendicularly to the plane of the mirror, divides the angle formed by the incident and reflected rays into two equal angles. Excerpt from the class diary:

["The light comes (reflected light) the same way as it goes (incident light)" (Ângelo). Some fail to understand Ângelo's reasoning. Bruna clarifies her colleague's idea: "when the light is hitting at an angle, it is reflected with the same angle". "It's like an axis of symmetry, it is here, dotted on the diagram" – adds Eva. "The line on one side must be equal to the line on the other side, because they are symmetrical" (Daniela). "It is like there is always a symmetry in the middle of the two lines, the one that goes and the one that comes" (Lara). Joel, in conclusion, states: if the light goes in a diagonal, it also comes back in a diagonal, with the same angle. Their ideas are already quite mature and so I mention the fact that the angle formed between the symmetry axis (normal) and the incident ray is equal to the angle formed between the symmetry axis (normal) and the reflected ray].

- *The students build a periscope and make observations.*

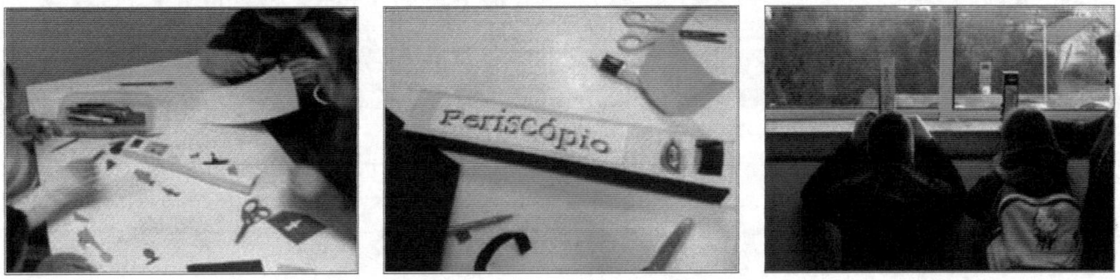

Figure 3. Periscopes built by the students.

3.2. Assessment of the student's learning

Three weeks after the pedagogical intervention, students answered individually to a questionnaire with true or false items. The following table shows the results obtained.

Table 2. Lessons subject and duration.

	Items	Correct answers (%)
1.	We can only see an object if there is light.	15 (94%)
2.	Our eyes are light sources.	16 (100%)
3.	A window is a light source.	15 (94%)
4.	All objects send to our eyes the light they receive.	9 (56%)
5.	I can see a pencil because it emits its own light.	15 (94%)
6.	In a completely dark room I could not see anything.	16 (100%)
7.	The light propagates in all directions in "curves".	16 (100%)
8.	Objects that let some light pass through are called translucent.	16 (100%)
9.	All materials let light passes through.	16 (100%)
10.	Shadows are formed when light falls on opaque materials.	11 (69%)
11.	A mirror does not form shadows.	7 (44%)
12.	Light is reflected when it strikes a mirror.	14 (88%)
13.	The light reflected by a mirror spreads in all directions.	13 (81%)
14.	Light curves when it travels from one transparent substance to another.	12 (74%)

These results suggest that the learning acquired by the students was meaningful because it is long-lasting, as opposed to memorized learning, which is soon forgotten [26].

4. Final considerations

The analysis of the class diary shows that the children are capable of overcoming complex challenges of a cognitive nature when these are approached in a collaborative context, of stimulation and freedom to express their thoughts. Thus, they become active and reflective subjects in the learning process [27]. In the previously discussed learning case, students: a) begin by elaborating predictions about what happens when the light of a flashlight hits a mirror and a cardstock; b) suggest ways to test their predictions; c) in group, test the predictions and make observations; d) reflect on their observations and identify differences between light reflected from the mirror and from the cardstock - regular and diffuse reflection; e) predict the path of the light between two mirrors; f) test their predictions and interpret the observations made; g) record the light path on their individual record sheets; h) infer and communicate to the class the path of light between the two mirrors; i) identify regularities, recognising that light is reflected according to a certain rule - the law of reflection; j) transfer the knowledge developed and apply it in the construction of a periscope; l) with the periscope, observe outside of the classroom. All throughout this learning process, the meanings that students gradually construct are subject to collective discussion and reflection, in a context of social interaction [28], aiming at the construction of socially enriched meanings and shared by a growing number of students. In this way, the social interaction generated among the children and between them and the teacher, promotes higher levels of learning. This is consistent with Harlen, when he refers that "interactions among students and between students and teachers are needed for inquiry-based learning, with the teacher having a key role" [29, p. 2]. The teacher, through a process of questioning, stimulator of thought and action in the students [3], supports their individual and collective cognitive activity [30, 31]. Through this process of questioning, guided by the teacher, students are able to reach higher levels of comprehension and develop better reasoning skills, which they would not be able to achieve without support. All this, as supported by Sá [3] and Harlen [2, 29], entails great personal and intellectual involvement by the student and is closely dependent on an intervention intentionally guided by the teacher, which aims at promoting in students both the construction of meanings that are more consistent with reality and the development of scientific process skills. The intentionality with which the research/teacher conducts their educational action is in line with a reflective practice, in the way they regulate and provide feedback to the children's joint cognitive activity, through continuous questioning, which stimulates reflection and action.

The data contained in the class diaries takes on the nature of a sample of the learning acquired by the children and, therefore, does not allow drawing any inference on the individual level of learning achieved by each one. However, the results of the questionnaire administered 3 weeks after the lessons shows that most of the children in the class acquired a solid knowledge about light and its interaction with the materials. According to Coll and Martín [26], an evaluation that is based on the consideration of an instant situation is unreliable, as it does not take into account the dynamic nature or the temporal dimension of the meaning construction process. In this sense, the results obtained in the questionnaire also allow claiming that this learning was significant, as opposed to memorisation, which is quickly forgotten.

Lastly, we would like to point out that the initial and in-service training of primary school teachers should be able to endow them not only with scientific knowledge in elementary physics, but also with specific didactic knowledge on how to explore the different curricular topics with the children. The development of this knowledge should be based on the data and tools that emerge from research undertaken with children in the classroom context. Research in science education should offer fruitful elements to support the educative action of the teachers. In this sense, the analysis presented in this article, about the activity "Reflection of light" may constitute both a didactic resource for teacher training and an element of support for those teachers, so that, in similar contexts, they are able to promote the same teaching and learning process with their students.

Journal of Physics: Conference Series **605** (2015) 012041 doi:10.1088/1742-6596/605/1/012041

5. References

[1] Duschl A, Schweingruber A and Shouse A 2007 *Taking science to school: Learning and teaching science in grades K-8* (Washington, DC: National Academy Press)

[2] Harlen W 2007 *Enseñanza y aprendizaje de las ciencias* (Madrid: Ediciones Morata)

[3] Sá J 2002 *Renovar as Práticos no 1º Ciclo Pela Via das Ciências da Natureza* (Porto: Porto Editora)

[4] Furman M 2009 *O ensino de Ciências no Ensino Fundamental: colocando as pedras fundacionais do pensamento científico* (Vila Siqueira: Sangari Brasil)

[5] Varela P 2012 *Experimental Science Teaching in Primary School: Reflective Construction of Meanings and Promotion of Transversal Skills* (Saarbrücken: Lambert Academic Publishing)

[6] Charpack G 2005 *As Ciências na Escola Primária: Uma Proposta de Acção* (Mem Martins: Editorial Inquérito)

[7] National Research Council 2000 *Inquiry and the national science education standards: A guide for teaching and learning* (Washington, DC: National Academy Press)

[8] Rocard, M 2007 *Science education now: A renewed pedagogy for the future of Europe* (European Commission Report: Brussels)

[9] Osborne J and Dillon J 2008 *Science Education in Europe: Critical Reflections* (London: Nuffield Foundation)

[10] Linn, M Clark D and Slotta J 2003 WISE Design for knowledge integration *Sci. Educ.* **87** 517-538

[11] Drayton B and Falk J 2001Tell-tale signs of the inquiry-oriented classroom *NASSP Bulletin* **85** 24-34

[12] Eshach H and Fried M 2005 Should science be taught in early childhood? *J. Sci. Edu. Tec.* **14** 315-336

[13] Cuevas P, Lee O, Hart J and Deaktor R 2005 Improving Science Inquiry with Elementary Students of Diverse Backgrounds *J. Res. Sci. Teach.* 42, 337-357

[14] Dyasi H 1999 *Inquiry: Thoughts, Views and Strategies for the K-5 Classroom* ed. I Rankin (Arlington: National Science Foundation) pp 9-13

[15] Dyasi, H 2006 *Linking Science & Literacy in the K-8 Classroom* ed R Douglas et al (Arlington: NSTA Press) pp 3-16

[16] Partridge J 2006 Conducting a science investigation in a primary classroom *Teach. Sci.* **52** 44-5

[17] Minner D, Levy A and Century J 2010 Inquiry-based science instruction: what is it and does it matter? Results from a research synthesis years 1984 to 2002 *J. Res. Sci. Teach.* **47** 474-496

[18] Wilson C, Taylor J , Kowalski S and Carlson J 2010 The Relative Effects and Equity of Inquiry-based and Commonplace Science Teaching on Students' Knowledge, Reasoning and Argumentation *J. Res. Sci. Teach.* **47** 276-301

[19] Abd-El-Khalick F, Baujaoude S, Duschl R, Lederman N, Mamlok-Naaman R and Hofstein A 2004 Inquiry in science education: International perspectives *Sci. Educ.* **88** 397-419

[20] Kask K and Rannikmäe M 2009 Towards a model describing student learning related to inquiry based experimental work linked to everyday situations," *J. Sci. Educ.* **10** 15-19

[21] Appleton K 2002 Science Activities That Work: Perceptions of Primary School Teachers *Res. Sci. Educ.* **32** 393-410

[22] Kang N 2008 Understanding Teachers' Conceptions of Classroom Inquiry with a Teaching Scenario Survey Instrument *J. Sci. Teach. Educ.* **19** 337-354

[23] Ministério da Educação 2004 *Organização Curricular e programas Ensino Básico – 1º Ciclo Ministério da Educação* (Mem Martins: DEB)

[24] Zabalza A 2004 *Diarios de clase: un instrumento de investigación* (Madrid: Narcea)

[25] Sá J 2002 Diary Writing: An Interpretative Research Method of Teaching and Learning *Educ. Res. Eval.* **8** 149-168

[26] Coll C and Martín E 2001 *O construtivismo na sala de aula. Novas perspectivas para a acção pedagógica* (Porto: Edições ASA) pp 196-221

[27] Varela P 2012 The reflexive experimental construction of meanings on the shape of the Earth and the alternation of day and night *Int. Elec. J. Elem. Educ.* **5** 5-26
[28] Ibáñez V and Alemany I 2005 La interacción y la regulación de los procesos de enseñanza-aprendizaje en la clase de ciencias: análisis de una experiencia *Ens. Cienc.* **23** 97-110
[29] Harlen W 2004 Evaluating Inquiry-Based Science Developments (Available online)
[30] Chin C 2006 Classroom Interaction in Science: Teacher questioning and feedback to students' responses *Int. J. Sci. Educ.* **28** 1315-46
[31] Kawalkar A and Vijapurkar J 2011 Scaffolding Science Talk: The role of teachers' questions in the inquiry classroom *Int. J. Sci. Educ.* **35** 2004-27.

Journal of Physics: Conference Series **605** (2015) 012042 doi:10.1088/1742-6596/605/1/012042

Implementation of high concentration photovoltaic grid connected power plant for training, research, innovation and solar electricity production

A Barhdadi[1]*, A Benazzouz[1], B Fabrizio[2] and D Verdilio[2]

[1]Semiconductors Physics and Solar Energy Research Team, École Normale Supérieure, Mohammed V University, PO Box: 5118-Takaddoum, Rabat, Morocco
[2]BECAR s.r.l, Beghelli Group, Viale della Pace, 1 Monteveglio, Bologna, Italy
*Senior Associate at Inter. Centre for Theoretical Physics, Trieste, Italy,

*E-mail: barhdadi@ictp.it

Abstract. In this paper, performance monitoring program is applied to a grid-connected high concentration photovoltaic power plant in order to identify any operational problem and to make sure of its optimal and continuous power generation working conditions. A preventive maintenance plan was also established and proposed for the whole system.

1. Introduction

CPV is far the leading technology and the appropriate solution for high Direct Normal Irradiation (DNI) regions as in the south of Morocco. It delivers the outstanding efficiencies at all levels (cell, module, and system) and the highest energy yield with comparison to the other solar technologies. It is also the best technology for solar systems operating at high temperatures with consistent energy production [1]. CPV has also a compelling cost advantage because of its leading LCOE (levelled cost of energy) in the target market [2]. It is also environmentally advantaged (no permanent shadowing, minimal impact to land, dual land usage, flexible layout sites, best cradle-to-cradle footprint) and does not need any water consumption for electricity generation. Moreover, CPV is a proven, certified, predictable technology with rapid deployment, scalability, and a strong solar market forecast [1].

Recently, the European Beghelli Group has developed and patented an innovative High Concentration PhotoVoltaic (HCPV) power generation system baptized "Life Tree". A demonstration prototype of this new system has been lately installed and implemented at the Teachers College (Ecole Normale Supérieure, ENS) of Mohammed V University (UM5) [3] for training, research and innovation purposes. The installation is well furnished by the necessary meteorological equipments, locally grid-connected, and operating as any small PV power plant. This achievement has been performed in the framework of a scientific and technical cooperation agreement aiming at improving the system performances in terms of energy efficiency, technology, and cost-effectiveness. Researchers and PhD/Master's students from Semiconductors Physics and Solar Energy Research Team (PSES) are now developing training sessions and performing proper research activities on the system to reach the agreement objectives. They are also focusing on designing and developing innovative integrated and reliable components that should improve the system and boost its

Journal of Physics: Conference Series **605** (2015) 012042 doi:10.1088/1742-6596/605/1/012042

competitiveness in the PV market. The purpose of this paper is to present some of the main results we obtained from the study of system performance monitoring after one year of effective and continuous operating.

2. Description of the system

The power generation system is composed by: (a) a sun tracker (heliostat), which hosts a set of 48 HCPV single modules of 64 cells each, (b) the DC/AC converters (inverters) and (c) the control electronics. The system's nominal peak power is 8160 W at 1 kW/m² DNI and standard cells temperature of 25 °C. The system total weight is 3900 kg and its size is 7.670 x 5.560 m².

The tracker is based on a galvanized steel frame mounted on top of a steel pole which can rotate from east to west (azimuthal rotation) and from bottom to top (zenithal rotation) according to the two angular degrees of freedom. The tracker is specifically designed for the HCPV applications. It is a high accuracy bi-axial solar tracker, with typical tracking error lower than 0.1°. The azimuthal and zenithal rotations are driven by two electric brushless motors. Each motor is driven by an electronic PWM circuit embedded in the tracker's control box. The tracking control is based on specifically developed hybrid algorithm. It is an astronomic algorithm which uses the sun ephemeris data, corrected by the sun disc's position identified by the solar camera mounted on the top of the tracker. The tracker is equipped with an anemometer, which is mounted on its top nearby the solar camera. When the wind speed exceed 50 km/h the control box drives the zenithal motor to recover the tracker in 90° home position (fully horizontal) to minimize the surface exposed to the wind force and reach a safe position. The tracker's home position can withstand 190 km/h wind speed.

The high concentration single module is composed by 64 (8 x 8) III-V, triple junction cells, assembled on 8 aluminium fins which are facing the module's front side to release the heat in excess. The 64 triple junction cells are connected in series. Each cell has its own anti-parallel silicon diode to bypass the cell current in case of single cell's shadowing. Each cell outputs typically 2.65 V, while the photocurrent at Maximum Power Point (MPP) is about 1 A. The technical characteristics of each HCPV single module are shown in Table 1.

Table 1. The technical characteristics of the HCPV modules

Size (cm³)	Active Area (m²)	Weight (kg)	Operating Temperature (°C)	Conversion Efficiency	Open Circuit Voltage (V)	Short Circuit Current (A)	MPP Voltage (V)	Peak Power (W)
91.2 x 91.2 x 17.4	0.732	32	- 40 to + 85	22 %	198	1.1	176	170

The sun radiation is concentrated on the cells by parabolic high reflection mirrors. Fig. 1 shows the optical converter's section structure associated to each cell. The optical system is a single reflection type. The sun rays cross the frontal glass and hit the parabolic reflector which concentrates on the secondary optic element (SOE). The SOE homogenizes the light on the cell's surface, thanks to its kaleidoscope behaviour and also enhances the angular performance, allowing a wide angular acceptance of the optical system. The angular acceptance is defined as the maximum angle misalignment which keeps the output power greater than 95% of the maximum. The concentrated sun coming out of the prism lights the solar cell. The geometric concentration factor is about 1350 X (the ratio between the paraboloid's front area (110 mm²) and the cell's area (9 mm²)). Basically, the system is designed to operate at very high efficiency under sunlight intensities more than 1000 times higher than in the case of traditional PV [4].

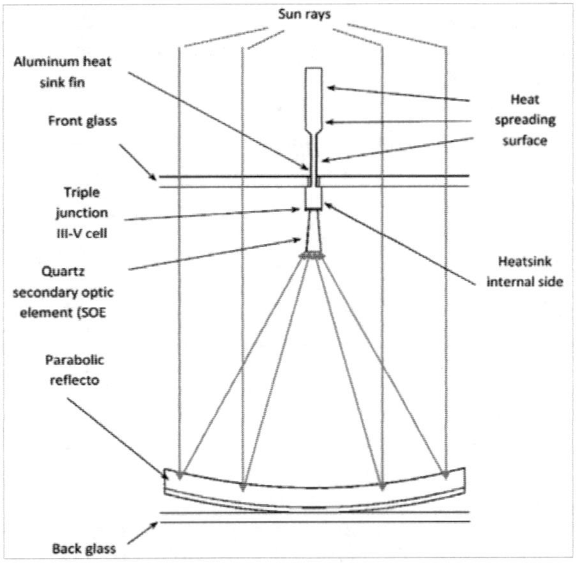

Figure 1. Optical system diagram

Each HCPV single module is connected to its own inverter which converts the 170 V DC power input into 230 V AC 50 Hz single phase power output. The input is electrical insulated from the output, keeping a high level of electrical safety at system level. In fact the maximum voltage at module level is the module's cell open circuit voltage, lower than 200 V DC. This feature is quite important because it minimizes the problems related to the lack of insulation at module level, a known trouble of all PV systems.

The inverter is made of a double high frequency power converting stage: the input inverter operates the MPPT (Maximum Power Point Tracking) algorithm which continuously looks for the maximum power point. The inverter's output stage is the inverter which injects the current into the grid synchronously with the grid voltage, in sinusoidal shape at the 50 Hz grid frequency. The grid injected current is proportional at any time to the available input DC maximum power extracted by the MPPT algorithm. The inverter, controlled by a specific microprocessor, keeps synchronized with the grid frequency by means of a digital PLL, and implements all the output protections to suddenly interrupt the current injection if the grid parameters are not compliant with the safety operating conditions. Each inverter allows the maximum possible energy yield of the HCPV generator, since every single module is independent from the others and there is no problem of module shadowing, mismatching or misalignment. Any aging effect of the system which could eventually lead to mismatching conditions is completely eliminated, since each module will always inject into the grid its maximum available power in any condition. The inverters are connected to the three phase junction box for the connection to the HCPV generator to the 400 V grid.

The control system is based on a specific electronic board (COVECO) included in the metal sealed box fixed to the HCPV generator. The box is visible in fig. 18, which also shows the two brushless motors which manage the two tracker movements. The COVECO board embeds a LINUX CPU which runs the proprietary SW driving the tracker and monitoring the HCPV generator's operation. The COVECO board is also equipped with a simple 2 rows characters display with 4 keys as a basic man-machine interface for simple installation start up procedures, such as running the motors to move the tracker. The more complete control interface is implemented on a remote computer connected to the COVECO via Ethernet or DSSS radio.

The solar camera device is mounted on the top of the tracker and embeds the CCD camera, the wide angle irradiation sensor, and the narrow angle irradiation sensor. The irradiation sensors are monitored by the control box. The sensors are calibrated at the moment of installation and are used as reference sensors to measure the irradiation levels and estimate the generator efficiency.

A very important system's function is the "tracker scan". This function allows to automatically analyzing the angular performance of the whole tracker (and obviously of every single module). Starting the scan procedure, the tracker automatically, while it continues to track the sun position, adds a variable angular offset which is changed in small steps, one at a time; each offset step is kept for the time needed to complete an I-V acquisition (embedded in each inverter) from all the modules; at each step the I-V data is stored and a new step is defined; a complete matrix is scanned of all the angles around the sun pointing position. For each offset angle the maximum power output of each module is calculated and the maximum power output of the whole generator is then calculated.

3. Procedure

The optimal operating and continuous productivity of the system is threatened by uncontrollable criteria such as random failures and climatic conditions. This is why an appropriate and systematic performance monitoring is necessary to identify any operational problem before implementing the fitting solution to make sure of an optimal and uninterrupted activity of the system. To do this, we applied the International Electro-technical Commission Standard IEC 61724 for the performance monitoring of photovoltaic plants [5]. In the beginning, we defined our monitoring strategy according to the above standard. This strategy, detailed in our recent published report [6], is based on the following three steps: (1) Purpose of the performance monitoring and therefore the end use of data recorded (2) Selection of system parameters that should be monitored and the level of accuracy required. (3) Definition of monitoring period or time monitoring (T_m) and maximum recording intervals (MRIs) required. In our case, MRIs is set at 10 min and T_m is calculated from the date when the power plant starts operating for the first time at ENS (April 4th, 2013) until the date when we started doing its performance monitoring (Marsh 27th 2014). T_m is then set at 8640 hours (360 days). Hence, our strategy is in perfect conformity with the international standards and recommendations for monitoring the performance of photovoltaic installations.

Before starting the performance analysis, it was necessary to check the measurement quality of all system sensors. The checking includes power and energy supplied by each inverter, global normal irradiation, ambient and cell temperatures, humidity, and tracker angles. A reasonable set of limits was defined for each recorded parameter based on its known characteristics, the PV plant and the environment. The limits define the maximum and minimum allowable values for the parameter. Data that falls outside these limits or are otherwise inconsistent with other data are not included in the subsequent analyses.

To evaluate the quality of our checking process, two indicator parameters have to be calculated: The checking duration (T_d) in hours and the data availability (D_a) defined by the ratio between T_m and T_d. In our case, T_d = 1488 hours and $D_a = T_m/T_d = 5.8$

4. Performance analysis and results

4.1. Performance analysis in sunny days

To identify the performance problems in the system, we focus mainly on the very well sunny days of the year. Those can be easily identified by historical measurements of DNI and daily power supplied by the system. For these sunny days, the measurements recorded between 11 h and 15 h are collected. During this period, PV modules are in similar operating conditions to standard test conditions in laboratory. Indeed, in this period, there is no shading, the air mass is not so high, and there is no cloud. In such conditions, sources of losses in term of energy production are at the minimum and operational performance problems of the system are easily identifiable.

In this analysis, we evaluate PV cells temperature and their coefficient of thermal degradation. We calculate also the coefficient of performance of each module for each sunny day identified. The cell temperature T_{cell} is estimated by the following equation according to the CEI 82-25 Italian Electro-

Journal of Physics: Conference Series **605** (2015) 012042 doi:10.1088/1742-6596/605/1/012042

technical Commission Standard [7]. There are several other relationships, but Italian manufacturers recommend the use of this formula.

$$T_{cell} = T_{amb} + (NOCT - 20) \times E_{DNI} \times (4 \times I_p)^{-1} \quad (1)$$

T_{amb} is the mean temperature measured between 11 h and 15 h. E_{DNI} is the Direct Normal Irradiation (in kWh) received between 11 h and 15 h. $I_p = 0.9$ kWp.

The performance thermal degradation coefficient $K_{thermic}$ is calculated by the following relationship normalizing the performance coefficient relative to the temperature since we mainly consider operational issues.

$$K_{thermic} = 1 \text{ si } T_{cell} \leq 25 \text{ ìC} \quad (2)$$

$$K_{thermic} = 1 - (T_{cell} - 25) \times \sigma \text{ if } T_{cell} > 25 \text{ ìC} \quad (3)$$

σ depends on the cell technology. In our case, σ = 0.106 %

To calculate the performance ratio PR (also called coefficient of performance), we used the following formula which takes into account the effective power supplied by the system (P_f), its peak power (P_{STC}), the measured irradiation (H) and irradiation H_{STC} corresponding to the peak power ($H_{STC} = 0.9$ kWp/m²).

$$Rp = \frac{P_f}{P_{STC}} \times (K_{thermic} \times \frac{H}{H_{STC}})^{-1} \quad (4)$$

Fig. 2 provides a graphical analysis of the system performance in the sunny days considered in this exercise. The figure shows the evolution of PR for each sunny day and identifies the coefficient of nominal performance in green line. This coefficient is about 85%, when the performance ratio is below 85% the power plant is not operating at the top of its nominal performance. The graphical analysis reveals the presence of a serious operational problem during the summer months and significant measurements instability from the pyrheliometer (area circled in read).

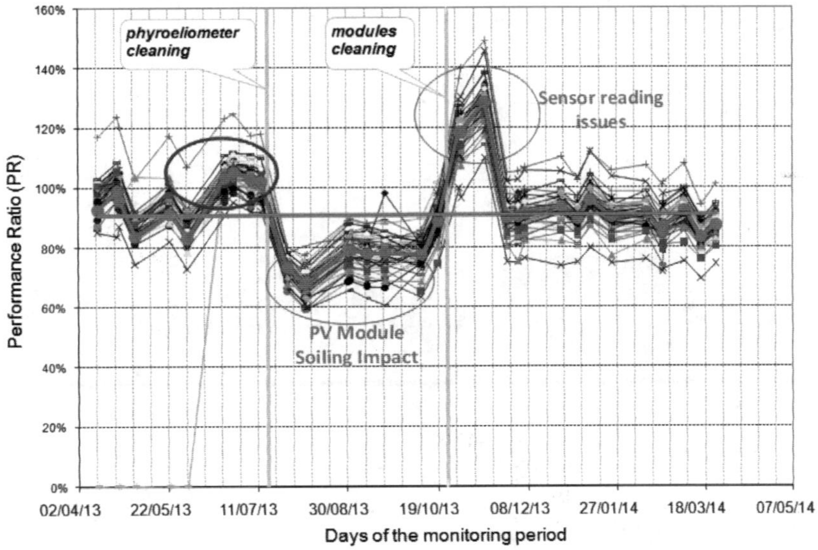

Figure 2. Graphical analyses of system performance in sunny days

In order to identify the source of the above problem (in the summer months), a monitoring of the maximum power supplied by the system was conducted.

4.2. Monitoring of the maximum power output

The maximum power measured at each inverter was reported on Fig. 3, which shows the evolution of this parameter during the monitoring period.

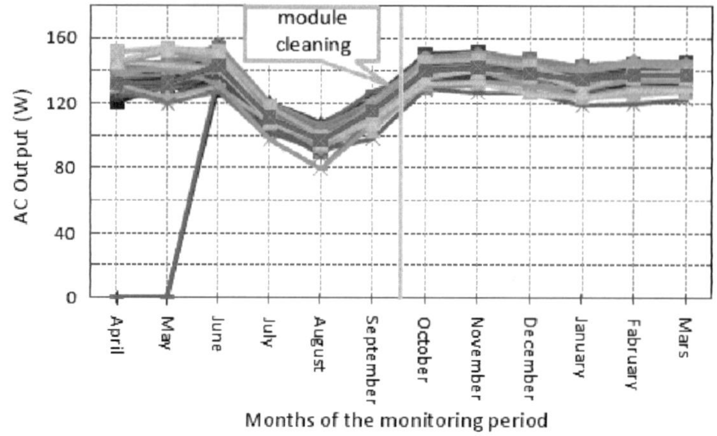

Figure 3. Graphical analysis of the maximum AC power provided by each PV module

The graph shows the presence of limitation in the power in the same period where the coefficient of performance of the system has had its lowest levels (area circled in blue). This allows us to deduce that the operational problem is coming either from the inverters or from the PV modules. However, since the maximum monthly power has increased after cleaning the modules, we can conclude that the operational problem exists because of the accumulation of dust on the modules surface. During the summer, the probability of precipitation is low and therefore the likelihood of accumulation of dust on the modules is high.

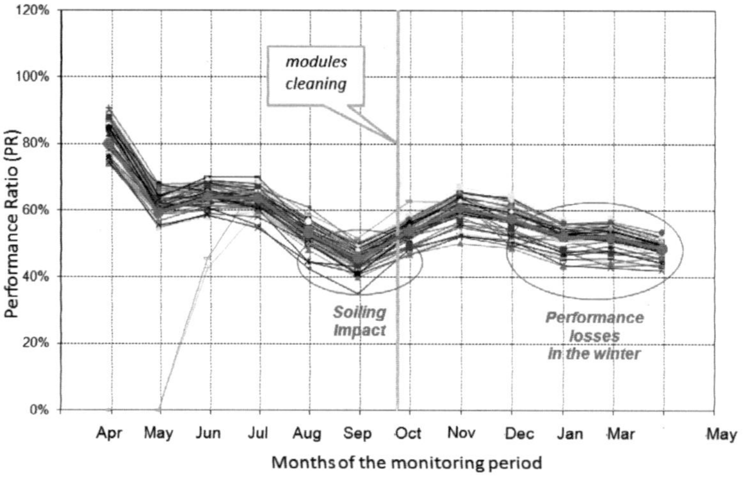

Figure 4. Graphical analysis of the monthly performance ratio of each PV module

4.3. Monthly performance coefficient of each PV module

As we did in the case of performance analysis in sunny days, an additional analysis of the monthly performance was performed in order to detect other sources of losses in the system. The results of this investigation are shown in Fig. 4 where the curves clearly reveal the impact of cleaning on the system performance and the presence of a performance problem in the winter season. This last behaviour can be explained by both shading and air mass effect. In the following section, we examine these issues.

4.4. Sources of losses in winter

4.4.1. Shading effect. Our analysis revealed that the system installation is suffering from some partial shading. The shading effect appears especially in the winter when the sun elevation in the sky is relatively the lowest. Fig. 5 shows the impact of shading on the system production in a typical winter day.

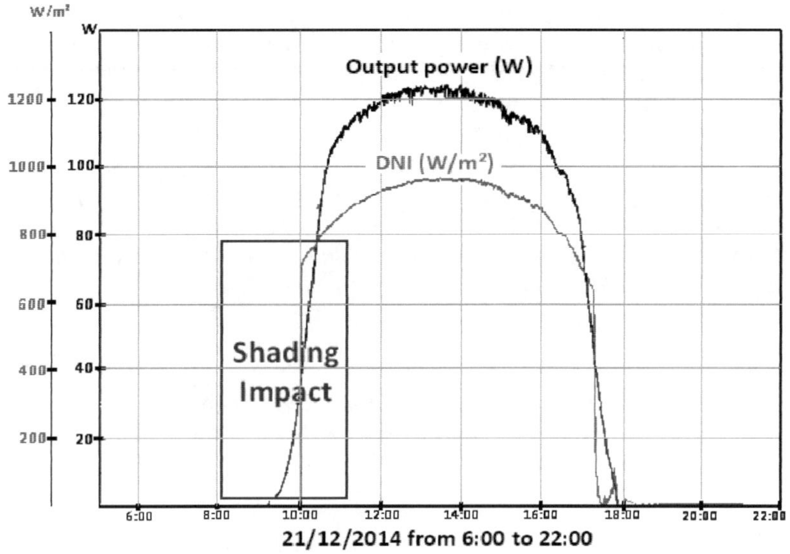

Figure 5. Impact of shading on the system power generation in the winter

Inside the blue frame, the curves show that the power production starts late in the morning after 10 am due to more than one-hour partial shading on the modules surface. This shading comes from the Solar Energy Laboratory building that has been under construction when the system installation was achieved.

4.4.2. Air Mass effect. PV modules are usually tested under STC characterized by an Air Mass fixed at 1.5 (AM 1.5). This parameter is among the factors that have a significant effect on PV module efficiency, particularly in the case of CPV technology. In winter, because the sun is lower on horizon than in summer, solar radiations is going through a larger atmosphere thickness and then undergo more significant changes and mitigations. AM is then obviously higher in winter, therefore PV modules efficiency is lower especially for CPV that produce only from direct sunrays.

5. Conclusion

The performance monitoring of the grid-connected CPV power plant shows that, apart from uncontrollable problems related to climate and physical conditions of the installation site, there are completely controllable issues, which are due only to a lack of cleaning and regular maintenance. Indeed, accumulation of dust on the surface of modules, instability of pyrheliometer measurements, and the long time response to the possible operational problems, are all things that can regularly be verified and avoided. For such problems, we have proposed appropriate solutions to eliminate them or at least reduce their impact. These solutions are integrated into a maintenance plan that we have established and whose implementation will optimize the energy production of the plant and ensure measurement accuracy of its sensors. The proposed plan will also allow a performance prediction and prevent possible malfunctions. To increase the continuity of production in response to different possible anomalies, we have also proposed the installation of an alarm system. All these issues will be presented in details in our next publication, which is still under progress.

6. References

[1] Mark C 2012 *Why CPV? The CPV value proposition* CPV Consortium Report
[2] Brett P 2011 *Cost and LCOE Generation Technology* GTM Research Report
[3] Barhdadi A, Jaziri H, Carpanelli M, Borelli G, Verdilio D 2013 *Implementation of a HCPV grid connected power system for training and research at the university Int. Conf. on Renewable Energy* (Amman, Jordan)
[4] Beghelli SpA Magazine *2011 High concentration Photovoltaic System* (Italy)
[5] International Electrotechnical Commission 1999 *Photovoltaic system performance monitoring – Guidelines for measurement, data exchange and analysis, International Standard IEC 61724*
[6] Benazzouz A 2014 *Surveillance de performance et mise en place de la maintenance préventive d'une centrale photovoltaïque à très haute concentration connectée au réseau basse tension Master thesis ENSA Kenitra* (Morocco)
[7] Italian Electrotechnical Commission 2010 *Guide for design and installation of photovoltaic (PV) systems connected to MV and LV networks* (Italy)

Acknowledgment

This work has been supported by IRESEN (Institut de Recherche en Énergie Solaire et Énergies Nouvelles) Moroccan Institute, in the framework of PROPRE.MA Inno-Project.

AUTHOR INDEX

AUTHOR INDEX